MATHEMATICAL MILESTONES

Dr. Clement E. Falbo

The Reading Glass
BOOKS

MATHEMATICAL MILESTONES
Copyright © 2023 by Dr. Clement E. Falbo

ISBN: Paperback   :   978-1959151104
         Hardback   :   978-1959151753
         e-book   :   978-1959151111

Library of Congress Control Number: 2023909027

The Reading Glass
BOOKS

CONTENTS

3

I am mathematics.

I sing of my Equations.

They rival music.

And painting, and literature and poetry.

My culture is defined by challenges.

As rich and deep as any human endeavor.

Unveiling infinities beyond infinities.

Asking if truth and beauty can be found

Outside of my assumptions and beliefs?

Harvesting understanding from knowledge.

I am delighted when Nature embraces my answers!

What happens if I assume such instead of so?

Imagination begets everything, including, imaginary numbers.

I see connections that surprise all, except Nature, of course.

If you love mathematics, I needed to write you this book.

To help you count the ways.

To see how creativity conquers obstacles.

But, what am I to make of you who say you hate mathematics?

I never heard you say you hate music.

Nor painting, nor literature nor poetry.

I needed to write you this book.

You need to read this book.

CHAPTER 1

WHERE DOES MATHEMATICS COME FROM?

Stone Age tools dating back to about 40,000 years ago were found at a site called the Nwya Devu in Tibet and at earlier dates in the Blombos Cave in South Africa. Also, in many places, drawings and symbols on cave walls depicted human knowledge. There can be no doubt that language and technology had their start in these early prehistoric times. It is safe to speculate that arithmetic, agriculture, and art began to emerge in such places at such times.

Imagine human beings trying to survive by acquiring food and shelter in wild environments. They had to confront major life-threatening problems. Every day they had to think about ways to overcome challenges in a better way today than they did yesterday. The creative human mind is no doubt, the source of our accumulated knowledge, including mathematics, science, literature, art and music that we pass down from one generation to the next.

But what about the stuff we study today in school? We might ask: How did we get to this place? Well, there is plenty of evidence that sophisticated mathematical activities were practiced in civilizations that existed before 3000 BCE. And this happened, independently, in the Eastern World (China), the Near East (Persia), Mesopotamia(Greece), and in the Western World (Europe).

For example, around 500 BCE in China, architects wanted to build a square wall around their city, so they invented the right triangle independently from the Pythagoreans in Greece. Surveyors used formulas to find the areas of the cities enclosed by those walls. Merchants used the abacus to keep track of their purchases, inventories and sales of millet and fish. Warriors wanted to learn the best strategies for winning a battle and gamblers wanted to compute the best odds in betting on the outcomes of wars and games. Thus, we have the beginnings of geometry, computers, matrix algebra and probability in the East. Meanwhile, these same sort of things were happening in the West..

In general, let's say that mathematics comes from human attempts to solve hard problems. Time and time again, new concepts and new branches of mathematics arose when we humans encountered a tough nut to crack. When such a problem was finally solved, it became apparent to the solver and others, even years later, that the same technique would work for totally unrelated problems. Next thing you know, these new mathematical methods would be confronted by some new challenge. Ironically, the new problems could even arise from the previous state of the art itself. This created new and seemingly impossible obstacles, needing new and deeper methods.

Not to belabor this "chicken and egg" theory, let me invite you into my house of mathematics.

There is a sense in which mathematics can be defined as a house with a library of collected problems, solved and unsolved. In addition, this library contains treatises on formulas and methods that might be worth applying to future problems. It also includes alcoves where visitors can come and sit and try to get insights for attacking problems of their own makings or those found in the library. They can write up their results and keep them secret, as Carl Friedrich Gauss did in the 1800's with some of his work, ("Few, but ripe." was his motto) or they could put them up on the shelves (publish them) for future visitors. Everyone is welcome to the house of mathematics.

As we enter the house, we realize that much of modern physics, engineering and economics contributed to the mathematics created in the last 3 or 4 centuries. When we study prime numbers we will be discovering beautiful, but hidden, truths that exist in the counting numbers. In our study of abstract algebras, we will have the pleasure of taking stock of powerful techniques and applications that have been discovered in these relatively recent, times. In some of the back rooms, looking at dusty old records of mathematical knowledge dating back beyond thousands of years, we discover the early beginnings of recent events, specifically number theory, geometry and algebra.

Almost all mathematical concepts are derived from the counting numbers, algebra, and geometry. In geometry, we can construct figures, consisting of points, lines and curves, marked on paper, or as done in Athens by the ancient Greek mathematicians, scratched in sand with a stick. Alternatively, geometric entities can exist mentally defined only by a set of assumptions called *axioms.*

In 300 BCE, Euclid brought together three hundred years of Western mathematical knowledge and organized it into thirteen books called, *The Elements.* Euclid's books laid the foundations of plane and solid geometry, number theory, trigonometry, and the beginnings of algebra. Later, around 240 BCE, Archimedes did much more with geometry and numerical computations, applying them to his inventions of engineering tools, war machines, and the solutions of problems requiring calculus, nearly two millennia before the dates usually attributed to the invention of calculus.

The first axiom in Euclidean geometry is "A straight line can be drawn from any point to any point." Scholars over the centuries have been critical of the imprecision of Euclid's treatment of geometry. He failed to meet the modern, more stringent, requirements later applied to evolving standards in mathematics. Consider Figure 1.1. Is this an example of a straight line being drawn from any point to any point?

A

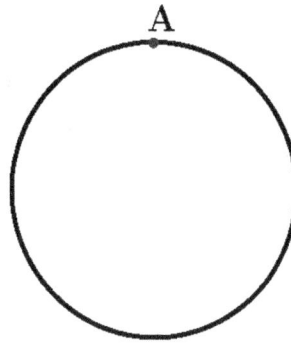

FIGURE 1.1 IS THIS A STRAIGHT LINE?

It could be. As a matter of fact, there is no way to settle this question from just this one axiom. The whole geometric system might consist of just this single point, and lines could be defined like this. If Euclid meant (and it is clear that he did) his geometry to contain more than one point, then the axiom is faulty. In the 18th and 19th centuries, mathematicians re-wrote the axioms and definitions in more precise language; they required this axiom to say that a *straight line can be drawn from any point to any other point,* stating clearly that the line is defined by two distinct points, and not by one point alone. Sometimes this axiom is shortened to say "Two points determine a line." What is required for the construction of a line is simply the identification of two points. So that in any geometric figure if we are given that A and B are two (distinct) points then we can justify saying "Construct the line AB through the two given points A and B." The original Euclidean definition of circles and angles also suffered from the same kind of unconscious assumptions that something more was meant than what was said. Never-the-less, if you are willing to forgive such gaffs, Euclid's work was really a substantial achievement, a testament to the greatness of Greek mathematicians.

Algebra

Actually, the seeds of algebra were sown by Euclid himself when he introduced the idea of a unit length, creating a "metric" (measurement) in geometry in order to add, multiply, subtract and divide lengths, areas and volumes. This type of metric geometry dealt with numbers, and was, eventually, the beginning of algebra, started by Diophantus, in Greece and much further developed in Persia by Al-Khwarizimi around 800 CE. Our word "algebra" came from the title of his book *Al-jabr,* which, apparently, meant it was a book about balancing equations.

Analytic geometry

For the most part, Euclidean Geometry is *synthetic geometry* because when you solve problems in it you are "synthesizing" (building up) the geometric figures. In 1630, René Descartes introduced the notion of *analytic geometry,* solving problems by "analyzing" them, (breaking them down). He started solving geometric problems by use of algebraic equations and a *coordinate system,* which we call the *Cartesian Coordinates.*

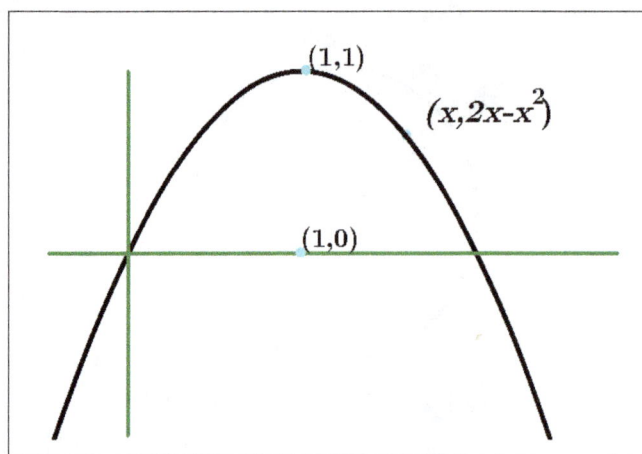

$(1,1)$

$(x, 2x - x^2)$

$(1,0)$

FIGURE 1.2 GRAPH OF THE EQUATION $y = 2x - x^2$.

In analytic geometry a curve, such as a parabola can be defined by an *equation,* for example the equation, $y = 2x - x^2$, can be represented by the set of all points (x, y), in the coordinate plane where x is any number and y is twice x minus the square of x in the coordinate plane as shown in Figure 1.2. Every point on the graph has coordinates (x, y), where x is any number and $y = 2x - x^2$.

Analytic geometry lead to a flurry of activities over the 60 years between 1630 and 1690 when other mathematicians such as Pierre de Fermat from France used the coordinate system to start solving problems concerning instantaneous rates of growth, and when Bonaventura Cavalieri from Italy used these coordinates to find areas of geometric regions enclosed by curves. Finally, Isaac Newton from England and, independently, Gotfried von Leibnitz, from Germany harvested all the bits and pieces being produced over these years and amalgamated them into a coherent body of mathematics called the calculus. Newton himself proclaimed in 1675: "If I have seen further, it is by standing on the shoulders of giants."

In Chapter 2, we will see that, at first, algebra was a practical tool used to solve problems in measurements, accounting and construction work. But, astronomy, physics, other sciences and pure mathematical curiosity, itself, moved both algebra and geometry into more advanced forms. Over the centuries, these subjects evolved into the many facets of mathematics today.

Perhaps Euclid's most significant and longest-lasting contribution to mathematics is what really defines formal mathematics. He organized mathematical statements about geometry into a list of things, called the *postulates* or *axioms* which we are willing *to assume* as true (for the sake of argument) versus another list of things, called *propositions, or theorems,* that we *can prove* to be true. Throughout history, any mathematical statement that is claimed to be true in geometry or algebra or any other system, can have that status of being true only if: a) it is one of the assumed postulates of that system *or* b) it is one of the statements that has been proved to be true in that system. But, just what do we mean by "proved to be true"?

Development of modern mathematics
What is a proof?

Put simply, a proof of some assertion, in algebra, geometry (or any other mathematical system) is a sequence of steps that can be used to logically derive that assertion from the axioms of that system. An equally valid proof would be a sequence of steps showing that a *denial* of the assertion leads to a contradiction of the axioms.

In other words, a proof is a method that lets us justify saying that the assertion is *true* – a logical consequence of the axioms. We cannot prove that the axioms themselves are true because we have already assumed them to be true. Rather than asking the question "are the axioms true?" we need to ask, "are the axioms sufficient for some purpose and do they result in theorems of some substantial consequence?" The study of axioms, themselves in this way is an interesting examination of questions about mathematics and questions about how and why it works. In a sense, it is looking at mathematics from a bird' eye view, and is known as *metamathematics*.

Rapid progress

In the years between 1630 and 1800 mathematicians concentrated on solving practical problems and it is fair to say that they paid little or no attention to metamathematics. With the discovery, however, of non-euclidean geometry, and imaginary numbers, they recognized the need to firm up the foundations by examining the axioms. This was especially true since many practical problems in gravity, electricity, magnetism, and sub-atomic forces became more abstract, in physics, and new concepts such as group theory and topology became more abstract in mathematics.

Relativity, space-time continua, various forces in nature, and quantum physics stimulated increasingly new demands on mathematical systems. Around the 1880s mathematicians, and physicists, finally after a lot of introspection, established almost all of the axioms needed to define the real number system.

Even so, one major problem left to be solved was that of the continuity of a function, especially any function that was defined as the limit of an infinite series. The need to fix this problem was "in the air", so to speak, and several mathematicians invented their own axioms to complete the real number line. These various axioms: the *Least Upper Bound Axiom*, the *Nested Sequence Axiom* and others, turned out to be equivalent to each other in the sense that if you assumed any one, you could prove all the others. One of the earliest and most widely accepted such axiom was the one posed by the German mathematician, Richard Dedekind in 1880; his axiom is called the *Dedekind Cut Axiom.*

In any case, the addition of any one of these completeness axioms became the crowning achievement that would insure that the real number line was continuous. Now, the scientific community was sanguine about the most fundamental questions regarding axioms, theorems and logic in mathematics. They could confidently say that "truth" in a system meant "provable" from the assumptions of that system. It is relative and not an absolute.[1]

Disturbing questions

At the end of the nineteenth and beginning of the twentieth centuries, the study of sets became central to almost all mathematical ideas. It was the most natural logical conse-

---

[1]You will find all of these axioms, including the Dedekind cut axiom in the Chapter 4.

quence in the development of a unified mathematical science.   Unfortunately, however, a crisis occurred.   When the axioms came under closer scrutiny, serious questions arose concerning their logical consistency, their completeness and even their value to humanity. What contradictions arise when we work with sets of numbers or sets of other kinds of elements?   Can a set contain itself as an element?   We know what we mean by limits, but what do we mean by infinity?   Why does it appear that there are two different sizes of infinity; one for the rational numbers and one for the irrational numbers?   Are there more infinities?   And does *truth* itself exist as an absolute construct?   Are all axiom systems equally "good"?   And what about Kurt Gödel's proof, upsetting the apple cart in 1930?   We will tell you this story in Chapter 12 of this book.

Practical resolution

Aside from these disturbing questions, what can we properly say about the nature of mathematics and about proofs?   Abstractly, mathematics is a collection of non-contradictory logical statements based upon previously laid-out sets of axioms.   Even so, we must recognize that mathematics is much more than just that   The axioms are based upon real attempts to solve real problems faced by real human beings.   For that reason mathematics has become very effective in describing what is going on and what the problems are and what to do about them.   Thus, everyday observations and thoughtful guesses find their way into formulating whatever assumptions you want to make as the basis for a mathematical system.   Following this path we will see that new mathematical systems are created not only by new axioms, but by new meanings assigned to the operators of *multiplication,* and *addition* as well as to new binary relations.   We will see how the axioms of the real number line apply to vector analysis, complex variables, quaternions, and various fields of technology such as computer sciences and engineering.

But let us begin at the beginning.   We want to introduce the language of mathematics, specifically mathematical sentences, called *equations*.   Believe it or not, the only verb needed in a mathematical sentence is the infinitive form *to be,* and its conjugations: *is, are, equals, was, were*, all denoted by the equal sign, $=$.   We also use negations such as: *is not, are not, will not be, not equals, less than*, and *greater than,* denoted by $\neq$, $<$ and $>$.

CHAPTER 2

THE EQUAL SIGN

Mathematical notation

Some people say mathematics is a language, but I believe it is more accurate to say that mathematics, what ever it is, *has* its own language and it is universal. Of course, every field has its jargon. So does mathematics, but when I say mathematics has its own universal language, I mean its sentences are expressed in symbols understood by any mathematician, in any country in the world.

The notation, usually the Roman alphabet $a, b, c...$ and Arabic numerals, $1, 2, 3...$ used in mathematics today may not even be the letters or numbers used anywhere in the native language of that person. Mathematical symbols are not in the Cyrillic, nor the Hebrew, nor the Arabic alphabet. And the Chinese, Japanese, and other characters are not used world-wide as mathematical symbols. If the mathematical equation

$$x^2 + y^2 = z^2$$

appears in the middle of a French document, then a translation of that document into Russian or Chinese, that same expression

$$x^2 + y^2 = z^2$$

will appear in the translated document, but the rest of the page will be in the Cyrillic alphabet or in Chinese characters.

This is not surprising since the symbols used in mathematics today evolved over the centuries by mathematicians working in the Western World: Greece, Mesopotamia, Germany, Italy, France, and Persia. In fact, we use the first half $a, b, c, ...m, n$ of the alphabet to stand for *constants*, and letters from the second half $p, q, ...y, z$ to stand for *variables*. This convention was permanently established world-wide in 1630 by the French mathematician René Descartes. The plus and minus signs have been used universally for the past 600 years and the equal sign for the past 460 years. In earlier times, we used language-dependent words or abbreviations for plus, minus, division, multiplication, exponents and equals. But the universally accepted notation used today has been developed by "trial and error" during attempts by mathematicians and scientists to communicate

internationally over several centuries.   Table 2.1 presents a brief summary of the notation
as it developed over the centuries:

Who	Where	When	What	Meaning
Euclid	Greece	300 BCE	$A, B, C, \ldots$	Points, lines,...
Diophantus	Greece	300 CE	abbreviations	Number, operation
Al-Khwarizmi	Persia	825 CE	$1, 2, \ldots, 10, \ldots$	Numerals
Al-Kashi	Persia	1400 CE	$+, -, \sqrt{}$	plus, minus, roots
Recorde	Britain	1557 CE	$=$	Equal sign
Bombelli	Italy	1572 CE	$ax^n + bx^{n-1} +$	Polynomials
Descartes	France	1630 CE	$(x, y), a, b, x, y$	Coord., const. vars.

TABLE 2.1 THE ORIGINS OF SOME MATHEMATICAL NOTATION.

To get an idea of how extensive was the development of mathematical notation see
Joseph Mazur's book, *Enlightening Symbols*, Princeton University Press, 2014.

This is not to say that all of mathematics originally developed in the Western World.
In Asia, especially in China, as early as 250 BCE, the Eastern World had advanced mathematical concepts, including decimal and binary place value systems, negative numbers
and many theorems in algebra, geometry, and number theory.   Some concepts in China
were studied even earlier than the same ideas in the Western World.   Pascal's Triangle
is an example we will discuss as it relates to infinite series in Chapter 6.   Although
some of the concepts in Asian mathematics preceded the same topics in the West, the
mathematical notation did not survive, internationally, in competition with that of the
West.

The current notation has become so very efficient, we might even say that it is the
best notation possible.   A notable exception might be that the juxtaposition of two letters
such as ab which is often used to mean multiplication of a by b.   In some cases, however,
juxtaposition does not mean multiplication, such as: $3\frac{1}{2}$ means add 3 to $\frac{1}{2}$.   Or 34 does
not mean multiply 3 by 4, but it may represent the base *ten* number *thirty-four,* or the
base *five* number, *nineteen,* and so forth.   This notation adds to confusion for all but
the anointed mathematics student.   In general, however, the world-wide mathematical
notation has been, pretty much, unchanged for the last couple of centuries.

In 1910, two British mathematicians Alfred North Whitehead and Bertrand Russell,
using standard notation for set theory, wrote a three volume work, *Principia Mathematica,*
purporting to create a foundation of logic and set theory that would be a complete system
capable of answering all mathematical questions with no possibility of any contradictions.
In Figure 2.1 we show part of a page from that book.

Dem.

$$\vdash . *56{\cdot}101 . *37{\cdot}106 . \supset$$
$$\vdash : R \,\epsilon\, \dot{2} . \equiv . R \,\epsilon\, \breve{D}``1 . R \,\epsilon\, \breve{\Game}``1 .$$
$$[*22{\cdot}33] \quad \equiv . R \,\epsilon\, \breve{D}``1 \cap \breve{\Game}``1 : \supset \vdash . \text{Prop}$$

*56·103. $\vdash : R \,\epsilon\, \dot{2} . \supset . \dot{\Game}! R$

Dem.

$$\vdash . *56{\cdot}101 . \supset \vdash : R \,\epsilon\, \dot{2} . \supset . D`R \,\epsilon\, 1 .$$
$$[*52{\cdot}16] \qquad\qquad \supset . \Game! D`R .$$
$$[*33{\cdot}24] \qquad\qquad \supset . \dot{\Game}! R : \supset \vdash . \text{Prop}$$

*56·104. $\vdash : R \,\epsilon\, 0_r . \equiv . R = \dot{\Lambda} \quad [(*56{\cdot}03)]$

*56·11. $\vdash : R \,\epsilon\, 2_r . \equiv . (\Game x, y) . x \neq y . R = x \downarrow y \quad [*20{\cdot}3 . (*56{\cdot}02)]$

*56·111. $\vdash : R \,\epsilon\, 2_r . \equiv . D`R, \Game`R \,\epsilon\, 1 . D`R \cap \Game`R = \Lambda$

Dem.

$$\vdash . *51{\cdot}231 . *55{\cdot}16 . \supset$$
$$\vdash : x \neq y . R = x \downarrow y . \equiv . \iota`x \cap \iota`y = \Lambda . D`R = \iota`x . \Game`R = \iota`y .$$
$$[*13{\cdot}193] \qquad \equiv . D`R \cap \Game`R = \Lambda . D`R = \iota`x . \Game`R = \iota`y \qquad (1)$$
$$\vdash . (1) . *56{\cdot}11 . *11{\cdot}11{\cdot}341 . \supset$$
$$\vdash :. R \,\epsilon\, 2_r . \equiv : (\Game x, y) . D`R \cap \Game`R = \Lambda . D`R = \iota`x . \Game`R = \iota`y :$$
$$[*11{\cdot}45] \qquad \equiv : D`R \cap \Game`R = \Lambda : (\Game x, y) . D`R = \iota`x . \Game`R = \iota`y :$$
$$[*11{\cdot}54] \qquad \equiv : D`R \cap \Game`R = \Lambda : (\Game x) . D`R = \iota`x : (\Game y) . \Game`R = \iota`y :$$
$$[*52{\cdot}1] \qquad \equiv : D`R \cap \Game`R = \Lambda . D`R, \Game`R \,\epsilon\, 1 :. \supset \vdash . \text{Prop}$$

Figure 2.1 PAGE FROM PRINCIPIA

After a brief introduction in English, almost every page of those books contained only mathematical notation with no words at all. Everyone could read this book, and even though the book represented a monumental work, Whitehead and Russell's plan was an utter failure. This was not because of the notation but for another reason all together. We shall see why later in this book.

A notable exception to the unchanging mathematical notation comes from the Twenti-eth Century revolution in mathematics, namely *computer science*. For example, computer languages use the equal sign in a way that is quite different from the usual mathematical way. The computer statement $x = 7$ means that the number 7 is *stored* in an address, (a place in the computer's memory) called x. After you have written $x = 7$ in your program, if you write $x = x + 1$ later in that same program, you are not saying $7 = 7 + 1$; that would make no sense. What $x = x + 1$ means is that, now, you want to place into the address x, what was previously in x increased by 1.

Equations
Equal sign

Normally, the mathematical symbol $=$ means "is" or "are". Equations are *sentences* using this verb and mathematical inequalities are sentences using various negations of this verb, not equal $\neq$, greater than or less than, $>$ or $<$. That's it. In this sense, mathematics is simpler than any natural language.

Grammar

Furthermore, most mathematical statements are in the predicate-nominative case, where the subject is the object. Much as you would respond on the telephone, "This is she", when some one asks to speak to you. It is the same thing as "She is this". Because the sentence $1 + 1 = 2$ means the same thing as $2 = 1 + 1$. In other words, if x and y are any numbers, then $x = y$ means the same thing as $y = x$. This is called the *symmetric law of equals*.

The reason the equal sign represents the predicate nominative is that it represents a state of **being**, not a state of action. In Chemistry, you may say "Two molecules of Hydrogen gas combined with one molecule of Oxygen gas will yield two molecules of water." This can be written symbolically as:

$$2H_2 + O_2 \longrightarrow 2H_2O$$

But it would *not* be proper to use an equal sign because it is not a sentence about a state of *being*; it is a statement about a chemical reaction between two elements *yielding* a molecule. So, the correct verb to use is not the verb *to be*, but rather the verb *to yield*. The usual symbol for yield is "$\longrightarrow$". When the process is reversible we could use the double arrow $\rightleftarrows$.

Adjectives and nouns

A teacher in the lower grades will try to keep students from misusing the equal sign by making up rules, such as "you can't add apples to oranges." When they do this they are trying to prevent students from mistakenly writing $x + y = 2x$ or $x + y = 2y$. Neither of which is true unless $x = y$. In desperation, a teacher might say "$1 + 1$ is not always 2". What they mean is 1 orange plus 1 apple is neither 2 apples, nor 2 oranges.

This problem occurs when you say that $1y$ stands for "one orange" and $1x$ stands for "one apple". Here "orange" and "apple" are nouns and the "1" is the adjective modifying each noun. It would be better to think of each of x and y as a *number,* used as an adjective modifying the number 1. Doing this, $1y$ becomes $y1$, or y *ones,* and $1x$ becomes $x1$, or x *ones.* Therefore, we could write $1x + 1y$ as $x1 + y1$, meaning x **ones** plus y **ones**; thus, x and y are not oranges and apples, but rather **numbers**, telling you *how many* 1's are to be added to *how many* other 1's.

   Example: What is $2x + 3y$?

   Well, it depends on what the **numbers** x and y are, and not on any fruity ideas.

   We can re-write this as a slightly more awkward expression: "What are x **twos** added to y **threes**?",

$$x2 + y3$$

We can know this as soon as we find out what the numbers x and y are, otherwise, we just leave it in the form: $2x + 3y$.

Substitutions

Substitutions are not only important in team sports such as basketball and football, but they are just as important in mathematics. In football you substitute by replacing a player in a given position on the team by another player into that same position. You may think of substitution in math as replacing some part of an expression by something else.

Example: If you are given that: $y = 7x + 9$, read as "y is seven x's plus nine" and you want to substitute 2 in for the x, that is, $x = 2$, then you can say $y = 23$. How did we get this mysterious answer? By substitution! We start with the original expression (in words)

"y is seven x's plus nine".

Now if $x = 2$, you *substitute* 2 in for x, then

"y is seven *twos* plus nine".

"y is fourteen plus nine".

"y is 23"

You can substitute any thing you want in for x, (another number, a fraction another letter or even another mathematical expression) then multiply it by seven and add nine to get another answer for y.

Examples: Let $y = 7x + 9$,

$$
\begin{aligned}
\text{If } x &= 3, \text{then } y = 21 + 9 = 30 \\
\text{If } x &= \frac{1}{2}, \text{then } y = \frac{7}{2} + 9 = \frac{25}{2} \\
\text{If } x &= t, \text{then } y = 7t + 9 \\
\text{If } x &= \sin(\theta), \text{then } y = 7\sin(\theta) + 9
\end{aligned}
$$

Substitution is not a new idea in mathematics, the basic rule was stated in 300 BCE by Euclid, when he wrote, as a common notion, that: *Things equal to the same thing are equal to one another.*

Symbolically, we write this as

If $x = y$ and $y = z$, then $x = z$

which is called the *transitive law of equals.* This really is a substitution because we substituted x in for y in the expression $y = z$ to get $x = z$.

Solving an equation for its unknown

A theme in this book is that mathematics is beautiful and powerful. An example of both its beauty and power is our ability to substitute not only *expressions,* such as $\sin(\theta)$ for x, as in the previous example, but also the ability to substitute rules in one mathematical system as rules in another system. A specific case of this would be the ability to replace one equation by another equation. "Things equal to the same thing are equal to each other" as Euclid said. For example, if we know that the equation $r = \frac{d}{t}$ is true then we may multiply this equation by the equation $t = t$, getting $r \cdot t = \frac{d}{t} \cdot t$, which says the same thing as $r \cdot t = d$. You might say "equals times equals are equal", but this is too vague for my taste.

Problem in one variable   Consider the following riddle posed by a fictitious Persian Sheik in 800 CE.

"The five fold of Abdul's sword decreased by 6 cubits is its double increased by 9 cubits. He who can tell me the length of Abdul's sword will receive a handsome reward."

Al-Khwarizmi, the mathematician speaks up, "Your Highness, the five fold of the sword diminished by its double is six increased by nine cubits, My Sublime Leader. The threefold of the sword is 15 cubits, Oh, Honored One. Abdul's sword is 5 cubits."

"Say what? How did you get that?"

"Well, you see, Sire, your riddle is a linear equation in one variable, the sword length. If we let x stand for the sword length, then your pronouncement can be written as

$$5x - 6 = 2x + 9$$

which no doubt your royal intelligence can follow."

Then the wise Al-Khwarizimi, who was the inventor of algebra, went on to say, "Subtract $2x$ from $5x$ and add 6 to 9 to get

$$
\begin{aligned}
5x - 2x &= 6 + 9 \\
3x &= 15
\end{aligned}
$$

as you can see by logic."

"But, tell me, young Persian, why can you subtract $2x$ from $5x$?"

"Oh, Exalted One, the great Greek mathematician, Euclid, told me that *Equals added to equals are equal,* so I start with Your riddle $5x - 6 = 2x + 9$ and add a new equation $-2x + 6 = -2x + 6$, which everyone in the Sheikdom agrees is the veritable truth.

$$
\begin{aligned}
5x - 6 &= 2x + 9 \\
-2x + 6 &= -2x + 6 \\
5x - 2x + (-6 + 6) &= (2x - 2x) + 9 + 6 \\
3x &= 15
\end{aligned}
$$

and that, Your Eminence, is the nature of Al-jabr." (Algebra). The Sheik was impressed and rewarded Al-Khwarizmi with a fine Arabian horse.

In today's world, adding equals to equals is a step in the modern method, called *row reduction*, for solving linear equations. You may have learned it in grade school as gathering all the *unknowns* on one side of the equation and all the *knowns* on the other side. Or sometimes it is taught as *when you take a quantity across the equals, you change its sign.*

Problems in two variables   We make a significant step up from this problem to one in which we try to solve a *system of equations* (two or more equations). For example, if we have two equations

$$5x - 2y \;\; = \;\; 6, \text{ and} \tag{2.1}$$

$$\frac{1}{2}x + y \;\; = \;\; 15 \tag{2.2}$$

We want to find the the two numbers x and y that make both of the these equations true.

Looking at these two equations, Al-Khwarizmi might say, "Hmm, if we had a $2y$ instead of just y in the second equation, we could add the two equations and the $-2y$ and the $+2y$ would cancel out leaving us with a new equation with no y term at all. Then we could just solve the resulting equation for x, the only variable left."

So, let us multiply the second equation (on both sides) by 2, getting:

$$x + 2y = 30 \tag{2.3}$$

The two Equations (2.1) and (2.2) can be written as

$$5x - 2y \;\; = \;\; 6 \text{ and} \tag{2.4}$$

$$x + 2y \;\; = \;\; 30 \tag{2.5}$$

Equations (2.4) and (2.5) are equivalent to Equations (2.1) and (2.2), so they will have the same solutions, x, and y. Add Equations (2.4) and (2.5), then solve for x:

$$6x + 0y \;\; = \;\; 36, \text{ so}$$

$$x \;\; = \;\; 6$$

Now, substitute $x = 6$ into either Equation (2.1) or (2.2) to find that

$$y = 12$$

In these problems we use substitution of one equation for another with the goal of reducing the number of variables in an equation. This process is a topic in a field called *matrix algebra* in which we concentrate on how to manipulate the coefficients of the variables.

What is a coefficient?

What do we mean by manipulating the coefficients of a variable? What is a coefficient? It is the constant that is multiplied by the unknown variable. For example, if x and y are variables, in the expression $5x - 2y$, then 5 is the coefficient of x and -2 is the coefficient of y. If we have an expression with only the variable such as just x, then the coefficient of x is 1, because $x = 1x$. Sometimes, in a problem, the coefficients are an unknown constants, such as $ax + by$, but the context of the problem will help us see what to do with them.

In the following problem, we will set up a small 2 equation, 2 variable problem, a two by two system. But in many applications in industry and business, there may be as many as 50 or more variables in that many equations, which would require powerful computers to solve. Never-the-less we can illustrate the actual computer algorithm for solving these large systems with our smaller example. In fact, you can probably install programs in your own home computer, and even in your calculator to solve n by n systems of linear equations for small n.

Let us now consider how to set up some of the mathematical terminology used in everyone's favorite subject, NOT!, – *word problems*.

Word problems

Of course, mathematics must have a vocabulary.   Usually a problem is stated in your native language and then converted to mathematical symbolism.

HOW TO SET UP A WORD PROBLEM
Most word problems can be set up for solution by pretending that you know the answer to the question, and then assigning a name (a variable) to be that answer. For example, when you are asked to find some quantity that satisfies some condition, start by quoting these words from a Beatle's song: "There will be an answer; let it be ..." x. Then you start requiring x to have the properties outlined in the problem.

Problem: Let us say that a young entrepreneur is selling fat wood[1] door-to-door in Joseph, Oregon. She plans to sell each small bundle (ten sticks) for five dollars. It costs her \$25 to buy an axe. Then she has to go to the forest to collect logs and the cost of the wood turns out to about 16 cents per stick (\$1.60 per bundle). How many bundles must she sell in order to make enough money to equal her cost in producing that many bundles? Of course, she wants to eventually sell more than that to make a profit.

Solution:

We will put these words into algebraic equations. We start by pretending that we have an answer to the question "how many bundles must she sell?" and that answer is x. What is the revenue, r, from selling x bundles? It is \$5 times x, so $r = 5x$. What is the cost, c, of producing those x bundles to be sold? It is \$25 (for the axe) and \$1.60 times the number x, of bundles sold. Thus, $c = 25 + 1.6x$. We have two equations:

$$r = 5x, \text{ the revenue equation.}$$
$$c = 25 + 1.6x, \text{ the cost equation.}$$

The problem is asking what x do you think will make the cost equal the revenue. This is called the beak-even point. Just set them equal.

$$\begin{aligned} \text{revenue} &= \text{cost} \\ 5x &= 25 + 1.6x \\ -1.6x &= -1.6x \text{ equals subtracted from equals} \\ 3.4x &= 25 \\ x &= 25/(3.4) \\ x &= 7.3529... \end{aligned}$$

The answer is a fraction between 7 and 8, but she is selling whole bundles, not fractions. Let us look at her revenue and cost in two cases. One where she sells 7 bundles, and the

---

[1] Small kindling sticks with a high resin content used to start a quick fire in a home fireplace.

other where she sells 8 bundles. For seven bundles, her revenue is

$$r = \$5 \times 7 = \$35$$

But the cost is

$$c = \$25 + \$1.60 \times 7 = \$36.20$$

She will lose \$1.20.

If she sells eight bundles, her revenue is

$$r = \$5 \times 8 = \$40$$

And her cost is:

$$c = \$25 + 8 \times \$1.60 = \$37.80$$

giving her a profit of \$2.20.

In order to get a complete picture of this enterprise, let us look at Figure 2.2 where we plotted the straight line graphs of the cost and revenue equations for the sales of *any* number of bundles.

$$y = 5x$$
$$y = 25 + 1.6x$$

These are the same revenue and cost equations given above, but written in terms of lines in the xy coordinate system (which we will study later). You can see that the revenue starts out below the cost but climbs up faster (the revenue line has a steeper slope than the cost line) and finally catches up with the cost at about $x = 7$, then after that it stays above the cost line showing profit after $x = 8$.

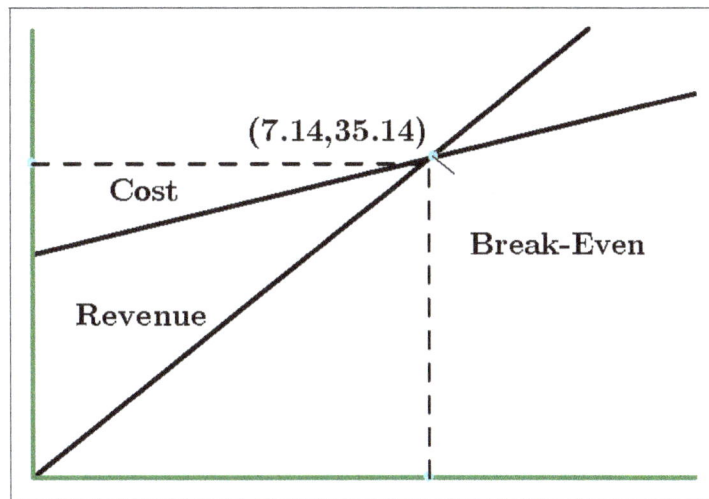

FIGURE 2.2 BREAK-EVEN POINT

Note: The revenue starts out at zero; no sales, no revenue. The (fixed) cost starts out at \$25, no sales, the cost is still \$25.

The reason that there is a break even point at all is that the revenue line has a greater *slope* than the cost line. The revenue goes up faster because you gain \$5 per bundle while the cost goes up at \$1.60 per bundle. The slope is called the *rise to run ratio*; here the rise in revenue is 5 when the run (bundles) is 1. In the cost equation the rise is 1.6 for the run of 1 bundle, so the revenue is catching up with the cost. If she sold the bundles at only \$1.60 the slope of both lines would be the same and the lines would be parallel; they would never cross and there would be no break even point.

Any time we are trying to solve a pair of linear equations in two variables, there will be no solution if both lines have the same slope. In fact, when the equations are for lines with the same slope, it may be that there are not really two lines, but the same line. In which case, there is no unique solution because all points are break even points. Later, we will devise a test to see whether or not two or more linear equations have a solution before we even start trying to solve the system.

Systems of linear equations

Let each of a, b, and c be a constant. If x and y are variables, then a *linear equation* in x and y is: $ax + by = c$.

If, in addition, we had constants, d, e, and f and two linear equations in x and y, then, we would have a "two by two" system because there are 2 variables and 2 equations for which we want to find to numbers x and y that satisfy both equations simultaneously.

$$
\begin{aligned}
ax + by &= c \\
dx + ey &= f
\end{aligned}
$$

An example of this was the system of two equations, one for cost and the other for revenue discussed in the fat wood problem shown in Figure 2.2

A 3 by 3 linear system has three equations and three variables, x, y, and z, and we want to find values for those three variables that satisfy all three equations simultaneously.

$$
\begin{aligned}
a_1 x + b_1 y + c_1 z &= d_1 \\
a_2 x + b_2 y + c_2 z &= d_2 \\
a_3 x + b_3 y + c_3 z &= d_3
\end{aligned}
$$

where all of the a, b, c, and d 's are constants.

When we change over to the matrix format for these systems, a common vernacular has been informally adopted. The equations are called "rows" and the coefficients (the a, b, c, etc.) are called "columns." When we are dealing with a large system, we use a double subscript notation a_{ij} on the constants, the first to denote the row and the second to denote the column.

Row reduction method for solving a linear system

The method we are about to introduce will double the pleasure you can obtain from solving equations. It is like getting a new Alpha-Romeo to zip around town; it is a sleek vehicle that can be applied to solve n by n systems (n equations with n variables) for any positive whole number n. It was first rolled out of the showroom and presented to the world about 2300 years ago.[2] Here it is for $n = 2$.

---

[2]This procedure was invented by Chinese mathematicians as early as 300 BCE. We will discuss the historical context later in Chapter 9.

Two equations with two unknowns

Problem:

Find the values of x and y that satisfy the two equations:

$$3x - 5y = 8 \tag{2.6}$$
$$4x + 7y = 11 \tag{2.7}$$

We are going to simplify this formulation of the problem by "ignoring" the x and y and concentrating on the constants 3, 5, 8, and 4, 7, 11. The numbers 3, -5 are the coefficients of x and y in the first equation and 4 and 7 are the coefficients of x and y in the second equation. We arrange the coefficients of x and y into a 2 by 2 array of numbers, called *a matrix* as follows:

$$\begin{pmatrix} 3 & -5 \\ 4 & 7 \end{pmatrix}$$

The numbers 8 and 11 on the other side of the equation we will arrange into the 2 by 1 matrix:

$$\begin{pmatrix} 8 \\ 11 \end{pmatrix}$$

To see a relation between these matrices (plural of matrix) and the original problem, Equations (2.6) and (2.7), we will also write out the 2 by 1 matrix for the variables, x and y, called a *column matrix*

$$\begin{pmatrix} x \\ y \end{pmatrix}$$

Any matrix that has only one column is called a column matrix. Now, the system (2.6), (2.7) is

$$\begin{pmatrix} 3 & -5 \\ 4 & 7 \end{pmatrix} \times \begin{pmatrix} x \\ y \end{pmatrix} = \begin{pmatrix} 8 \\ 11 \end{pmatrix} \tag{2.8}$$

Whoa! What is Equation (2.8)? How is it related to the system (2.6) and (2.7)? and why do we want to write it this way?

It is a *matrix equation* and is one of the most powerful tools for solving systems of equations. The idea is that we have simplified the problem enough for it to be programmed into a series of steps, called row reduction, that a computer would mindlessly go through and get the answer. Here, finally, is an even more bare bones way to write this system, omitting the x and y, altogether. It is called the *augmented matrix* for this system.

$$\begin{bmatrix} 3 & -5 & 8 \\ 4 & 7 & 11 \end{bmatrix}$$

RULES FOR ROW REDUCTION:
1. We can replace any row by a multiple of that same row.
2. We can add or subtract any row from a given row and replace that given row by the result of the subtraction or addition.

Here we go: Replace Row 2 by $\frac{3}{4}$ times Row 2

$$\begin{bmatrix} 3 & -5 & 8 \\ 3 & \frac{21}{4} & \frac{33}{4} \end{bmatrix}$$

Replace Row 2 by Row 2 minus Row 1

$$\begin{bmatrix} 3 & -5 & 8 \\ 0 & \frac{41}{4} & \frac{1}{4} \end{bmatrix}$$

Replace Row 2 by $\frac{4}{41}$ times Row 2

$$\begin{bmatrix} 3 & -5 & 8 \\ 0 & 1 & \frac{1}{41} \end{bmatrix}$$

Replace Row 1 by Row 1 plus 5 times Row 2

$$\begin{bmatrix} 3 & 0 & \frac{333}{41} \\ 0 & 1 & \frac{1}{41} \end{bmatrix}$$

Replace Row 1 by $\frac{1}{3}$ times Row 1

$$\begin{bmatrix} 1 & 0 & \frac{111}{4} \\ 0 & 1 & \frac{1}{41} \end{bmatrix}$$

We are done!

Restore this augmented matrix to the 2 by 1 variable matrix form as in Equation (2.8)

$$\begin{pmatrix} 1 & 0 \\ 0 & 1 \end{pmatrix} \times \begin{pmatrix} x \\ y \end{pmatrix} = \begin{pmatrix} \frac{111}{41} \\ \frac{1}{41} \end{pmatrix} \tag{2.9}$$

Now, restore the matrix equation to the original form in the system (2.6), (2.7)

$$\begin{aligned} 1x + 0y &= \frac{111}{41} \\ 0x + 1y &= \frac{1}{41} \end{aligned}$$

In other words, the solution to the system is $x = 111/41$ and $y = 1/41$
Check:

$$\begin{aligned} 3x - 5y &= 333/41 - 5/41 = 328/41 = 8 \\ 4x + 7y &= 444/41 + 7/41 = 451/41 = 11 \end{aligned}$$

In Equation (2.9) the matrix $\begin{pmatrix} 1 & 0 \\ 0 & 1 \end{pmatrix}$ is called the *identity matrix* for all 2 by 2 matrices. We always deliberately apply the rules for row reduction toward getting the identity matrix. We make this our target because it gives you two equations one in x only and the other in y only.

When row-reduction fails

Row reduction does not always work; it fails when there is no solution or there is no unique solution. What this means, geometrically, is that it fails if there is no intersection point for the two lines or if both equations are of the same line.

Consider the two by two system:

$$\begin{aligned} ax + by &= r \\ cx + dy &= s \end{aligned}$$

which we can write in matrix form

$$\begin{pmatrix} a & b \\ c & d \end{pmatrix} \begin{pmatrix} x \\ y \end{pmatrix} = \begin{pmatrix} r \\ s \end{pmatrix}$$

Now, we need to know whether the two lines are parallel or not. They might even be two equations for the same line, which would mean we cannot find a unique solution.

Here is the test we want to apply: If the numbers a and b are in the same ratio as c and d,

$$\frac{a}{b} = \frac{c}{d} \tag{2.10}$$

then the two lines have the same slope; so, there is no answer or there is no unique answer.

Note: Equation (2.10) can be written as

$$ad - cb = 0$$

The expression $ad - cb$ has a name; it is called the *determinant of the matrix*

$$M = \begin{pmatrix} a & b \\ c & d \end{pmatrix}$$

and is denoted by $\det(M)$; so,

$$\det(M) = ad - cb$$

A visual way to compute the determinant is to draw an arrow on the *main diagonal*, from the upper left corner to the lower right and multiply these two numbers. Then draw an arrow on the other diagonal (sometimes called the "off" diagonal) from the lower left to the upper right and multiply these, then subtract the second product from the first. See Figure 2.3.

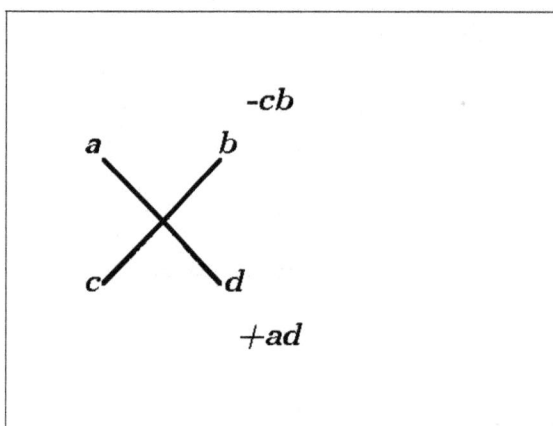

FIGURE 2.3 $\det \begin{pmatrix} a & b \\ c & d \end{pmatrix} = ad - cb$

Problem: Find the determinants of the following 2 by 2 matrix

$$M = \begin{pmatrix} 5 & -3 \\ 1 & 2 \end{pmatrix}$$

Solution:

$$\begin{aligned} \det(M) &= (5 \times 2) - (1 \times (-3)) \\ &= 13 \end{aligned}$$

Problem: Verify that $\det(A) = 0$, for the following matrix A.

$$A = \begin{pmatrix} 3 & 2 \\ 6 & 4 \end{pmatrix}$$

Problem: Verify that

$$\det \begin{pmatrix} -1 & 2 \\ 7 & -3 \end{pmatrix} = -11$$

Problem: Without trying to solve it, does the following system have a solution?

$$\begin{aligned} 5x - 2.5y &= 11 \\ 4x - 2y &= 9 \end{aligned}$$

Solution: No, because

$$\det \begin{pmatrix} 5 & -2.5 \\ 4 & -2 \end{pmatrix} = 0$$

Three equations with three unknowns.

Things get a little tougher for a 3 by 3 systems, but row reduction works in just the same way as it did for a 2 by 2 system.

Example: Given that the following system has a solution, for x, y and z find it by row reduction.

$$1x + 1y + 1z = 6$$
$$0x + 3y + 2z = 9$$
$$4x + 0y - z = 5$$

What we want to do is write out the augmented matrix, then start adding and subtracting multiples of rows as needed to move toward changing all of the coefficients of x, y, and z to look like the following identity matrix for a 3 by 3.

$$\begin{pmatrix} 1 & 0 & 0 \\ 0 & 1 & 0 \\ 0 & 0 & 1 \end{pmatrix}$$

Let us begin. Start with the augmented matrix

$$\begin{bmatrix} 1 & 1 & 1 & 6 \\ 0 & 3 & 2 & 9 \\ 4 & 0 & -1 & 5 \end{bmatrix}$$

Replace Row3 by Row 3 reduced by 4 times Row 1

$$\begin{bmatrix} 1 & 1 & 1 & 6 \\ 0 & 3 & 2 & 9 \\ 0 & -4 & -5 & -19 \end{bmatrix}$$

Replace Row 1 by Row 1 minus $\frac{1}{3}$ of Row 2.

$$\begin{bmatrix} 1 & 0 & \frac{1}{3} & 3 \\ 0 & 3 & 2 & 9 \\ 0 & -4 & -5 & -19 \end{bmatrix}$$

Replace Row 3 by Row 3 plus $\frac{4}{3}$ of Row 2

$$\begin{bmatrix} 1 & 0 & \frac{1}{3} & 3 \\ 0 & 3 & 2 & 9 \\ 0 & 0 & -\frac{7}{3} & -7 \end{bmatrix}$$

Replace Row 3 by $-\frac{3}{7}$ times Row 3

$$\begin{bmatrix} 1 & 0 & \frac{1}{3} & 3 \\ 0 & 3 & 2 & 9 \\ 0 & 0 & 1 & 3 \end{bmatrix}$$

Replace Row 1 by Row 1 minus $\frac{1}{3}$ of Row 3

$$\begin{bmatrix} 1 & 0 & 0 & 2 \\ 0 & 3 & 2 & 9 \\ 0 & 0 & 1 & 3 \end{bmatrix}$$

Replace Row 2 by $\frac{1}{3}$ of Row 2

$$\begin{bmatrix} 1 & 0 & 0 & 2 \\ 0 & 1 & \frac{2}{3} & 3 \\ 0 & 0 & 1 & 3 \end{bmatrix}$$

Replace Row 2 by Row 2 minus $\frac{2}{3}$ time Row 3

$$\begin{bmatrix} 1 & 0 & 0 & 2 \\ 0 & 1 & 0 & 1 \\ 0 & 0 & 1 & 3 \end{bmatrix}$$

Done!

$$\begin{pmatrix} 1 & 0 & 0 \\ 0 & 1 & 0 \\ 0 & 0 & 1 \end{pmatrix} \times \begin{pmatrix} x \\ y \\ z \end{pmatrix} = \begin{pmatrix} 2 \\ 1 \\ 3 \end{pmatrix}$$

The solution is: $\{x = 2, y = 1, z = 3\}$.

Systems with more variables than equations

We have been solving "square systems," like the 2 by 2 or the 3 by 3 systems; those that have the same number of variables as equations In a real problem we might encounter systems that have more variables than equations[3].

$$\begin{aligned} 1x + 2y + 3z &= 10 & (2.11) \\ 4x + 5y + 6z &= 15 & (2.12) \end{aligned}$$

What good would it be for us to set up a problem like this? There are logistic, or inventory problems in military and business applications in which the extra variable stands for an unrealized asset or a potential customer. It may be that there had been a third equation and this was, originally, a three by three system, but some circumstance required us to drop one of them as being infeasible for some reason, say a new unreliable supplier. Now we may consider the option of setting any one of the current variables, x, y, or z equal to zero, thereby dropping an item from our inventory. This would leave us with a square system that can be solved for the other two variables.

There *is* a cost to doing this because it means that we are ignoring some market, or failing to access some supplier. This cost is called an "opportunity cost." We may wish to minimize that quantity in our decision making by devising a way to test the resulting system in order to determine whether or not one of the other variables should have been the deleted one.

For the simple 2 by 3 system, Equations (2.11) and (2.12), we can experiment with deleting one variable at a time, and solving a square, 2 by 2 "subsystem."

There are exactly 3 ways to pick one of three things. We can set $z = 0$ and solve the two equations for x and y. Or, we can set $y = 0$ and solve the resulting system for x and z, or we can set $x = 0$ and solve the resulting two by two system for y and z.

---

[3]Or , it could be the other way around. We might have a problem with more equations than variables.

If we set $z = 0$, then

$$
\begin{aligned}
1x + 2y &= 10 \\
4x + 5y &= 15
\end{aligned}
$$

If we set $y = 0$, then

$$
\begin{aligned}
1x + 3z &= 10 \\
4x + 6z &= 15
\end{aligned}
$$

If we set $x = 0$, then

$$
\begin{aligned}
2y + 3z &= 10 \\
5y + 6z &= 15
\end{aligned}
$$

In each case we get a solution for the two variables and add zero as the value for the dropped variable, to complete the solution. This is not too hard to do for a 3 by 2 system, but what about a 8 by 5 system? There are 56 ways to take 8 things 5 at a time, so there are that many five by five square subsystems, a pretty formidable task.

Some modest, you might even say small, real-life applications may have 20 variables and 14 equations. Even so, such a system would have us solving more than $38,000$ square subsystems. This would be a difficult to do, even with a computer. Furthermore, it would be a very inefficient way to handle such a problem. Fortunately, a new method called the *simplex* method was devised in the 1940's that greatly reduces the number of square systems you need to check out. For every square system that has been solved, the simplex method tells you what variables you need to replace and which variables you should use to replace them in order to get a better answer. In other words, it keeps tack of the opportunity cost for any current square system and calculates what, if any, replacement variable would improve the solution. We will discuss this method in sufficient detail in our Chapter on Operations Research, Chapter 10. Before we can take advantage of this wonderful technology, we need to learn more about matrices.

Under certain conditions, matrices can be added, multiplied, subtracted, divided, and raised to powers. When matrices are used in this manner, they form a practical applied subject called *matrix algebra*, which has applications in economics, physics, and other sciences as well as in business and mathematics. In order to carry out these operations, we need to learn the *rules*, actually *definitions*, for addition, subtraction, multiplication, and division. Here, we stick to matrix algebra for small systems.

Introduction to matrix algebra

Converting systems of equations into matrix equations was invented in 300 BCE, then again in the 1500's CE, and then again in 1800's CE. There can be no doubt, that it must have been considered to be an important tool. But, today, its strength comes from our ability to hand it over to a computer.

It is equivalent to the invention of the washing machine. In 1900, when he or she got weary of the tedium of beating the dirt out of clothes, someone invented the first electric washing machine.

This is not an easy subject, especially learning how to multiply two matrices and how to find the multiplicative inverse of a larger than 2 by 2 matrix. It takes some effort on the part of the reader to learn, in general, how to multiply matrices, but it is a tool worth having.

Let us begin at the beginning. Since this chapter is about the equal sign, we will ask, when are two matrices equal?

Equal matrices

The first rule we need is the *rule of equals*. When is the matrix $\mathbf{A}$ equal to the matrix $\mathbf{B}$?

Matrix $\mathbf{A}$ and $\mathbf{B}$ are equal if

1. They are both m by n matrices, and

2. Every element in the *ith* row and *jth* column of $\mathbf{A}$ is equal to the corresponding element in the *ith* row and *jth* column of $\mathbf{B}$

Example: Let $\mathbf{A}$ and $\mathbf{B}$, be the following two 4 by 3 matrices:

$$\mathbf{A} = \begin{pmatrix} 1 & 7 & x \\ -2 & 0.5 & 8 \\ y & w & 0 \\ 0 & 1 & 9 \end{pmatrix}$$

and

$$\mathbf{B} = \begin{pmatrix} 1 & 7 & 22 \\ -2 & p & q \\ 4 & 18 & 0 \\ 0 & 1 & t \end{pmatrix}$$

then $\mathbf{A}=\mathbf{B}$

$$\begin{pmatrix} 1 & 7 & x \\ -2 & 0.5 & 8 \\ y & w & 0 \\ 0 & 1 & 9 \end{pmatrix} = \begin{pmatrix} 1 & 7 & 22 \\ -2 & p & q \\ 4 & 18 & 0 \\ 0 & 1 & t \end{pmatrix}$$

If and only if

$$x = 22, \; p = 0.5, \; q = 8, \; y = 4, \; w = 18, \text{ and } t = 9$$

Addition and subtraction of matrices

Now we move on to the addition of matrices. To study matrix algebra we need to learn the definitions and rules for adding and multiplying matrices.

Adding matrices is pretty simple; just add the elements in the corresponding row and column positions.

Example:

Add the following 2 by 4 matrices:

$$\begin{pmatrix} 1 & 1 & 5 & 0.143 \\ -0.7 & \pi & 29 & 101 \end{pmatrix} + \begin{pmatrix} -8 & 0 & 13 & 3 \\ -1.3 & 14 & 2 & -4 \end{pmatrix} = \begin{pmatrix} -7 & 1 & 18 & 3.143 \\ -2 & 14+\pi & 31 & 97 \end{pmatrix}$$

For larger matrices, you do the same thing, adding numbers in one matrix to the corresponding numbers in the other matrix, then placing that sum in the same corresponding position in the new matrix.

Example: The entries in following matrices tabulate the deaths in U.S. Armed forces in three wars. The numbers in the *Other* column represents deaths from diseases. Data are from the 1971 *Information Please Almanac*[4]

Spanish-American	Battle	Other
Army	369	2,061
Navy	10	0
Marines	6	0

World War I	Battle	Other
Army	50,510	55,868
Navy	431	6,858
Marines	2,461	390

World War II	Battle	Other
Army	234,874	83,400
Navy	36,950	25,664
Marines	19,733	4,778

We convert these tables to matrices, $\mathbf{W}_S$, $\mathbf{W}_I$, and $\mathbf{W}_{II}$ respectively. For example, the Spanish-American table is the matrix:

$$\mathbf{W}_S = \begin{pmatrix} 369 & 2,061 \\ 10 & 0 \\ 6 & 0 \end{pmatrix}$$

Problems:
(a) Find the sum of the three matrices $\mathbf{W}_s$, $\mathbf{W}_I$, and $\mathbf{W}_{II}$.
(b) How many Army personnel died from diseases in the three wars?
Solutions:
(a)

$$\mathbf{W}_S + \mathbf{W}_I + \mathbf{W}_{II} = \begin{pmatrix} 285,753 & 141,329 \\ 37,391 & 32,522 \\ 22,200 & 5,168 \end{pmatrix}$$

(b) 141,329

The additive identity matrix (the zero matrix).

There is an n by n matrix, $\mathbf{0}$, such that if $\mathbf{M}$ is any n by n matrix, then $\mathbf{M} + \mathbf{0} = \mathbf{M}$. All of the elements of $\mathbf{0}$ are 0's. For example, if $n = 2$, then the additive identity matrix is,

$$\mathbf{0} = \begin{pmatrix} 0 & 0 \\ 0 & 0 \end{pmatrix} \tag{2.13}$$

Thus for any 2 by 2 matrix, $\mathbf{M} + \mathbf{0} = \mathbf{M}$

$$\begin{pmatrix} a & b \\ c & d \end{pmatrix} + \begin{pmatrix} 0 & 0 \\ 0 & 0 \end{pmatrix} = \begin{pmatrix} a & b \\ c & d \end{pmatrix}$$

[4]The U. S. Bureau of Census *Statistical Abstract of the United States* 1969 (Washington, DC: GPO)

just like $x + 0 = x$, in the number system.

In general, the m by n matrix with nothing but zero elements is the additive identity element for m by n matrices. For example if $\mathbf{A}$ is a 3 by 4 matrix, then $\mathbf{0}$ is the 3 by 4 matrix with all zeros.

The additive inverse matrix (the negative matrix)

The additive inverse of a matrix $\mathbf{M}$ is the matrix $-\mathbf{M}$, in which every element of $-\mathbf{M}$ is the negative of the corresponding element in $\mathbf{M}$.

$$\text{If } \mathbf{M} = \begin{pmatrix} a & b \\ c & d \end{pmatrix}, \text{ then } -\mathbf{M} = \begin{pmatrix} -a & -b \\ -c & -d \end{pmatrix} \text{ then}$$

$$\mathbf{M} + (-\mathbf{M}) = \mathbf{0}$$

To subtract a matrix $\mathbf{Y}$ from a matrix $\mathbf{X}$, add the negative of $\mathbf{Y}$ to $\mathbf{X}$.

$$\mathbf{X} - \mathbf{Y} = \mathbf{X} + (-\mathbf{Y})$$

Now we know how to add and subtract matrices. In the next section, we will learn how to multiply matrices. As we shall see multiplication is much more difficult, but it gives us powerful results in applied problems, such as economics, and game theory.

Multiplication of matrices, $\mathbf{A} \times \mathbf{B}$

What makes multiplication of two matrices difficult to learn? It is *not* just a matter of multiplying corresponding elements of the two matrices. Why do we want take the trouble to learn how to do it? It gives us a well-organized method for keeping track of data from several sources. It also gives us an overall picture of what is going on in a complicated situation.

Example: Let us look back at a previously discussed system of equations:

$$\begin{aligned} 3x - 5y &= 8 \\ 4x + 7y &= 11 \end{aligned}$$

We re-wrote this as Equation (2.8).

$$\begin{pmatrix} 3 & -5 \\ 4 & 7 \end{pmatrix} \begin{pmatrix} x \\ y \end{pmatrix} = \begin{pmatrix} 8 \\ 11 \end{pmatrix}$$

This provides a hint as to how matrix multiplication is defined. Think of the matrix

$$\begin{pmatrix} 3 & -5 \\ 4 & 7 \end{pmatrix}$$

as consisting of two rows, $(3, -5)$ and $(4, 7)$, called "row matrices", and the column $\begin{pmatrix} x \\ y \end{pmatrix}$ as a column matrix. We will show you how to get the product

$$\begin{pmatrix} 3 & -5 \\ 4 & 7 \end{pmatrix} \begin{pmatrix} x \\ y \end{pmatrix} = \begin{pmatrix} 3x - 5y \\ 4x + 7y \end{pmatrix}$$

Multiply each row matrix by the column matrix as follows:

In, Row 1, we multiply the first row-number 3, times the first column-number x, getting $3x$, then multiply the second row-number, -5, times the second column-number y, getting $-5y$, then add these two products getting $3x + (-5y)$.

Put this answer, $3x + (-5y)$, actually $3x - 5y$, into the first row of the final product.

$$\left(\boxed{3}, \boxed{-5}\right) \times \left(\boxed{\begin{matrix} x \\ y \end{matrix}}\right) \;=\; \left(\boxed{3} \times \boxed{x} + \boxed{-5} \times \boxed{y}\right)$$

$$= \;(3x - 5y)$$

$$\begin{pmatrix} 3 & -5 \\ 4 & 7 \end{pmatrix} \begin{pmatrix} x \\ y \end{pmatrix} = \begin{pmatrix} 3x - 5y \\ \dots \end{pmatrix}$$

Next repeat this for the second row matrix, namely $(4, 7)$ times $\begin{pmatrix} x \\ y \end{pmatrix}$, getting $4x + 7y$.

$$\left(\boxed{4}, \boxed{7}\right) \times \left(\boxed{\begin{matrix} x \\ y \end{matrix}}\right) \;=\; \left(\boxed{4} \times \boxed{x} + \boxed{7} \times \boxed{y}\right)$$

$$= \; 4x + 7y$$

This answer goes in the second row to complete the multiplication

$$\begin{pmatrix} 3 & -5 \\ 4 & 7 \end{pmatrix} \begin{pmatrix} x \\ y \end{pmatrix} = \begin{pmatrix} 3x - 5y \\ 4x + 7y \end{pmatrix}$$

Set this equal to the column matrix $\begin{pmatrix} 8 \\ 11 \end{pmatrix}$ and you get

$$\begin{pmatrix} 3 & -5 \\ 4 & 7 \end{pmatrix} \begin{pmatrix} x \\ y \end{pmatrix} = \begin{pmatrix} 8 \\ 11 \end{pmatrix}$$

In the product of any 2 by 2 matrix

$$\begin{pmatrix} a & b \\ c & d \end{pmatrix}$$

by any 2 by 1 column matrix,

$$\begin{pmatrix} x \\ y \end{pmatrix}$$

the multiplication looks like this:

$$\begin{pmatrix} a & b \\ c & d \end{pmatrix} \times \begin{pmatrix} x \\ y \end{pmatrix} = \begin{pmatrix} ax + by \\ cx + dy \end{pmatrix}$$

This gives away the whole secret of matrix multiplication. If you want to multiply a 2 by 2 matrix times any other matrix that has two rows and not, necessarily, just one column, but any number of columns, you just multiply the rows in first matrix times each column of the second matrix and put all the resulting answers into their proper row and column place.

Example: Multiply the following two by two matrix **M** times the two by five matrix **N**. (Hint: you just repeat the row column multiplication process five times in each of the two rows).

$$\begin{pmatrix} a & b \\ c & d \end{pmatrix} \times \begin{pmatrix} x & u & r & p & j \\ y & v & s & q & k \end{pmatrix}$$

Answer

$$\begin{pmatrix} ax+by & au+bv & ar+bs & ap+bq & aj+bk \\ cx+dy & cu+dv & cr+ds & cp+dq & cj+dk \end{pmatrix}$$

This same row by row multiplication of column by column works for any two matrices in which the number of columns of the first matrix match up with the number of rows for the second matrix.

Example: Any 3 by 2 matrix times a 2 by 1 matrix is similarly computed (multiplying each row of the first matrix by each column of the second); it is:

$$\begin{pmatrix} a & b \\ c & d \\ e & g \end{pmatrix} \times \begin{pmatrix} x \\ y \end{pmatrix} = \begin{pmatrix} ax+by \\ cx+dy \\ ex+gy \end{pmatrix}$$

Farmers' market example:

Here is an interesting problems that uses matrix multiplication

Problem:

At a Farmer's Market in Eastern Oregon a vendor finds that 45% of all the melons he sells are from Hermiston, 36% are from Imnaha and 19% are from Cove. Also, 40% of all the tomatoes he sells are from Hermiston and 30% are from Imnaha and 30% from Cove. One week he sold \$200 worth of melons and \$440 worth of tomatoes. Find out how much of the revenue was for produce from Hermiston, how much from Imnaha and how much from Cove.

Solution:

We can set this up as a matrix multiplication problem.

Let **A** be the matrix

$$\mathbf{A} = \begin{array}{c} \\ \text{Hermiston} \\ \text{Imnaha} \\ \text{Cove} \end{array} \begin{array}{cc} \text{Melons} & \text{Tomatoes} \\ \begin{pmatrix} 0.45 & 0.40 \\ 0.36 & 0.30 \\ 0.19 & 0.30 \end{pmatrix} \end{array}$$

Look at matrix **A**. Row 1 tells you that 45% of the melons and 40% of the tomatoes he sold were from Hermiston, and so forth, as stated in the problem. Now we will write out a matrix, **B**, that gives us the revenue data for melons and tomatoes as follows:

$$\mathbf{B} = \begin{array}{c} \\ \text{Melons} \\ \text{Tomatoes} \end{array} \begin{array}{c} \text{Revenue} \\ \begin{pmatrix} \$200 \\ \$440 \end{pmatrix} \end{array}$$

You will be delighted to see that the answer to the question raised in this problem comes from just multiplying matrix **A** times matrix **B**.

$$\mathbf{A} \times \mathbf{B} = \begin{pmatrix} 0.45 & 0.40 \\ 0.36 & 0.30 \\ 0.19 & 0.30 \end{pmatrix} \times \begin{pmatrix} \$200 \\ \$440 \end{pmatrix}$$

$$= \begin{pmatrix} 0.45 \times \$200 + 0.40 \times \$440 \\ 0.36 \times \$200 + 0.30 \times \$440 \\ 0.19 \times \$200 + 0.30 \times \$440 \end{pmatrix}$$

$$= \begin{pmatrix} \$266 \\ \$204 \\ \$170 \end{pmatrix} \begin{matrix} \text{Hermiston} \\ \text{Imnaha} \\ \text{Cove} \end{matrix}$$

We can get a more nuanced analysis of this problem if we were to make the second matrix, **B**, show us another detail, namely, the profit. We can do this by writing **B** as a 2 by 2 matrix instead of a 2 by 1 We just learned how to do such multiplications, using the row times column method for each row and column and putting the answers into their correct row-column places.

Thus, taking the Farmers' market example a step further, assume the same matrix **A**, but now suppose **B** is a matrix with two columns, one for revenue from the produce and the other profit, as follows.

$$\mathbf{B} = \begin{matrix} \text{Melons} \\ \text{Tomatoes} \end{matrix} \begin{matrix} \text{Revenue} & \text{Profit} \\ \begin{pmatrix} \$200 & \$105 \\ \$440 & \$310 \end{pmatrix} \end{matrix}$$

Now when we multiply **A** × **B**, we will be multiplying the three rows of matrix **A** times the 2 by 2 matrix **B**. We will get two columns in the answer, one for revenue (the same one we got before) *and* one for profit. Just concentrating on the second column, here is the multiplication.

$$\mathbf{A} \times \mathbf{B} = \begin{pmatrix} 0.45 & 0.40 \\ 0.36 & 0.30 \\ 0.19 & 0.30 \end{pmatrix} \times \begin{pmatrix} \$200 & \$105 \\ \$440 & \$310 \end{pmatrix}$$

$$= \begin{matrix} \text{Revenue} & \text{Profit} \\ \begin{pmatrix} \$266 & \$171.25 \\ \$204 & \$130.80 \\ \$170 & \$112.95 \end{pmatrix} \end{matrix} \begin{matrix} \text{Hermiston} \\ \text{Imnaha} \\ \text{Cove} \end{matrix}$$

Raising a matrix to a power

If we wanted to find the square of a matrix, we can just multiply it times itself. For the cube, we just multiply it by itself 3 times and so forth.

Example: If **R** is the following matrix, find $\mathbf{R}^2$.

$$\mathbf{R} = \begin{pmatrix} 3 & 1 \\ 0.5 & 2 \end{pmatrix}$$

$$\begin{pmatrix} 3 & 1 \\ 0.5 & 2 \end{pmatrix}^2 = \begin{pmatrix} 3 & 1 \\ 0.5 & 2 \end{pmatrix} \times \begin{pmatrix} 3 & 1 \\ 0.5 & 2 \end{pmatrix}$$
$$= \begin{pmatrix} 9.5 & 5 \\ 2.5 & 4.5 \end{pmatrix}$$

When we multiply matrices that are not square, the number of columns of the first matrix must match the number of rows of the second matrix. We just follow the same pattern of multiplying rows by columns.

Example: Find

$$\begin{pmatrix} 1 & 0 & -2 \\ 3 & 1 & 1 \\ 7 & 0 & 8 \\ -5 & 6 & 9 \end{pmatrix} \times \begin{pmatrix} -11 & 4 \\ 2 & 1 \\ 0 & -10 \end{pmatrix} = \begin{pmatrix} -11 & 24 \\ -31 & 3 \\ -77 & -52 \\ 67 & -104 \end{pmatrix}$$

Even if the row column match-up is correct, for both the resulting matrices, the products $\mathbf{M} \times \mathbf{N}$ and $\mathbf{N} \times \mathbf{M}$ might not be the same.

Problem: If $\mathbf{X}$ and $\mathbf{Y}$ are the two matrices given below,

$$\mathbf{X} = \begin{pmatrix} 1 & 2 & 1 \\ 2 & 0 & 3 \end{pmatrix}, \mathbf{Y} = \begin{pmatrix} 2 & -1 \\ 1 & 4 \\ 7 & 0 \end{pmatrix}$$

Show that $\mathbf{X} \times \mathbf{Y} \neq \mathbf{Y} \times \mathbf{X}$.

Solution:

First let us compute these two products:

$$\mathbf{X} \times \mathbf{Y} = \begin{pmatrix} 1 & 2 & 1 \\ 2 & 0 & 3 \end{pmatrix} \times \begin{pmatrix} 2 & -1 \\ 1 & 4 \\ 7 & 0 \end{pmatrix}$$
$$= \begin{pmatrix} 11 & 7 \\ 25 & -2 \end{pmatrix}$$

but

$$\mathbf{Y} \times \mathbf{X} = \begin{pmatrix} 2 & -1 \\ 1 & 4 \\ 7 & 0 \end{pmatrix} \times \begin{pmatrix} 1 & 2 & 1 \\ 2 & 0 & 3 \end{pmatrix}$$
$$= \begin{pmatrix} 0 & 4 & -1 \\ 9 & 2 & 13 \\ 7 & 14 & 7 \end{pmatrix}$$

$$\begin{pmatrix} 11 & 7 \\ 25 & -2 \end{pmatrix} \neq \begin{pmatrix} 0 & 4 & -1 \\ 9 & 2 & 13 \\ 7 & 14 & 7 \end{pmatrix}$$

They do not satisfy the definition of what we mean by saying two matrices are equal, therefore, $\mathbf{X} \times \mathbf{Y} \neq \mathbf{Y} \times \mathbf{X}$.

Example: Lamb birthing

The data in this example are based on a study of lambs born to ewes on two consecutive years: 1952 and 1953, in Tallis, G. M. "The Maximum Likelihood Estimation of Correlation from Contingency Tables." *Biometrics* Vol. 18(1962).

↗			1953	
		None	Single	Twins
	None	58	26	8
1952	Single	52	58	12
	Twins	1	3	9

The birth history of lambs to ewes over a two year period is shown in the table above. See the arrow in the upper left corner indicating the switch from births in 1952 to births in 1953.

Here is how you can read this table: In the first row we have those ewes who were barren in 1952. Of these, 58 were also barren in 1953, but 26 others had singles in 1953, and 8 others, happily, had twins. Read the next two rows in this same fashion.

Let us write the numbers in this table as the following matrix:

$$\mathbf{A} = \begin{pmatrix} 58 & 26 & 8 \\ 52 & 58 & 12 \\ 1 & 3 & 9 \end{pmatrix}$$

We can represent the no births (0), single birth (1) and twin births (2) by the column matrix.

$$\mathbf{v} = \begin{pmatrix} 0 \\ 1 \\ 2 \end{pmatrix}$$

Problem:

Find the product **Av** and interpret the meaning of the entries in that product.

Answer:

$$\mathbf{Av} = \begin{pmatrix} 58 & 26 & 8 \\ 52 & 58 & 12 \\ 1 & 3 & 9 \end{pmatrix} \begin{pmatrix} 0 \\ 1 \\ 2 \end{pmatrix} = \begin{pmatrix} 42 \\ 82 \\ 21 \end{pmatrix}$$

The number of lambs born in 1953 to the ewes that had been barren in 1952 was 42 because: $58 \times 0 + 26 \times 1 + 8 \times 2 = 42$. The number of lambs born in 1953 to the ewes having single births in 1952 was 82, and the number of lambs born in 1953 to the ewes having twins in 1952 was 21.

Problem: How many lambs were born in 1953?

Answer: 145; add up the entries in **Av**.

Problem: Find the number of lambs born in 1952.

Answer: You could solve this problem by *transposing* the matrix **A** (making its rows be columns and its columns be rows), creating a new matrix **B**. Now the rows are 1953 and the columns are 1952.

$$\mathbf{Bv} = \begin{pmatrix} 58 & 52 & 1 \\ 26 & 58 & 3 \\ 8 & 12 & 9 \end{pmatrix} \begin{pmatrix} 0 \\ 1 \\ 2 \end{pmatrix} = \begin{pmatrix} 54 \\ 64 \\ 30 \end{pmatrix}$$

There were $54 + 64 + 30 = 148$ lambs born in 1952 because 54 can come from the 1953 barren ewes, 64 came from the 1953 single birthers and 30 came from the 1953 twin bearers.

So, now we know how to add, subtract, and multiply matrices. How about *division*? Actually we don't use that word, we say finding the *multiplicative inverse* of a matrix. But first we need to tell you what the multiplicative identity matrix is.

Multiplicative identity matrix

For any positive integer m, if $\mathbf{I}_m$ is a square m by m matrix, in which all of the elements on the main diagonal are 1's and all of the other elements are zeros, then $\mathbf{I}$ is the multiplicative identity matrix for all m by m matrices. Thus, for example:

$$\mathbf{I}_4 = \begin{pmatrix} 1 & 0 & 0 & 0 \\ 0 & 1 & 0 & 0 \\ 0 & 0 & 1 & 0 \\ 0 & 0 & 0 & 1 \end{pmatrix}$$

is the multiplicative identity matrix for all 4 by 4 matrices. Frequently, when we know the dimensions of the square matrices we are working with, the subscript m, is left off the notation for an identity matrix. So, we could write $\mathbf{I}$, instead of $\mathbf{I}_4$.

Problem:

Show that the matrix $\mathbf{I} = \begin{pmatrix} 1 & 0 \\ 0 & 1 \end{pmatrix}$ is the multiplicative identity for all 2 by 2 matrices.

Solution:

Let $\mathbf{M} = \begin{pmatrix} x & y \\ u & v \end{pmatrix}$ be any two-by two matrix, and multiply it by I.

$$\begin{pmatrix} 1 & 0 \\ 0 & 1 \end{pmatrix} \cdot \begin{pmatrix} x & y \\ u & v \end{pmatrix} = \begin{pmatrix} 1x + 0u & 1y + 0v \\ 0x + 1u & 0y + 1v \end{pmatrix} = \begin{pmatrix} x & y \\ u & v \end{pmatrix}$$

So,

$$\mathbf{I} \times \mathbf{M} = \mathbf{M}$$

Multiplicative inverses.

If we are given a matrix $\mathbf{A}$, can we find a matrix $\mathbf{B}$ such that $\mathbf{A} \times \mathbf{B} = \mathbf{I}$? That is, can we find the multiplicative inverse of a matrix? (Which is the same thing as asking can we divide by a matrix?) We can, in a limited way.

In ordinary arithmetic, we know that we cannot divide by zero. There is no number x that is $\frac{1}{0}$. Here is why. If there were to be a number, $x = \frac{1}{0}$, then we could multiply x by 0 and since any number times zero is zero, then $x \times 0 = 0$. But on the other hand, any number, x times its reciprocal, $\frac{1}{x}$ is 1, so $x \times 0 = \frac{1}{0} \times 0 = 1$. Thus $x \times 0$ is both 1 and 0, which is impossible by the axioms of the real number system—which says, among other things, $1 \neq 0$. Actually, if we allow 1 to be 0, the whole number system collapses to a single entity, namely zero and all mathematics will disappear from the Earth and the rest of the universe.

A similar thing happens when we try to find the reciprocal of a matrix. Not only can we not divide by the zero matrix, we *cannot divide by any matrix that has a zero for its determinant.*

Problem: If

$$\mathbf{M} = \begin{pmatrix} \frac{1}{3} & 7 \\ 1 & 21 \end{pmatrix}$$

show that there is no multiplicative inverse of $\mathbf{M}$.

Solution:

This looks easy, just look at the determinant of $\mathbf{M}$. It is $\frac{1}{3} \cdot 21 - 1 \cdot 7 = 7 - 7 = 0$, so M has no inverse. But you don't have to take my word for it. Just try to find an inverse of $\mathbf{M}$.

Let $\mathbf{X} = \begin{pmatrix} a & b \\ c & d \end{pmatrix}$ be the multiplicative inverse of the matrix $\mathbf{M}$ given above, then $\mathbf{X} \cdot \mathbf{M} = \mathbf{I}$, therefore

$$\begin{pmatrix} a & b \\ c & d \end{pmatrix} \cdot \begin{pmatrix} \frac{1}{3} & 7 \\ 1 & 21 \end{pmatrix} = \begin{pmatrix} 1 & 0 \\ 0 & 1 \end{pmatrix}$$
$$\begin{pmatrix} \frac{1}{3}a + b & 7a + 21b \\ \frac{1}{3}c + d & 7c + 21d \end{pmatrix} = \begin{pmatrix} 1 & 0 \\ 0 & 1 \end{pmatrix}$$

Now, since matrix equality tells us that these matrices are equal if and only if they are equal element by element, that gives us four equations

$$\frac{1}{3}a + b = 1 \tag{2.14}$$
$$7a + 21b = 0 \tag{2.15}$$

and

$$\frac{1}{3}c + d = 0 \tag{2.16}$$
$$7c + 21d = 1 \tag{2.17}$$

But we do know that neither one of these two system can have a solution because each one has a coefficient matrix whose determinant is zero. This is a problem because it means they are parallel lines. There are *no* unique numbers a and b that satisfy Equations (2.14) and (2.15). Also there cannot be a unique solution c and d to Equations (2.16) and (2.17). Therefore, the purported matrix, $\begin{pmatrix} a & b \\ c & d \end{pmatrix}$ does not exist.

Problem: If

$$A = \begin{pmatrix} 2 & 1 \\ 3 & 5 \end{pmatrix}$$

Show that $\det(A)$ is not zero, and find the multiplicative inverse of A.

Solution: First, let us find the determinant. It is

$$\det(A) = 2 \cdot 5 - 3 \cdot 1 = 10 - 3 = 7$$

All right, the determinant is not zero. Let us find a matrix B such that $A \cdot B = I$

$$\begin{pmatrix} a & b \\ c & d \end{pmatrix} \cdot \begin{pmatrix} 2 & 1 \\ 3 & 5 \end{pmatrix} = \begin{pmatrix} 1 & 0 \\ 0 & 1 \end{pmatrix}$$

$$\begin{pmatrix} 2a+3b & a+5b \\ 2c+3d & c+5d \end{pmatrix} = \begin{pmatrix} 1 & 0 \\ 0 & 1 \end{pmatrix}$$

So, now we must solve a pair of 2 by 2 systems:

$$\begin{aligned} 2a + 3b &= 1 \\ a + 5b &= 0 \end{aligned}$$

$$\text{and}$$

$$\begin{aligned} 2c + 3d &= 0 \\ c + 5d &= 1 \end{aligned}$$

Solving the first system, by any method you want, you will get $a = \frac{5}{7}$, $b = \frac{-1}{7}$. And for the second system, you get $c = \frac{-3}{7}$, $d = \frac{2}{7}$. Putting these numbers in for a, b, c, and d, we get the inverse:

$$A^{-1} = \begin{pmatrix} \frac{5}{7} & -\frac{1}{7} \\ -\frac{3}{7} & \frac{2}{7} \end{pmatrix} \tag{2.18}$$

Proof:

$$\begin{pmatrix} \frac{5}{7} & -\frac{1}{7} \\ -\frac{3}{7} & \frac{2}{7} \end{pmatrix} \cdot \begin{pmatrix} 2 & 1 \\ 3 & 5 \end{pmatrix} = \begin{pmatrix} \frac{10}{7} - \frac{3}{7} & \frac{5}{7} + \frac{-5}{7} \\ \frac{-6}{7} + \frac{5}{7} & \frac{-3}{7} + \frac{10}{7} \end{pmatrix} = \begin{pmatrix} 1 & 0 \\ 0 & 1 \end{pmatrix}$$

Scalar multiplication

We need to stop with the matrix multiplication right now and ask what happens if you just multiply a matrix by an ordinary number that is not a matrix? For example, we would like to write

$$4 \times \begin{pmatrix} 1 & 1 \\ 3 & 2 \end{pmatrix} = \begin{pmatrix} 4 & 4 \\ 12 & 8 \end{pmatrix}$$

This type of multiplication is called "scalar multiplication"; we multiply every element in the matrix by the scalar. Also if any number (scalar) is a multiple of every element of a matrix, we can factor that number out of the matrix and it will be a scalar multiplying the whole matrix.

$$3 \times \begin{pmatrix} 1 & 5 & 7 \\ 2 & 8 & -4 \\ 0 & \frac{1}{3} & 9 \end{pmatrix} = \begin{pmatrix} 3 & 15 & 21 \\ 6 & 24 & -12 \\ 0 & 1 & 27 \end{pmatrix}$$

$$\frac{1}{2} \times \begin{pmatrix} 2 & 4 \\ 10 & 50 \end{pmatrix} = \begin{pmatrix} 1 & 2 \\ 5 & 25 \end{pmatrix}$$

This comes from the fact that addition, especially addition of the same quantity over and over becomes a sort of multiplication. For example if you add $x + x + x + x$, your are really multiplying 4 times x.

Look back at the matrix A^{-1} in Equation (2.18). It can be written as

$$A^{-1} = \begin{pmatrix} \frac{5}{7} & -\frac{1}{7} \\ -\frac{3}{7} & \frac{2}{7} \end{pmatrix} = \frac{1}{7} \times \begin{pmatrix} 5 & -1 \\ -3 & 2 \end{pmatrix}$$

Cool way to get the multiplicative inverse

Just think of that! The inverse of A is got by switching the 2 and 5 on the *main* diagonal, and changing the signs of the 3 and 1 (without switching them) on the *off* diagonal and multiplying by the reciprocal of the determinant. How cool is that!

But is this always true for any 2 by 2 matrix that has a non-zero determinant? Would I ask if it weren't?

It is! And you can work it out for yourself.

Problem: Given the following matrix, C, use the cool method just discussed to find the multiplicative inverse C^{-1}

$$C = \begin{pmatrix} 6 & 2 \\ 2 & 1 \end{pmatrix}$$

Solution: First, find the determinant of C. It is $\det(C) = 2$. So yes, the inverse exists. Next, reverse the elements on the main diagonal, 6 and 1 and change the signs on the off diagonal, 2 and 2. Then multiply by $\frac{1}{\det(C)}$. Therefore,

$$C^{-1} = \frac{1}{2} \begin{pmatrix} 1 & -2 \\ -2 & 6 \end{pmatrix}$$

Now, check this to see if $C \times C^{-1}$ really is I.

$$C \times C^{-1} = \begin{pmatrix} 6 & 2 \\ 2 & 1 \end{pmatrix} \times \frac{1}{2} \begin{pmatrix} 1 & -2 \\ -2 & 6 \end{pmatrix} = \frac{1}{2} \begin{pmatrix} 2 & 0 \\ 0 & 2 \end{pmatrix} = \begin{pmatrix} 1 & 0 \\ 0 & 1 \end{pmatrix}$$

Some applications of small matrices

Although 2 by 2 and 3 by 3 matrices are rather simple and do not have the power that larger matrices have, they still do have a couple of interesting applications. One is a simple matrix *game theory* model and the other can be used to create simple *unbreakable codes*

A matrix game

An n by n matrix can be used to study game theory. We can let the rows of the matrix represent payoffs to the two players. A row player secretly selects a certain row and the column player secretly selects a certain column, if there is a positive number (called the payoff) in *that* row and *that* column, then Row wins that amount and Column loses that amount. If there is a negative number in that row and column, the row player loses and the column player wins that amount. This is called a zero-sum game. In other games, the selected row and column might have two different amounts, one for the row and one for the column player. In this case it is called a non-zero sum game.

Let us consider the following 3 by 3 matrix that models the Rock-Paper-Scissors game:

	Rock	Paper	Scissors
Rock	0	−1	+1
Paper	+1	0	−1
Scissors	−1	+1	0

Call the two players Row and Column.   There is a hierarchy in which rock beats scissor (a rock can break scissors), paper can cover a rock and scissors can cut paper. Row does not know which column is being picked by Column and Column does not know which row will be picked Row.   They simultaneously reveal their choices.   Let us say several games are played.

Game 1: Row plays Rock, Column plays Paper.   Column wins 1 point.

Game 2. Row plays Rock and Column plays Rock; neither side wins a point

Game 3 Row plays paper and Column plays Rock; Row wins 1 point.

It doesn't take long to realize that neither player has the advantage here, and that luck determines the long run outcomes.

Larger n by n games (with n greater than 3) with more serious payoffs involving probabilities provide us with interesting strategies, in which luck is not a significant factor.   We will look at these in a later chapter in this book.

A matrix and its inverse as a coder and decoder.

Matrices can be used to establish "unbreakable codes." Well, maybe not 100% unbreakable because more and more powerful computers are always being created to try to break codes that are always being made more and more difficult. It is an *Arms Race*.

One method a cryptographer uses in trying to break a coded message is based upon *letter frequency* in any given language. Some letters appear more frequently than others, (in English, it is the letter "e") so sometimes a coded message will give the code breaker a clue because the more frequently used letter will correspond to the more frequently used symbol in the coded message.   This can become a dead give-away and provides a start to unraveling the message.

We can use a simple two by two matrix to encode a message in such a way that the frequency of letters is completely distorted making the frequency completely misleading with the same symbol being used for entirely different letters.

Example: Let's say you want to encode the following message

Rosebud number 356

This is called the plain text message.   First, we want to translate it to a *string* consisting of only numbers.   This not actually a coded message yet.   It is called the *translated* text and it is totally transparent.

Take the simplest, translation of this message into numbers by assigning each letter of the alphabet, A,B,C, ..Z, a positive integer, $1, 2, 3, ..26$.   Also, let us encode a *space* by the number 27, and encode any digit $x = 0, 1,...9$ in the plain text message by $28 + x$.   In other words, 0 is encoded as 28, 1 is 29, etc. We show this translation in table 2.2.

A	B	C	D	E	F	G.	H	I	J	K	L	M
1	2	3	4	5	6	7	8	9	10	11	12	13
N	O	P	Q	R	S	T	U	V	W	X	Y	Z
14	15	16	17	18	19	20	21	22	23	24	25	26
_	0	1	2	3	4	5	6	7	8	9	..	..
27	28	29	30	31	32	33	34	35	36	37	..	..

Table 2.2 Translation Table

Using this table, the plain text message would be the following string of 18 characters

$$18\ 15\ 19\ 5\ 2\ 21\ 4\ 27\ 14\ 21\ 13\ 2\ 5\ 18\ 27\ 31\ 33\ 34$$

Which is totally inadequate as a code. Anyone looking at it can see that the 27's are spaced out at about one word length apart, so they are likely to be word breaks. Here 21 and 5 are frequent, so one of these might be an E, and the other another vowel, etc. making this easily deciphered. But if we set up the numbers in the translated text into a 2 by 9 matrix

$$\mathbf{M} = \begin{pmatrix} 18 & 15 & 19 & 5 & 2 & 21 & 4 & 27 & 14 \\ 21 & 13 & 2 & 5 & 18 & 27 & 31 & 33 & 34 \end{pmatrix}$$

and use a secret 2 by 2 matrix we can get a coded message in which the word frequency pattern cannot be uncovered.

Let us use the matrix $\mathbf{C} = \begin{pmatrix} 6 & 2 \\ 2 & 1 \end{pmatrix}$ just discussed above. We can encode the matrix $\mathbf{M}$ by finding $\mathbf{C} \times \mathbf{M}$.

$$\mathbf{C} \times \mathbf{M} = \begin{pmatrix} 6 & 2 \\ 2 & 1 \end{pmatrix} \times \begin{pmatrix} 18 & 15 & 19 & 5 & 2 & 21 & 4 & 27 & 14 \\ 21 & 13 & 2 & 5 & 18 & 27 & 31 & 33 & 34 \end{pmatrix}$$

$$\mathbf{F} = \begin{pmatrix} 150 & 116 & 118 & 40 & 48 & 180 & 86 & 228 & 152 \\ 57 & 43 & 40 & 15 & 22 & 69 & 39 & 87 & 62 \end{pmatrix}$$

Where we denoted the final coded message, $\mathbf{C} \times \mathbf{M}$, as $\mathbf{F}$. Notice that the letter frequency is completely destroyed. In fact, the only repeated number in $\mathbf{F}$ is 40, and we know it to stand for two different things in the plain text message, E and B. Also each obvious space, 27, between words in the original message, $\mathbf{M}$, has been totally obliterated, becoming 228 and 69 in the final message, $\mathbf{F}$. There is only one way to decode this final message, $\mathbf{F}$ in order to recover the translated message $\mathbf{M}$ and that is to multiply the inverse $\mathbf{C}^{-1}$ times $\mathbf{F}$.

$$\begin{pmatrix} \frac{1}{2} & -1 \\ -1 & 3 \end{pmatrix} \times \mathbf{F} = \mathbf{M}$$

Try it for yourself and you will see that you get the matrix $\mathbf{M}$ back, then use the translation table to recover the secret plain text message. Thus, if your fellow spy knows what matrix you used to code the message, she only needs to find the inverse of that matrix to decode it.

Problem: You are on a vacation in Maui and your friend has constructed a plain text message and translated it to a matrix $\mathbf{A}$ according to Table 2.2. Then she coded it with the matrix $\mathbf{C}$, (the one we just used) and has sent it to you as a coded message $\mathbf{B}$. Here is that coded message

$$\mathbf{B} = \begin{pmatrix} 102 & 18 & 174 & 58 \\ 43 & 8 & 65 & 24 \end{pmatrix}$$

What was the plain text message she sent you?

Solution: Left to the reader. Hint, $\mathbf{B} = \mathbf{C} \times \mathbf{A}$, find $\mathbf{A}$, then translate that back to the plain text message.

Linear models in mathematics

What else can matrices do? Well, in the past fifty years, matrices have been important in helping economists win Nobel Prizes; this is because a matrix can be used to describe a complex system in terms of its interacting parts. Some examples are: buyers and sellers in a market, intergenerational (parent-child) relations, economic dynamics, game theory, and models in labor-management negotiations.

An Assignment Problem

The following example is a problem involving the assignment of various personnel to various tasks. It is taken from the book by Richard Bronson, <u>Operations Research</u>, McGraw-Hill, 1997. We will learn how to solve such problems in a later chapter.

Example:

A legal firm has accepted five new cases, each of which can be handled adequately by any one of its five junior partners. Due to differences in experience and expertise, however, the junior partners would spend varying amounts of time on the cases. A senior partner has estimated the time requirements (in hours) as shown below:

	Case 1	Case 2	Case 3	Case4	Case 5
Lawyer 1	145	122	130	95	115
Lawyer 2	80	63	85	48	78
Lawyer 3	121	107	93	69	95
Lawyer 4	118	83	116	80	195
Lawyer 5	97	75	120	80	111

Determine an optimal assignment of cases to lawyers such that each junior partner receives a different case and the total hours expended by the firm is minimized.

This problem can be solved by trial and error; there are only 120 ways to assign the cases to the lawyers. This is because one of the five cases can be assigned to the first lawyer, then one of the four remaining cases can be assigned to the second lawyer, which makes 5×4, or twenty ways to assign the first two cases, then one of three remaining cases can be assigned to the next lawyer, and so on. This makes $5 \times 4 \times 3 \times 2 \times 1 = 120$ ways to assign all 5 cases. So, the senior partners could waste hours of their time by scheduling each of these 120 possibilities in order to determine the minimum time, or they could use an integer programming algorithm from *Operations Research* to solve the problem in about 5 minutes.

Other examples of word problems solved by computer algorithms include: Distillation of crude oil and distribution of products, Smart phone apps, and warehousing problems.

All of these problems are solved with the aid of computers, which use a modified version of the language and the logic and the symbols of mathematics. Before a computer can do these computations, we have to build in the mathematical rules it must follow to carry out such tasks. We shall see how we can design computer circuits to do this. It all depends upon clearly specifying the fundamental operations of mathematics. The next three chapters tell you exactly what these are.

CHAPTER 3

BREAKTHROUGHS

As we mathematicians ambled on down the street, complacent and happy, some people began to ask us questions. We answered them. Then they began to ask us how did we know that we really answered them? We decided to think more deeply about these answers. We began to realize that they were, indeed wonderful questions, for which we had to invent new answers. So we continued with our tool building.

"What are *numbers?*"

"What is *addition?*"

"What are the *laws* of addition and multiplication?",

"What is *algebra?*"

What are numbers?

An important field of mathematics is *number theory*. It is an area often referred to as *pure mathematics* and it deals with rich and varied problems with long histories. Solutions in this field are, usually, confined to those that can be expressed only by whole numbers, the *integers*. In 1850, a German mathematician, Leopold Kronecker, one of the world's foremost number theorists stated his philosophy of mathematics as follows:

> "Die ganzen Zahlen hat der liebe Gott gemacht, alles andere ist Menschenwerk."

In English this says:

> "God made the integers, all else is the work of man."

He did not take this position frivolously, he applied it in his mathematical work which rejected ideas such as irrational numbers. He can be thought of as one of the earliest constructivist mathematicians, demanding stringent adherence to the properties of whole numbers. He was even able to block the publications of some mathematical papers that did not follow his philosophy. But fortunately, for the progress of mathematics his intransigence did not prevail.

The Integers

Let us begin with the whole numbers, *the integers.* These are the positive numbers we count with $(1, 2, 3, ...$ going on forever$)$ and their negatives, and the number zero.

$$\{... - 5, -4, -3, -2, -1, 0, 1, 2, 3, 4, 5, ...\}$$

This set is the most fundamental concept in mathematics. Some of the hardest problems that can be posed deal with questions about the existence or non-existence of whole number solutions. For centuries, prior to 1994, it had been asserted that there are no whole number solutions to the equation $x^3 + y^3 = z^3$, nor to the equation $x^4 + y^4 = z^4$. But for, years, no one had been able to prove any of these special cases, to say nothing of the general case, which states that for any positive integer $n \geq 3$, there are no three integers x, y, z such that :

$$x^n + y^n = z^n$$

As early as 250 CE, Diophantus of Alexandria, a Greek mathematician, tried to show that there were not three integers x, y and z satisfying this equation for $n = 3$. He failed. The problem continued to be unsolved over the next 1400 years. Then, in 1637, Pierre de Fermat shocked the mathematical world by claiming that he actually had a proof for the general case, in which he proved that there were no solutions for any positive integer $n \geq 3$. This claim became known worldwide as Fermat' sLast Theorem (FLT).

He sometimes wrote his proofs in the margin of whatever book he was reading, and in this case he wrote, in the margin, that this one was too small to hold the proof, so he did not write it out. And, he never wrote it out elsewhere as far as anyone knows. Ironically, the book he was reading at that time was one by Diophantus. Everyone searched high and low for his proof and could not find it. Also, no one could produce their own proof based on the mathematical methods known at that time, nor even based on mathematics developed over the next three and a half centuries.

Between 1637 and 1993, great mathematicians, like Euler, and others tried to prove FLT and some of them did do so *for special cases,* but not in general. For example, in 1750, Leonard Euler and, independently, Fredrick Gauss proved there were no whole number solutions for the equation when $n = 3$, the case that Diophantus could not solve. Later, in 1825, Adrien-Marie Legendre solved it for $n = 5$, and in 1839, Gabriel Lamé proved it for $n = 7$. Substantial cash prizes were offered for a valid general proof, but they were never won.

By the late 1900's FLT had been proved, with help of computers, for all integers up to $n = 150,000$, but *not* for every n. Finally, in 1994, a British mathematician, Andrew Wiles, did prove that for every integer $n \geq 3$, the equation $x^n + y^n = z^n$ does not have a solution in integers.

The FLT saga is an example of attempts to find whole number solutions to problems. Some of these problems deal with *prime numbers.* A prime number is a positive integer that cannot be factored into integers other than itself and 1. One of the hardest such problems that is still unsolved today(2022) is the Goldbach Conjecture (1742), which is: *Every even number greater than 2, can be written as the sum of two primes, not necessarily distinct.* This statement has been shown to be true for every integer up to $2n = 4 \times 10^{17}$, the number 4 followed by 17 zeros, but it has not yet been proved for *every* even number.

Equations in which the objective is to find integer solutions, or to prove, without solving the equation, whether or not an integer solution even exists are called *Diophantine Equations.* These equations also include problems in which we just want to know whether or not a solution has some special property, such as, among others, is it even? Is it prime? We will show how to approach Diophantine Equations later in this chapter using a mathematical tool called *modular arithmetic*

All Else

As we move beyond the integers we encounter some of Kronecker's "all else," such as, non-integral rational numbers, irrational numbers, imaginary numbers, complex numbers, vectors, and other types, some of which don't seem to be numbers at all. As we expand our understanding of nature from subatomic particles to galaxies, black holes, and dark matter we should not be surprised to learn that the meaning of *number* and what we mean by the notion of adding two numbers, or multiplying two numbers has undergone substantial changes over recent years.

Babylonian Cuneiform tables show that as far back as 2000 BCE, people were attempting to solve quadratic equations. Later, attempts to solve cubic, quartic, and other polynomial equations became the tough problems eluding their grasps. In 300 BCE, Euclid, and in 200 BCE Chinese mathematicians, all were working on such equations. In, 1594 a Flemish mathematician, Simon Stevin came up with the formula solving the general quadratic equation, $ax^2 + bx + c = 0$, "by radicals" meaning a solution in terms of the coefficients and square roots of the coefficients.

$$x = \frac{-b \pm \sqrt{b^2 - 4ac}}{2a}, \text{ for } a \neq 0$$

Even earlier in the 16th century, Italian mathematicians Gerolamo Cardano, Scipione del Ferro, and Niccolo Tartaglia were engaged in a tangled web of who gets proprietary credit for inventing the formula for solving cubic equations. Apparently Tartaglia could solve the equation

$$t^3 + pt + q = 0 \tag{3.1}$$

by radicals, finding t in terms of cube roots of the coefficients p and q. He showed his method to Cardano under the agreement that it would not be published. Later, Cardano saw a formula invented by del Ferro for solving the more general case

$$x^3 + bx^2 + cx + d = 0 \tag{3.2}$$

Cardano then published del Ferro's formla but gave Tartaglia credit for the solution to Equation (3.1). So who gets credit for what? It turns out that if you can solve Equation (3.1), Tartaglia's equation, then you can convert (3.2) into (3.1) and get solutions to del Ferro's equation. All it takes is the substitution, letting $x = t - \frac{b}{3}$ in Equation (3.2) to turn it into

$$t^3 + \frac{3c - b^2}{3}t + \frac{2b^3 - 9bc + 27d}{27} = 0$$

Now, just find Tartaglia's solution for

$$p = \frac{3c - b^2}{3}, \text{ and } q = \frac{2b^3 - 9bc + 27d}{27}$$

It is still quite messy to get the three values of t that satisfy Equation (3.1).   Some historians are calling these formulas the Cardano-Tartaglia-del Ferro formula.

Similar unwieldy formulas for the general quartic (fourth degree) equation were found, but, prior to the 1800's no one had produced a formula that could solve the general fifth degree polynomial by radicals, (in terms of fifth roots of the coefficients). But, in 1820, a nineteen year old Norwegian mathematician Neils Abel *proved* that *no such general formula* can ever be found. I can hear mathematics students all over the world shouting "hooray," so they would not have to learn such formulas. Now-a-days we use numerical methods and computers to get approximate solutions of any polynomial equation to any degree of accuracy we want.

Evariste Galois

Then, in 1829 and 1830, another 19 year old mathematician, a Frenchman this time, Evariste Galois submitted two papers to mathematics journals tying the solutions of higher polynomial equations, to a concept he invented, called a *group of permutations*. In these papers, he outlined a method that completely settled all questions regarding polynomial equations. But, due to editorial ineptitude, neither of these were published at that time.

Galois, unfortunately, was involved in radical politics and in 1832, he was challenged to a duel in a political altercation.   On the night of May 29, 1832, he hastily wrote a "testamentary letter" to his friend Auguste Chevalier outlining his ideas on how to find all the conditions *any* fifth or higher degree polynomial equation would have to meet before there could be a general solution by radicals. This paper was hastily written because he was to fight the duel the next day, which he did and which he lost and in which he was mortally wounded.

When mathematicians read his works, they realized that other things could be treated as numbers and that multiplication could be treated as a new kind of operation on these new kinds of things.   All of a sudden numbers emerged as *permutations* and multiplication became *compositions* of permutations, and groups became abstract mathematical systems.   Early work in group theory continued to concentrate on the abstract nature of the subject, and not on concrete physical objects.

Scientists, on the other hand, saw that such sets of elements could be related to real things. About seventy years after the invention of group theory, in early 1900, physicists working with subatomic particles noticed that the elements of group theory could be interpreted as the ways in which particles *flip, rotate, jump, split, translate,* or *maintain symmetry.*   And, they discovered that addition could be interpreted as a *combination* of these activities, creating other flips, rotations, and symmetries.

So, today, what do we mean by numbers? The answer is that we can still use the old concepts that preceded Abel and Galois, such as, whole numbers, fractions, imaginary numbers, lengths, areas, and volumes. But now, with certain caveats, almost *anything* that can be added to, multiplied by, subtracted from or divided by each other is acceptable as a number. Numbers can now include not only the measurement of abstract quantities, but the quantities themselves. Today, we define the operations of addition and multiplication as *binary operations,* that is, any well-defined combinations of any two quantities.

In Chapter 2 we saw how to treat matrices as numbers that could be added, multiplied and so forth. Now, we will see how we can add geometric figures such as *squares* to get other *squares*. Or even more abstractly, adding a *spin* to a *jump* to get a *symmetry*, which means initiating a spin concatenated with a jump resulting in a symmetry. This says that when we have a collection of activities and we cause one of these activities to be followed by another, we can recognize the compound result to be one of the activities in the collection. For example, suppose we said that x is a rotation of 45°, and y is a rotation of 90° then we could say that $x + x = y$, that is, an x followed by another x is a y. A rotation of 45° followed by a rotation of 45° is a rotation of 90°.

Binary Operations (Addition or Multiplication)

The addition of two elements, x and y of a set S is called a *binary operation* (or a *binary relation*) on S, only if the result $x + y$ is an element in that set S. In general, let us denote a binary operation on a set S as $\circledast$, read "circle star", then if x and y are in S, the result $x \circledast y$ must be in S. That is, the binary operation $\circledast$ must have the *closure property*.

Examples:

Let $\mathcal{O}$ be the set of odd numbers, $\{... - 5, -3, -1, 1, 3, 5, 7...\}$. If $\circledast$ is defined to be ordinary multiplication, $\times$, then $\circledast$ is closed on the set $\mathcal{O}$ because when x and y are odd and $x \circledast y = z$, then z is odd. Odd times odd = odd.

Now suppose we have the same set $\mathcal{O}$, but this time define the operator $\circledast$ to be ordinary addition, $+$, then $\circledast$ is not closed on the set $\mathcal{O}$ because when x and y are odd and $x \circledast y = z$, then z is even, therefore, z is not in $\mathcal{O}$.

How to understand abstract notation

When we start solving problems in various abstract mathematical systems we must pay attention to the currency of any definition or notation specific to the particular system under discussion. For example, in a set of matrices, I might say let A and B be any two n by n matrices, and define $A + B$ as an n by n matrix obtained by adding corresponding elements. But later on, in a different problem I might say, let A and B be two circles and define $A + B$ to be a circle whose area is the sum of the areas of A and B. I mean just concentrate on the current meaning of $+$ not the meaning we had previously assigned.

This is the nature of *fresh definitions* when we say "define your terms," used in philosophy, mathematics and legal contracts. The words in any specific contract mean only what they were defined to mean in that contract, not what they may have meant in some other unrelated one.

What properties can binary operations have?
Closure Law

Here, we will be talking about abstract, undefined, generalized combinations of two things. The actual nature of the combination should emerge, or at least settle down to a comprehensible concept as we say more and more about it. At the outset, however, we will say two things: First that we are starting with a well defined *set* of identifiable elements, and second that a combination of elements can qualify as a binary operation on that set only if it satisfies the *closure* law, meaning that the *result* of the combination must be one of the elements in the original given set. Instead of using just the notations $+$ or

· or ×, we will subject you to an abstract operator, ⊛,and define what it stands for in different settings.

Associative Law

A law that may be satisfied by the addition of matrices and other types of things is $A + (B + C) = (A + B) + C$. This is called the "associative law." It lets us conveniently "group" elements that are being added together. You may remember this from your arithmetic class in elementary school.

$$3 + 7 + 5 + 4 \;=\; (3 + 7) + (5 + 4)$$
$$=\; 10 + 9 = 19$$

The associative law is important in problems in which you need to combine (by adding or multiplying) more than just two numbers at a time. If we tried to apply the associative law to adding odd numbers it would make no sense, because neither $B + C$, nor $A + B$ are numbers in the system of odds. (You might say, "they are not odd enough to belong to the odd fellows.")

In general, for any operator ⊛ the associative law would be: $x \circledast (y \circledast z) = (x \circledast y) \circledast z$. A binary operator without the associative law works only with two elements at a time. In such a case, we would not be able to infallibly determine the value of expressions such as $x \circledast y \circledast z$. This would severely limit the usefulness of the mathematical system with that operator.

Commutative Law

The commutative law: $x \circledast y = y \circledast x$, for example, $A + B = B + A$ says that if you add any two elements of a set you get the same thing regardless of the *order* in which you add them. There are examples in which the commutative law does not hold, for example as we learned in Chapter 2, matrix multiplication is not commutative.

Also, any operator, ⊛, is not commutative if it requires one process be completed before the other is begun and is such that these processes could not be reversed. For example, in a bottling plant, if you fill the bottle and then cap the bottle, you are OK, but it would be disastrous if you set up the machinery to cap bottles then tried to fill them.

Identity Law, neutral elements

Let ⊛ be a closed binary operator on a set S of things $\{x, y, ...\}$. If there is some element e (not to be confused with the famous mathematical constant $e = 2.71828...$, nor the famous physical constant $e = mc^2$) in S such that, for all x in S, $x \circledast e = x$, then the operator, ⊛ satisfies the *identity law* and the element e is called the *identity (or neutral) element* for ⊛ in S.

Examples: In ordinary arithmetic if the operator ⊛ is addition, $+$, then the identity element e is 0 because $x + 0 = x$. If, on the other hand, the operator ⊛ is multiplication, ×, then the identity element is 1 because $x \times 1 = x$. In the set $\mathcal{O}$ of odd numbers, the number 1 is the identity element for multiplication because 1 is odd and if x is any odd then $x \times 1 = x$. And, $1 \times x = x$, since multiplication is commutative in the integers.

Inverse Law

If S is a system that has a closed operator $\circledast$, and e is the neutral element for $\circledast$, and if for each element x in S, there is an element y in S such that $x \circledast y = e$, then y is called the *inverse* of x for the operator $\circledast$. In ordinary arithmetic, if the operator $\circledast$ is $+$ where the identity element is 0, and if x is any element in, S, then the inverse of x is denoted by $-x$ thus, $x + (-x) = 0$. Likewise, if the operator $\circledast$ is multiplication, $\times$, where the identity e is 1, and if x is in S and $x \neq 0$, then the inverse of x is denoted by x^{-1} or by $1/x$. That is, $x \times x^{-1} = 1$, or $x \times (1/x) = 1$. Notice, here, that since we have required that $x \neq 0$, we are saying that 0 has no multiplicative inverse. This brings up the question: can the two neutral elements (the one for addition and the one for multiplication) be the same? Is it possible that $1 = 0$? The answer is no unless $+$ and $\times$ are both the very same operator. Unfortunately, the reader must keep track of which operator, $+$ or $\times$ is being discussed as to whether the identity is 0 or 1 and which inverse is being discussed as to whether it is x or x^{-1}. In this regard, special care needs to be taken when reading about the identities and inverses for a general operator such as $\circledast$, in which the neutral elements may be neither 0 nor 1. This is related to the admonition to "define your terms" and keep track of your definitions mentioned above.

Addition

What is addition?

Before the invention (or discovery?) of group theory, we had a pretty good idea of what addition meant. Consider two line segments of lengths a and b joined at one end point and placed along the number line, we get a line segment of length $a + b$. Or, if we have a jar of marbles and drop another marble into the jar we have added it. But today, we can think of adding an element a and an element b of that set by combining them somehow and precisely describing the rule by which they are combined. In other words, it takes some *defining statement* of the relationship we assign to the resulting outcome that we are willing to call $a + b$.

Adding Geometric Figures

In geometry, if we concoct a definition for adding two squares we want the sum to be another square and not a triangle or a circle, or some other geometric figure. As we have previously said, this restriction is called the *closure law* for addition, meaning that the system is "closed" in the sense that adding two elements of a given collection will **not** produce a new object *outside* of that collection.

Example: *Adding Squares.*

Let A and B be two squares whose areas are x^2 and y^2, respectively. We define addition of these two squares to be *a square* whose area is $x^2 + y^2$.

As you can see in Figure 3.1 such a new square can be created by making the sides x and y be the perpendicular sides of a right triangle, then the result (by the Pythagorean Theorem) is the square on the hypotenuse, whose area is $x^2 + y^2$.

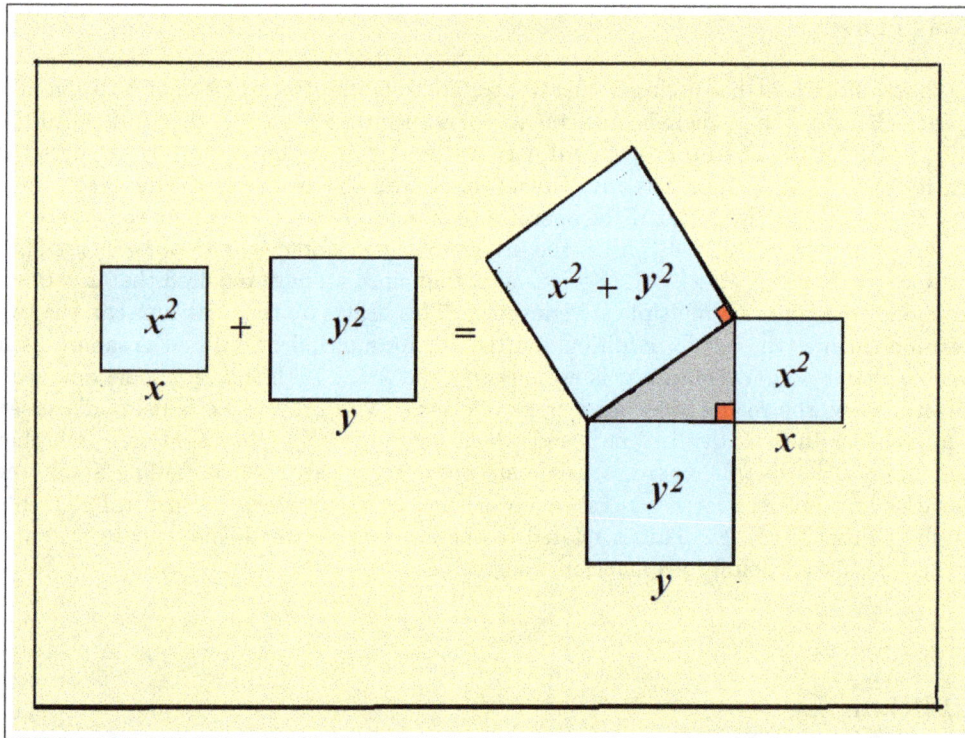

FIGURE 3.1 ADDING SQUARES

Using this definition, we can say that the addition is closed, associative, and commutative

Multiplication
What is multiplication?

If a system has an operator called addition as described above, we can also define other binary relations on the same system. Usually the second binary operator is called *multiplication*, but it could just as well be called something else. In order to really have two *different* binary relations we insist that the *neutral* elements (the identity elements) in the two systems must not be the same. For example, if 0 is the neutral element for addition and 1 is the neutral element for multiplication, we must require that $1 \neq 0$. Otherwise, we would have to conclude that 1 times any number is the same as 0. So all numbers would have to be zero. This actually could be a mathematical system alright, but it would would be very limited in its ability to solve many problems.

In some cases, one operator may have some property and the other not have that property. For example, addition may have an inverse, but multiplication does not, or addition may be commutative, but multiplication is not, etc. In fact, many useful systems exist in which we require addition to satisfy all of the properties, associative, commutative, identity and inverse, whereas multiplication need only satisfy one or two of the other laws.

When we do have two different operators + and ×, each satisfying all five properties of binary operators then we have almost all the tools we need to solve any algebra problem.

If, on the other hand, we deny some of these properties one at a time, we can still develop simpler mathematical systems to solve important problems in physics from Newtonian physics to quantum mechanics, to say nothing at all of the other problems in biology, chemistry, astronomy, economics, business, and everyday life.

Mathematical systems
Modular Arithmetic

An important set of numbers to study are the whole numbers

$$\{... -4, -3, -2, -1, 0, 1, 2, 3, 4, ...\}$$

there are infinitely many of them going off to the left with the negatives forever and going off to the right with positives forever.

The arithmetic of mod 12   Can you think of a rule for adding whole numbers that gives you the following results,

$$7 + 9 = 4, \text{ and } 11 + 6 = 5, \text{ and } 12 + 12 = 12$$

Hint, two words: Clock Arithmetic.   In a 24 hour clock, if it is 7 AM now, then 9 hours later it will be 16 (or 1600 hours), which translates to 4 PM in a twelve-hour clock, then $7 + 9 = 4$. Also, $11 + 6 = 17$ which translates to $11 + 6 = 5$ PM. The point of this little exercise is to show that even in a simple thing like addition, you need to the know the setting in which this operation on numbers is executed.

Rotating around a 12 hour clock like this is called *modular arithmetic* with 12 being the *modulus*, the place where the counting re-starts.   When 17 is changed to 5 it is because we re-started counting at 12, we say that 17 *is congruent to* 5 mod 12,

$$17 \cong 5 \bmod 12$$

The symbol $\cong$ is read "*congruent to.*"   It means that the difference between 17 and 5 is 12 or a multiple of 12.   Maybe the number 29 immediately came to your mind, after all, the difference between 29 and 5 is 24 also a multiple of 12.   And you are right! Since $29 - 5$ is 24, a multiple of 12, then

$$29 \cong 5 \bmod 12$$

You are also correct if you noticed that since 24 is a multiple of 6, then $29 \cong 5 \bmod 6$.

In general, we can derive all of the computations in modular arithmetic for any modulus from its formal definition:

Definition: (Congruence mod n)

Given an integer a, and a positive integer n, we say that an integer b is congruent to $a \bmod n$ (denoted $b \cong a \bmod n$) if and only if $b - a = n$ or a multiple of n.

For example, if $a = 0$, and $n = 12$, then the integer 60 is congruent to 0 mod 12, written as: $60 \cong 0 \bmod 12$, since $60 - 0$ is a multiple of 12.

This seems simple enough.   But why, for any unknown integer x, is the following congruence true?

$$109x \cong x \bmod 12$$

Answer: Because $109x - x = 108x$ which is 9 times $12x$, and no matter what integer x is, $108x$ is a multiple of 12.

We can use clock arithmetic as a powerful tool for determining whether or not equations have whole number solutions, without even solving the equation.

Example: Use clock arithmetic to show that there is no whole number, x, that satisfies the equation:

$$109x + 60 = 13x - 31$$

Answer: Observe that $31 \cong 7 \bmod 12$, because $31 - 7 = 24$. Now, use $\bmod 12$ on both sides of the equation to reduce it as follows:

$$
\begin{aligned}
(109x + 60) \bmod 12 \quad &\cong \quad (13x - 31) \bmod 12 \\
x + 0 \quad &\cong \quad (x - 7) \bmod (12) \\
x - (x - 7) \quad &= \quad \text{a multiple of } 12 \\
7 \quad &= \quad \text{a multiple of } 12
\end{aligned}
$$

Clearly, the number 7 cannot be a multiple of 12. So the equation has no whole number solution.

The arithmetic of mod 5 Of course, 12 is not the only number that can be used as a modulus. We can make a clock that has 5 numbers instead of 12. See Figure 3.2.

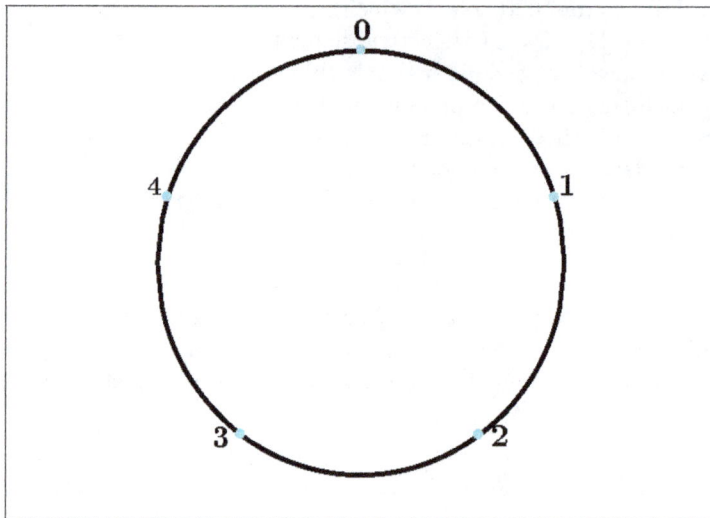

Figure 3.2 CLOCK FOR mod 5 ARITHMETIC

For any integer x, we can write $x \cong y \bmod 5$, where y is one of the integers, $0, 1, 2, 3$, or 4 or y is a multiple of 5 plus one of these numbers.

The definition of $\bmod n$, when applied to $n = 5$ is that $x \cong y \bmod 5$, means that $x - y$ is either 5 or a multiple of 5. How many numbers are congruent to $0 \bmod 5$? Infinitely many:

$$\ldots - 10 \cong -5 \cong 0 \cong 5 \cong 10 \cong 15 \cong \ldots 0 \bmod 5$$

How many are congruent to $1 \bmod 5$? Infinitely many

$$\ldots - 9 \cong -4 \cong 1 \cong 6 \cong 11 \cong 16 \cong \ldots 1 \bmod 5.$$

There are also infinitely many integers congruent to 2, 3, and 4, in mod 5. It may be easy to see that how the negative numbers -5, and -10, etc. are congruent to $0 \mod 5$, but not so easy to see how -4 or -9 is congruent to $1 \mod 5$. The negative numbers are a bit tricky. But just remember $-4 - 1$ is a multiple of 5 and so is $-9 - 1$, because subtraction is just addition of the negative

$$-4 - 1 = -4 + (-1) = -5$$

The following five lists show how we can separate *all integers* into the **five** classes of congruent integers in mod 5.

$$
\begin{aligned}
0 \bmod 5 &= \{-10, \ -5, \ 0, \ 5, \ 10, \ 15, 20, ...\} \\
1 \bmod 5 &= \{ \ -9, \ -4, \ 1, \ 6, \ 11, \ 16, \ 21, ...\} \\
2 \bmod 5 &= \{ \ -8, \ -3, \ 2, \ 7, \ 12, \ 17, \ 22, ...\} \\
3 \bmod 5 &= \{ \ -7, \ -2, \ 3, \ 8, \ 13, \ 18, \ 23, ...\} \\
4 \bmod 5 &= \{ \ -6, \ -1, \ 4, \ 9, \ 14, \ 19, \ 24, ...\}
\end{aligned}
$$

Each one of these five classes is called an *equivalence class* in mod 5. Thus, for example, $-7 - 2$ and 3 and 8, ... are all in the equivalence class congruent to $3 \mod 5$.

If you think that it is strange to see that -6 is congruent to 4, mod 5, just compute $-6 - 4 = -10$.

The arithmetic of mod 2 Now, let us break up all of the integers into **two** equivalence classes. We do this by considering all of the integers mod 2.

Example:

Using mod 2 we see that every integer is either congruent to $1 \mod 2$, or congruent to $0 \mod 2$. This simply says that every integer is either odd or even. There are two equivalence classes.

All the *odd* integers are congruent to $1 \mod 2$

$$... - 3 \cong -1 \cong 1 \cong 3 \cong 5 \cong 7 ... \bmod 2$$

All even integers are congruent to $0 \mod 2$

$$... - 4 \cong -2 \cong 0 \cong 2 \cong 4 \cong 6 ... \bmod 2$$

We can use mod 2 to solve problems such as the following:

Problem 1. Show, without actually solving the equation, that there is no integer x for which the following equation is true.

$$18x + 40 = 17$$

Solution: Look $(18x + 40) \bmod 2 = 0$, but $17 \bmod 2 = 1$, so this equation says $0 = 1$. Impossible!

Problem 2. Let x be an integer satisfying the following equation, show whether x is even or odd.

$$x + 4y - 19 = 17 - 8z + 9$$

Problem 3. Use mod 5 to show that there are no integers x and y that satisfies the following equation: $5x + 13 = 141 + 25y$.

Solutions to problems 2 and 3 left to the reader.

Problem 4. Show that if $a \cong b \bmod n$ and r is any multiple of n, then $a + r \cong b \bmod n$.

Solution: Let $r = mn$, then by definition if $a \cong b \bmod n$,

$$
\begin{aligned}
a - b &= kn, \text{ for some non-negative integer } k \text{ and} \\
a + r - b &= kn + mn = (k + m)n \\
a + r &\cong b \bmod n \text{ definition of congruence}
\end{aligned}
$$

An old problem

Historically, modular arithmetic has been around since the time of Euclid. A more recent instance, however, just 1600 years ago, was the following problem posed in a Chinese book in 400 CE.

> **Problem 5.** *We have a number of things, but we do not know, exactly, how many. If we count them by threes we have two left over. If we count by fives, we have three left over. If we count by sevens there are two left over. How many things are there?*

The problem is to find the first number x such that $x \cong 2 \bmod 3$ and $x \cong 3 \bmod 5$ and $x \cong 2 \bmod 7$.

Hint List a few positive numbers in each of the equivalence classes. $2 \bmod 3$; $3 \bmod 5$, and $2 \bmod 7$

$$
\begin{aligned}
2 \bmod 3 &= \{2, 5, 8, 11, 14, 17, 20, 23, 26, 29, ...\} \\
3 \bmod 5 &= \{3, 8, 13, 18, 23, 28, 33, ...\} \\
2 \bmod 7 &= \{2, 9, 16, 23, 30, 37, ...\}
\end{aligned}
$$

Find the first positive integer that is in all three classes. Once you know any value of x for which this is true, why is it that $x + 105$ would be another answer? There are infinitely many other answers, each one exactly 105 greater than the last one.

Congruence mod n is an Equivalence Relation

If n is any positive integer, it turns out that congruence $\bmod n$ is just like an equal sign, that is, it is a relation that has all the properties necessary to solve equations.

For any integers, a, b, c,

$$
a \cong a \bmod n
$$
$$
\text{If } a \cong b \bmod n, \text{ then } b \cong a \bmod n
$$
$$
\text{If } a \cong b \bmod n \text{ and } b \cong c \bmod n, \text{ then } a \cong c \bmod n
$$

These laws will help us develop the idea of addition and multiplication in modular arithmetic.

The numbers less than n and relatively prime to n are important to consider when we take a closer look at the two operators in modular arithmetic, *addition* and, especially, *multiplication*. They can be used to create *finite mathematical systems*, specifically, *finite groups*. Some finite groups can be associated with the concept of symmetry and used as models in studying molecular structures in chemistry.

Finite mathematical systems

If n is a positive integer, let $\mathcal{S}_n$ be a set containing exactly n elements, then $\mathcal{S}_n$ is *finite* set. We can list the elements in a finite set as $\{x_1, x_2, ..., x_n\}$. The subscripts on a single letter of the alphabet are being used, instead of just the ordinary letters of the alphabet $\{a, b, c, ...\}$; this way we will not run out of labels before we run out of numbers. The manner in which elements are labeled is <u>arbitrary</u> and often a matter of convenience. For example, if we wanted to make a list of the first 100 multiples of 7 starting with 0, we would write $\{0, 7, 14, ...7n, ..., 693\}$. The "general term", $7n$, is there to tell the reader what type of numbers (multiples of seven) are being listed, the 0 tells the reader that the first multiple of 7 in the list is 7×0, and $693 = 7 \times 99$ is the the 100th multiple of 7. The three dots ... (called an ellipsis) are there to tell the reader to fill in with all of the other missing multiples of 7 between 14 and 693. Depending upon how we wanted to use these data, we could have listed the first 100 multiples of 7 as: $\{7, 14, 21, ...7n, ...700\}$, starting with 7×1 and ending with 7×100. It is usually clear what kind of labelling we use by looking at the displayed numbers in the set brackets.

Finite Addition

Let $\mathcal{S}_n$ be the following finite set of integers.

$$\mathcal{S}_n = \{0, 1, 2, 3, ...n - 1\}$$

We will apply modular arithmetic addition by computing the sum of any two of these elements $\mod n$. If we take any two numbers x and y in $\mathcal{S}_n$, and we compute $x + y = z \mod n$, we will always re-write the sum, z, as one of the elements $\{0, 1, ...n - 1\}$ of $\mathcal{S}_n$. For example, in $\mod 4$ if we want $3 + 3$, we will write $3 + 3 = 2$, because $6 \cong 2 \mod 4$.

Let us not start with $n = 1$, because $\mathcal{S}_1 = \{0\}$ and the addition table $0 + 0 = 0$ is not very interesting. So, beginning with $\mathcal{S}_2$, which is the set $\{0, 1\}$, we have the following addition table in $\mod 2$.

+	0	1
0	0	1
1	1	0

Table 3.1

OK, in Table 3.1, $0 + 0 = 0, 0 + 1 = 1, 1 + 0 = 1$, but didn't you expect to see a 2 for $1 + 1$? Why does it show 0? Because $2 \cong 0 \mod 2$.

For $\mathcal{S}_3 = \{0, 1, 2\}$ and $\mathcal{S}_4 = \{0, 1, 2, 3\}$ the $\mod 3$ and $\mod 4$ addition tables are:

+	0	1	2
0	0	1	2
1	1	2	0
2	2	0	1

+	0	1	2	3
0	0	1	2	3
1	1	2	3	0
2	2	3	0	1
3	3	0	1	2

Table 3.2. $\mod 3$ and $\mod 4$

Notice that the values 0 appear where you might expect a 3 in the $\mod 3$ table and where you might expect 4 in the $\mod 4$ table. In the $\mod 4$ table why is $3 + 2 = 1$?

Two finite systems we want to study are: $\mathcal{S}_5$ and $\mathcal{S}_6$.

Since $\mathcal{S}_5 = \{0, 1, 2, 3, 4\}$, here is the table for finite addition in mod 5.

+	0	1	2	3	4
0	0	1	2	3	4
1	1	2	3	4	0
2	2	3	4	0	1
3	3	4	0	1	2
4	4	0	1	2	3

Table 3.3 mod 5

Here, the sum of 2 and 3 is written as 0 and not 5 because $5 \cong 0 \bmod 5$. In fact, every 0 you see in this table, where you might think a 5 belongs, there is a 0 because $5 \cong 0 \bmod 5$.

Next, we construct the finite addition table for mod 6 applied to the set $\mathcal{S}_6 = \{0, 1, 2, 3, 4, 5\}$.

+	0	1	2	3	4	5
0	0	1	2	3	4	5
1	1	2	3	4	5	0
2	2	3	4	5	0	1
3	3	4	5	0	1	2
4	4	5	0	1	2	3
5	5	0	1	2	3	4

Table 3.4 mod 6

The following rules hold for these two tables as well as for *any addition table* mod n for any set $\mathcal{S}_n$.

1. For any x and y in $\mathcal{S}_n$ the sum $x + y$ is in $\mathcal{S}_n$.

2. For any x, y, and z in $\mathcal{S}_n$, $x + (y + z) = (x + y) + z$.

3. For any x, y in $\mathcal{S}_n$, $x + y = y + x$.

4. The number 0 in $\mathcal{S}_n$ is such that for any x in $\mathcal{S}_n$, $x + 0 = x$.

5. If x is any number in $\mathcal{S}_n$, there is an inverse, y in $\mathcal{S}_n$, such that $x + y = 0$.

You may recognize these five rules as the closure law, associative law, commutative law, identity law, and the inverse law.

In ordinary *addition* the inverse of an element x is its negative $-x$. That is, for any x we want to find a number y such that $x + y = 0$. To see that this last rule holds, just notice that each row and each column has a 0 in it. If you pick a place with a 0 you will see that the number at the top of its column *plus* the number at left most side of its row add up to 0, or n, which is congruent to $0 \bmod n$. This means that every number has an (additive) inverse.

In the mod 5, table, we see that 4 is the inverse of 1 because $4 + 1 \cong 0 \bmod 5$. It may seem strange to say that 4 is the additive inverse of 1, $4 \cong -1 \bmod 5$, but it is because $4 - (-1) = 5 \approx 0 \bmod 5$. Also, $3 \cong -2 \bmod 5$. In the mod 6 table, we see that 3 is the inverse of 3 because $3 + 3 \cong 0 \bmod 6$.

Finite Multiplication

We can also define small mathematical systems using modular arithmetic *multiplication,* as follows. Let n be any positive integer and let x and y be any two integers in the set $M_n = \{0, 1, 2, 3, ...n - 1\}$. We want to construct a multiplication table showing the products $x \times y = z \bmod n$, always writing the number z as a number in the set M_n.

Let $n = 6$, so $M_6 = \{0, 1, 2, 3, 4, 5\}$ and construct the mod 6 multiplication table.

$\times$	0	1	2	3	4	5
0	0	0	0	0	0	0
1	0	1	2	3	4	5
2	0	2	4	0	2	4
3	0	3	0	3	0	3
4	0	4	2	0	4	2
5	0	5	4	3	2	1

Table 3.5 Multiplication, mod 6

Both the columns and the rows that are headed by 0 are superfluous in a multiplication table because every entry is 0. We can leave 0 out as one of the elements for a multiplication table. This table does have a multiplicative neutral element, and it is 1, so for any x in M_6, $x \times 1 = x$.

But there is something seriously wrong with Table 3.5, if we want it to be a system in which every number, x has a unique multiplicative inverse, y such that $x \times y = 1$. We can see this problem by noticing that it is **not true** that every row and column has a 1 in it. Thus, for example, there is no number y such that $2 \times y = 1$. In fact, none of the numbers 2, 3, or 4 have a multiplicative inverse. The fault is that the three positive numbers, 2, 3 and 4 are not relatively prime to 6. They each have a common factor with 6, other than 1. None of their products could be congruent to 1 mod 6. Let us omit 2, 3, 4, along with 0 and restrict the mod 6 multiplication table to just the positive numbers in M_6 that are relatively prime to 6, namely 1 and 5.

Thus, we will write $x \times y \cong z \bmod 6$, for $z = \{1, 5\}$ only. In this table, $5 \times 5 = 1$, and not 25, because $25 \cong 1 \bmod 6$; thus, 5 itself, is the multiplicative inverse of 5. In Table 3.6, below, we show the restricted multiplication table for mod 6 by leaving out the numbers $\{0, 2, 3, 4\}$ that were in Table 3.5.

$\times$	1	5
1	1	5
5	5	1

Table 3.6

Unlike the situation where n is composite, such as 6 in the previous example, when n is prime the situation is completely different. If n is prime, we can use all of the positive numbers $\{1, 2, 3, ...n - 1\}$ to construct the multiplication table because they are <u>all</u> relatively prime to n. Compare Table 3.5 above with the following mod 5 multiplication displayed in Table 3.7. table.

×	1	2	3	4
1	1	2	3	4
2	2	4	1	3
3	3	1	4	2
4	4	3	2	1

Table 3.7

These delightful finite arithmetic tables are not only interesting from a mathematical point of view, but they also have important applications in the sciences. They tell us that we need to know about the relationships between various types of entities such as molecules, galaxies, natural resources, computer networks, and subatomic particles.

When we gather together elements of a set and try to determine how they can be combined, we often find that these combinations may or may not obey some relationship. Take a binary relation $\circledast$, it may be associative, that is, $x \circledast (y \circledast z) = (x \circledast y) \circledast z$, or it may not be associative. Or $\circledast$ may or may not have a neutral element in the set, or $\circledast$ may or may not have an inverse, or $\circledast$ may be commutative: $x \circledast y = y \circledast x$, or may not be commutative: $x \circledast y \neq y \circledast x$.

These properties affect the behavior of the system. If you were managing such a system you would want to know the properties of the binary operator so as to make intelligent decisions about what kind of combinations you want to employ.

Example:

Let S be the set of instructions, actually *commands,* for making an apple pie:

$a = $ *preheat the oven to* $400°$

$b = $ *cut up apples*

$c = $ *make crust,*

$d = $ *put apples in crust*

Suppose the operator $\circledast$ is simply the word "next". This operator is closed because any two commands is a command.

It is a non-commutative operator, for example $b \circledast d \neq d \circledast b$. The resulting instruction $b \circledast d$ is OK, but the instruction $d \circledast b$ would do some serious damage to the pie crust.

What are some mathematical systems that we can get by specifying binary relationships?

Abstract algebras

In the various mathematical systems, the elements and the operators follow given rules; exactly which rules are obeyed will define what **type of mathematical system** we are working with such as: **semigroups, monoids, groups, Abelian Groups,** and **Rings,** and, as we will define later, **Fields.** Just like prime numbers, these entities exist not only for their truth and abstract beauty from which mathematicians derive a great deal of satisfaction, but also because they can be applied to the real world for many things: "shoes and ships–and sealing wax–and cabbages and kings.[1]"

In each case we are, at the very least, requiring that any binary relation $\circledast$ being discussed is closed.

---

[1]Charles Dodgson *aka* Lewis Carroll.

Semigroup

Definition: (Semigroup)

Any set that has an operation, such as addition, $+$, that is <u>closed</u> and <u>associative</u>, $x + (y + z) = (x + y) + z$, for all elements, x, y, z, of that set is called a <u>semigroup</u>.

This definition also applies to any other operator, $\circledast$.

$$x \circledast (y \circledast z) = (x \circledast y) \circledast z$$

Thus, if a set has a multiplicative operator $\times$ and $x \times (y \times z) = (x \times y) \times z$, then the set and its operator defines a semigroup.

Example Table 3.5, the first multiplication table for $\mathrm{mod}\, 6$, is an example of a semigroup. The multiplication operation on M_6 is closed (the result of multiplying any two numbers in the table is one of the numbers in the table) and the multiplication is associative, which can be verified by using this table to test all products $x \times (y \times z)$ and $(x \times y) \times z$. For instance, $5 \times (2 \times 4) = 5 \times 2 = 4$ and $(5 \times 2) \times 4 = 4 \times 4 = 4$. Therefore,

$$5 \times (2 \times 4) = (5 \times 2) \times 4$$

You can check out that this is true for any other set of three, not necessarily distinct, numbers. For example prove that:

$$3 \times (3 \times 4) = (3 \times 3) \times 4$$

The importance of knowing whether or not you have a semigroup is that if the binary relation that is *not associative*, then every operation between elements would be confined to no more than two elements at a time. You could never find out what would be true about combining any three elements.

Monoid

Definition: (Monoid)

A semi-group with a <u>neutral element</u> is called a <u>monoid</u>.

Example 1. Table 3.5 for $\mathrm{mod}\, 6$ is a monoid because it is a semigroup and has a neutral element 1 for multiplication.

Example 2. In our geometric example of the addition of squares earlier in this chapter, (See Figure 3.1) we have a monoid if we can imagine a square, e, whose sides are zero in length, then we can use that square (with area $= 0$) as the *neutral element*, so that if A is any square then $A + e = A$. Thus, our addition of squares also makes the set of squares a monoid. The neutral element is sometimes called the *identity* element.

Note: When we start with some system, like a semigroup for example, and we add a new independent axiom, such as requiring the existence of a neutral element, then new system still satisfies the rules of the original system. Thus, every monoid is still a semigroup, but not every semigroup is a monoid. This means that in the monoid we can still prove the theorems that were true in the semigroup. This hierarchical scheme is not unlike the concept of backward compatibility in technology when new updates are introduced. We can still do the old things with the new version, but we can not do the new things with the old version.

Group

Definition: (Group)

> A _group_ is a monoid with an <u>inverse.</u>

Let e be the neutral element and $\circledast$ be the operator in the monoid, and let x be any element, then the monoid is a group if and only if there is an element y such that $x \circledast y = e$.

We summarize this by saying that the binary relation $\circledast$ for a group satisfies four axioms:

(a) $\circledast$ is closed.

(b) $\circledast$ is associative.

(c) An element e exists such that for all x, $e \circledast x = x$.

(d) Every element has an inverse; for each x, there is a y such that $x \circledast y = e$.

Examples of Monoids that are not groups:

1. Table 3.5 for mod 6 is _not a group_ because there is at least one number, for example the number 2, which does not have a multiplicative inverse.

2 The addition of squares defined in Figure 3.1 is a monoid, but not a group because a non-zero square with area x^2, has no additive inverse square with area $-x^2$. That is, we cannot find a square whose area is $-x^2$ to add to a square whose area is x^2 in order to get $x^2 + (-x^2) = 0$. But wait! It is intriguing to ask what would happen if we were to use the imaginary unit $i = \sqrt{-1}$ to be the side of a square. Hmm? Naaa!

Example of a group: If we let Z be the set of all integers, positive and negative and 0, and let the binary relation be $+$ the usual addition operator, then Z and $+$, define a group.

This is true because $+$ is closed, associative, has an neutral element, 0 and for each integer n in Z, the integer $-n$, the (additive) inverse of n is in Z since $n + -n = 0$.

   Notation: If $\mathcal{S}$ is a set and $\circledast$ is a binary operator that satisfies the group requirements on the elements in $\mathcal{S}$, then we use the notation: $(\mathcal{S}, \circledast)$ is a group. So, in this example we say $(Z, +)$ is a group.

Abelian Group   If $(A, \circledast)$ is a group and $\circledast$ satisfies the commutative law:

$$x \circledast y = y \circledast x \tag{3.3}$$

then $(A, \circledast)$ is an _Abelian Group_, named for Abel; an operator which is not commutative can only be used to define non-Abelian groups.

   Example: Matrix addition is an operator that defines an Abelian group. Thus, if for some positive integer n, A_n is the set of all n by n matrices, and $+$ is the addition of matrices as defined before, then $(A_n, +)$ is an Abelian group. Also if A, B and C are n by n matrices, then $A + (B + C) = (A + B) + C$. There is a neutral n by n matrix O such that for all A, $A + O = A$ and if A is any n by n matrix, there is an n by n matrix $-A$ such that $A + -A = O$. Furthermore, matrix addition is commutative because $A + B = B + A$.

   On the other hand, however, matrices do not define an Abelian group under the multiplication operator, $\times$, because matrix multiplication is not commutative.

Example:

$$\begin{pmatrix} 1 & 3 \\ 1 & 2 \end{pmatrix} \times \begin{pmatrix} 5 & 1 \\ 6 & 6 \end{pmatrix} = \begin{pmatrix} 23 & 19 \\ 17 & 13 \end{pmatrix}, \text{ but}$$

$$\begin{pmatrix} 5 & 1 \\ 6 & 6 \end{pmatrix} \times \begin{pmatrix} 1 & 3 \\ 1 & 2 \end{pmatrix} = \begin{pmatrix} 6 & 17 \\ 12 & 30 \end{pmatrix}$$

A mathematical system may have two operators, addition and multiplication. If these have inverses then you might say they have *four* operators, addition, subtraction, multiplication and division. But we don't really count subtraction and division as separate operators because they are just the inverses of addition and multiplication.

In some cases, a system could be a group under addition but only a semigroup under multiplication, or it could be a monoid for one operator and a group under the other, or it could be an Abelian group for addition, but a non-Abelian group for multiplication, which is actually the case for matrix algebra. We define a mathematical system according to the axioms it satisfies.

An important such system is called a *Ring*. It has two binary operators $+$ and $\times$; it is a group with respect to one of them and semigroup with respect to the other. Furthermore, it satisfies a new axiom, called the *distributive law*, which combines the two operators as follows. For any triplet (x, y, z)

$$x \times (y + z) = x \times y + x \times z \tag{3.4}$$

Ring

Definition: (Ring) *A ring, $\mathcal{R}$, is a system of elements in which there are two binary operations, addition ($+$) and multiplication ($\times$), such that $\mathcal{R}$ is an Abelian group with respect to addition and a semigroup with respect to multiplication, and such that the distributive law.* Equation (3.4) *holds for all triplets in $\mathcal{R}$*

When molecular physicists needed to solve certain problems regarding *symmetry* in subatomic particles, they found that group theory and ring theory were the systems most suitable for explaining the conservation of various quantities.

Emmy Noether

In the 1920's, the German mathematician Emmy Noether, was an "unofficial" associate professor (meaning that she was neither paid nor assigned classes to teach) in the mathematics department in the University of Gottingen, Germany. Any class she taught was listed under the name of David Hilbert, but the students knew it was really one of Emmy's classes. She gave lectures and seminars on her research, *Ring Theory,* which was one of her inventions. In 1933, because of the rise of Hitler and his attempt to eradicate the Jews, she fled Germany and came to the America. She ended up as a professor (official, this time) at Bryn Mawr University. She is considered to be one of the greatest mathematicians in history. Among the major results she formulated and proved, was the following powerful theorem.

Theorem: *Every differentiable (smooth) symmetry in a physical system has corresponding conservation laws.*

The physical system is known as the *Noetherian Ring*. A non-physicist, would not be expected to understand the above theorem, but it is never-the-less impressive and is recognizable by many of us, as revealing a significant truth in a few words. It is often in the nature of mathematics to express deep ideas in such an economical fashion.

The ability to work with semi-groups, monoids, groups, rings and other algebraic structures is at once beautiful and powerful. The beauty lies in the great variety of different *elements* that are playing the same game–that is, obeying the same set of rules. The power lies in the fact that any problem solved in a given system is simultaneously solved in every interpretation of that system. Thus, regardless of what kind of *things* are being studied, all the theorems for that system are true about those things. So that any time a new scientific, technological, or business problem is encountered, one only needs to start working in the appropriate mathematical system that may have already addressed or even solved that problem.

Groups of permutations

Let M be an alphabet consisting of the four letters, $\{A, C, G, T\}$. Now, let us pick all of the possible three-letter words we can make from this alphabet. How many words are there and what are some of them? The answer is there are 64 words and some of them are

$$ACT,\ AAT,\ CAT,\ CCT,\ GGG,\ AGT, TGT, \ldots$$

Why are there 64 words? Any three letter word must start with one of the four letters in the alphabet, then it must be followed by one of these same four letters, possibly repeating the first letter, then it must end in any one of the same four letter, this gives you $4 \times 4 \times 4 = 64$ different words.

A biologist would call these "words" *codons*, and the letters A, G, C, T, of the alphabet are the *nucleotides* (chemical compounds). A = Adenine, C = Cytosine, G = Guanine, and T = Thymine. The codons are put together into the strands of DNA found in the bodies of anyone reading this book. These are facts that can be related to the readers by a mathematician whose wife is a biologist.

By-the-way, if we did not allow repetition of letters, then there would only be 24 words possible. In such as case, we could have 4 ways to pick the first letter, then there would only be three letters left in the alphabet to use for the second letter, and after that, there are only two letters to pick for the end of the word, that is $4 \times 3 \times 2 = 24$.

If we had an alphabet of only 3 letters $\{A, B, C\}$, and did not allow repeats, then there would only be 6 three-letter words.

$$ABC, ACB, BAC, BCA, CAB, CBA$$

These 6 words, are called *permutations* of the alphabet. They are all of the ways we can arrange and rearrange the letters in the alphabet. Note that with permutations we do not allow repetitions. It gets even simpler if we had only two letters, $\{A, B\}$, we could make only two 2 letter words, AB and BA. We want to show how the collection of all permutations can be used as elements of a set, and how to define a binary operator, $\circ$, called a *composition* to define a *group of permutations*.

Permutation group of order 2.

Let S_2 be the set $\{1,2\}$. There are two ways we can permute (arrange) these two elements (12) and (21). Let us define a set $A_2 = \{(12),(21)\}$ as being the collection of all permutations of S_2. Furthermore, we want to use the notation x_0 and x_1 to stand for these two permutations, defined as follows:

$$\begin{aligned} x_0(12) &= (12),\ \text{leaving (12)} \\ x_0(21) &= (21),\ \text{leaving (21)} \end{aligned}$$

and

$$\begin{aligned} x_1(12) &= (21),\ \text{reversing (12)} \\ x_1(21) &= (12),\ \text{reversing (21)} \end{aligned}$$

Then our set A_2 is $\{x_0, x_1\}$

When we say x_0 or x_1 is "applied" to some arrangement of 1 and 2, we mean that it is a permutation that works on <u>that</u> current arrangement by obeying what it is defined to do. That is x_0 leaving them alone and x_1 reversing them. If we *concatenate* two permutations, that is, apply one after the other, we say we are making a *composition* of the two permutations. Thus, $x_1(x_0(21)) = x_1(21) = (12)$ means first use x_0 on (21), leaving it as (21), then apply x_1 on this resulting (21), changing it to (12). The notation we use for concatenating two or more permutations is $\circ$. So now,

$$x_1(x_0(21)) = x_1 \circ x_0(21)$$

You may call the circle $\circ$ "times".

Example: $x_1 \circ x_1(12)$ means *first* apply x_1 to (12) getting a new arrangement (21), *then* apply x_1 to that arrangement.

$$\begin{aligned} x_1 \circ x_1(12) &= x_1(21) \\ &= (12) \end{aligned}$$

Example:
Find the composite permutation $x_1 \circ x_0 \circ x_1$ applied to (21). Here it is:

$$\begin{aligned} x_1 \circ x_0 \circ x_1(21) &= x_1 \circ x_0(12) \\ &= x_1(12) \\ &= (21) \end{aligned}$$

Table 3.8 below, is a multiplication table, where "multiplication" means compositions $\circ$ of permutations.

$\circ$	x_0	x_1
x_0	(12)	(21)
x_1	(21)	(12)

Table 3.8

In this system, we see that there is a neutral element, namely, $x_0 = (12)$ because

$x_0 \circ x = x$ for all x. Also in this system every element (that is every permutation) x in A_2 has an inverse.   In fact, in this case, each element *is* its own inverse.   Putting this table into the x_0, x_1 format, we have Table 3.9 below.

$\circ$	x_0	x_1
x_0	x_0	x_1
x_1	x_1	x_0

Table 3.9

This group, A_2, is the same as a group we could make by using addition of the numbers $\{0, 1\}$ in mod 2 arithmetic.

$+$	0	1
0	0	1
1	1	0

Table 3.10

The entry for $1 + 1$ is 0 because $1 + 1 = 2 \cong 0 \bmod 2$, but both $1 + 0$ and $0 + 1$ are $1 \bmod 2$, and again each element is its own inverse.

Because these two groups: (A_2), Table 3.9 and (mod 2), Table 3.10 can be matched one-to-one with their elements and their neutral elements and their inverses, they are called *isomorphic* groups.   Likewise, the group S_2 is isomorphic to the group defined on the set $\{-1, +1\}$ where the operator is $\times$, *ordinary multiplication*.

$\times$	-1	$+1$
-1	$+1$	-1
$+1$	-1	$+1$

Table 3.11

In physics, if a particle has two states, and these fit the properties of a group of order 2, then the particles behave as the elements of the groups isomorphic to those depicted in Tables 3.9, 3.10 or 3.11.   Any theorems that could be proved using these groups, would also be true in the physics setting.   Not only physics, but in any model, $\mathcal{M}$ that is isomorphic to some particular group, $\mathcal{G}$ then the solutions in $\mathcal{G}$ can be used to solve problems in the model, $\mathcal{M}$.   As Al Khwarizmi might say, "... and that, your Eminence, is the nature of Al-jabr."

Permutation group of order 6

Now let us move on to a more complicated example.   The set $S_3 = \{1, 2, 3\}$ has 6 permutations.   Define the collection of all of the permutations on S_3 as

$$A_6 = \{x_0, x_1, x_2, x_3, x_4, x_5, x_6\}$$

where each permutations is defined to rearrange the (123) as follows.

$$x_0 = (123), \; x_1 = (132), \; x_2 = (213)$$
$$x_3 = (231), \; x_4 = (312), \; x_5 = (321)$$

Note: In this list we are using a shorthand way to say what each of these do to the arrangement (123). For instance,

$$x_4(123) = (312)$$

moving the last number, 3, to the front and shoving the first and second numbers back into the second and third spots, but,

$$x_4(321) = (132)$$

because when x_4 does what it is supposed to do, it moves the last number to the front shoves the other two back.

Example:

The x_1 permutation keeps the *first* number in place but it switches the *second* and *third* ones around. Now working with x_1 applied to the permutation (213) saying these italicized words to yourself, find the permutation $x_1(213)$. It is

$$x_1(213) = (231)$$

Is (231) one of the x's? Yes it is x_3. This proves that the composition permutation, $x_1 \circ x_2$, is x_3.

Problem: Find $x_5 \circ x_2(123)$

Solution:

$$
\begin{aligned}
x_5 \circ x_2(123) &= x_5(213) \\
&= (312) \\
&= x_4
\end{aligned}
$$

Problem: Describe in words what x_2 does to (123), and apply x_2 to x_5.

Solution: In words, x_2, that is, (213), "keeps the last one in place and switches the first two."

Now, $x_2 \circ x_5 = x_2(321) = (231)$, which is x_3.

This is a good time to note that $x_5 \circ x_2 \neq x_2 \circ x_5$.

Table 3.12 is the table of all of the permutations of A_6 and their compositions. It is a times-table for $\mathcal{A}_6$. Where "times", in this case, means composition, and we have highlighted the identity permutations x_0.

$\circ$	x_0	x_1	x_2	x_3	x_4	x_5
x_0	x_0	x_1	x_2	x_3	x_4	x_5
x_1	x_1	x_0	x_3	x_2	x_5	x_4
x_2	x_2	x_4	x_0	x_5	x_1	x_3
x_3	x_3	x_5	x_1	x_4	x_0	x_2
x_4	x_4	x_2	x_5	x_0	x_3	x_1
x_5	x_5	x_3	x_4	x_1	x_2	x_0

Table 3.12 The group A_6

We can use this table to solve various equations in this system. For example, let us define x^2 to be $x \circ x$

Problems:

1. Solve the quadratic equation $x^2 = x_0$.

Solution: There are four solutions because we can find four numbers, x in the table, such that $x \circ x = x_0$. Namely: $x_0, x_1, x_2,$ and x_5. "Wait a minute!", you might ask, "How is $x_5 \circ x_5 = x_0$?" First notice what x_5 is defined to do to (123). When we say $x_5 = (321)$ we mean it has kept the 2 in the second place and reversed the first and third place numbers. So what will x_5 do to (321)? It will keep the middle one, 2, in place and reverse the first and third (here, the 3 and 1). That is $x_5 \circ x_5 = (123)$.

2. Show that the quadratic equation $x^2 = x_1$ has no solution.

Solution: If we look at all the products $x_n \circ x_n$, (the diagonal from the upper left to the lower right) we see that none of them is x_1.

3. Show that the group A_6 is not commutative.

4. Show that the three elements $\{x_0, x_3, x_4\}$, by themselves, also form a group by making up a times-table using this composition operator, $\circ$.

5. Show that $x_0^3 = x_0$.

Solution: We must, first figure out what we mean by x^3. It is $x \circ (x \circ x)$.

So, $x_0 \circ (x_0 \circ x_0) = x_0 \circ x_0 = x_0$.

Therefore $x = x_0$ is a solution to the cubic equation $x^3 = x_0$. There are two other less obvious solutions, $x = x_4$, and $x = x_3$.

Let us use Table 3.12 to check that x_4 is indeed a solution.

$$
\begin{aligned}
x_4 \circ x_4 \circ x_4 &= x_4 \circ (x_4 \circ x_4) \\
&= x_4 \circ x_3 \\
&= x_0
\end{aligned}
$$

Similarly we can show that $x = x_3$ is a solution.

Symmetric groups

Permutations of the elements $\{1, 2, 3, ..., n\}$ in the set S_n have an interesting geometric property. If we think of the numbers $1, 2, 3, .., n$ as vertices of a regular polygon (a geometric figure, such as triangle, a square, or a pentagon, etc.), then a permutation of these numbers can be an interchange of two vertices, obtained by rotating the polygon around a fixed point at the center, or by flipping it along a line of symmetry, so that one vertex replaces another. In the case where nothing is done to the polygon then each vertex is replaced by itself, meaning that this is the identity operation. As an example, let us consider an equilateral triangle whose vertices are labeled as $1, 2, 3$ in Figure 3.3, below:

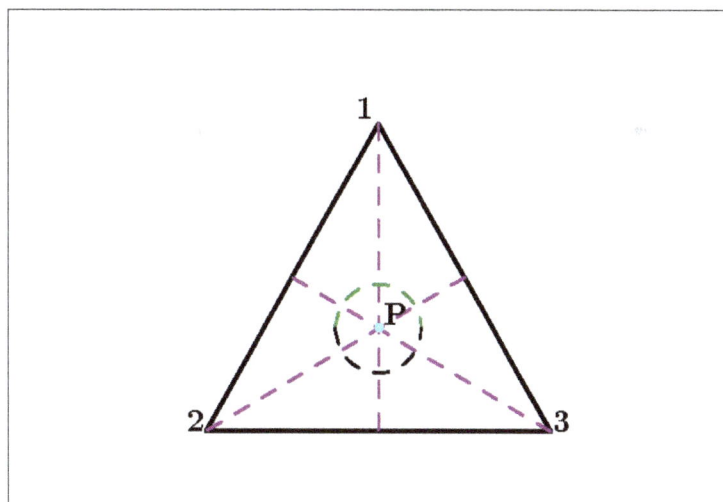

Figure 3.3 THE TRIANGLE 123

There are 6 ways we can interchange vertices because this triangle is symmetric about the point P and it can be <u>rotated</u> either $0°$ (the identity operation), or $240°$ or $120°$ about P, and this same triangle is symmetric along the dashed lines through the vertices 1, 3 and 2, so we can <u>flip</u> the triangle along these lines, keeping one vertex fixed and and replacing the other two vertices with each other. The following list shows how these moves correspond to the permutation group of order 6 that we constructed previously in Table 3.12.

1. No change. This means leave every vertex in its original place. This is called the identity symmetry $x_0 : (123) \to (123)$

2. Since the triangle is symmetric about the dashed line through the top vertex 1, we can leave 1 in place ("fix 1") and switch 2 and 3. Call this $x_1 : (123) \to (132)$

3. Fix 3 (keep 3 in its place) and switch 1 and 2 about the dashed line through 3. Call this $x_2 : (123) \to (213)$

4. Rotate the whole triangle $120°$ clockwise about P. This carries 1 to position 3, 2 to 1 and 3 to 2. Call this $x_3 : (123) \to (231)$

5. Rotate the whole triangle $240°$ clockwise about the point P. This carries 1 to position 2, 2 to 3, and 3 to 1; call this $x_4 : (123) \to (312)$

6. Fix 2 and switch 1 and 3 about the dashed line through 2, call this $x_5 : (123) \to (321)$

This example should make it clear why groups of permutations are called *symmetric groups*. Scientists have found uses for such groups in the structures of atoms and molecules–applications to atomic physics and molecular chemistry, all related to symmetry.

For example, molecules may have atoms such as Hydrogen and Oxygen arranged in various symmetric orientations around some other centrally positioned atom such as Carbon. Frequently, the arrangement of electron orbits, or the sharing of electrons or quantized energy levels, or atomic bonds, etc. can be determined from just the physical arrangements in the molecule, itself. Permutations of these symmetries can be interpreted as the elements of a symmetric group, and knowledge of their symmetric structure is useful in analyzing the atom, and predicting chemical reactions.

Number Theory

Sometimes the question comes up, "Is mathematics discovered or invented?" There is a case to be made for saying that the mathematical systems we just discussed are *inventions* as opposed to *discoveries*. We have artificially and deliberately defined a set of elements upon which we have arbitrarily assigned operators called addition and multiplication.

But there is one mathematical system in which (in my opinion) the beautiful truths are already in the system and just need to be uncovered by clever scrutiny. It is a more natural topic, namely the "natural numbers." These are the positive integers in Kronecker's quote given at the beginning of this chapter. Let us examine the counting numbers and see what kind of interesting patterns we can find there. The first hidden treasure we stumble upon is the path through the integers, called the *prime numbers.*

Prime numbers

Be reminded that the prime numbers are those positive integers greater than 1 that can't be factored into any factors other than 1 and the number itself. For example, 13 is prime. It is 1×13, and has no other prime factors. But 15 is *composite*–not prime, because it can be factored into two primes 3 and 5. The factors need not be all different, for example $48 = 2 \times 2 \times 2 \times 2 \times 3$. We say it has five factors, even though four of them are the same. Actually, it is more accurate to say that 48 has two prime factors, 2 and 3.

For an easy reference, here is a list of the first few prime numbers:

$$\{2, 3, 5, 7, 11, 13, 17, 19, 23, 29, 31, 37, 41, 43, 47, 53, 59, 61, 67, 71, 73, 79, 83\}$$

One interesting thing about prime numbers is that they cannot *evenly divide* two consecutive integers. For example, there is no prime number that can evenly divide both 34 and 35. Look, 34 can be divided by 17 and 2, but neither 2 nor 17 will go into 35, evenly (meaning with a zero remainder). And 35 can be divided evenly by 7 and 5, but neither of these will go into 34. You might think, "Ha! I can divide 34 and 35 by 1," well fine, but the trouble with that is 1 is *not* a prime number!

In general, if n is any integer and $n + 1$ is the next integer then n and $n + 1$ *cannot have any prime factors in common.* You can prove this for yourself by supposing that for some integer n, a prime number, p, can divide both n and $n + 1$ with 0 remainder then you would get

$$\frac{n+1}{p} = \frac{n}{p} + \frac{1}{p}$$

meaning that $\frac{n+1}{p}$, $\frac{n}{p}$, and $\frac{1}{p}$ would be whole numbers, but that would make $\frac{1}{p}$ be a whole number, so $p = 1$; but, fortunately, 1 is still not a prime number! This is a very important theorem (a valuable nugget) in the positive integers. One reason is that we can use it to prove that there is no largest prime number.

Using this fact in 300 BCE, Euclid gave a proof that there are *infinitely many prime numbers.* Here it is.

Euclid's proof that there are infinitely many primes

Suppose that there are only a finite number of primes and suppose that n is the last prime, that is, n is the largest possible prime number. Then let M be the *product* of all prime numbers less than or equal to n.

$$M = 2 \times 3 \times 5 \times 7 \times 11 \times ... \times n$$

Now, add 1 and look at $M + 1$. Either $M + 1$ is either prime or not prime. If $M + 1$ is prime then it is bigger than the supposed biggest prime, n. So, this is impossible. On the other hand, if $M + 1$ is not prime, then it must have prime factors, *none of which* can be factors of M, because, as we have just said, no two consecutive integers like M and $M + 1$ can have a common prime factor. Therefore, any prime factor, p of $M + 1$ cannot be less than or equal to n meaning that all of the prime factors of $M + 1$ are greater than n. This, also, is impossible. On the other hand, ...Wait! There is no other hand. Therefore, by the tradition of logic, we have proved that there is no largest prime.

Other theorems about primes were also proved two millennia ago. Around 275 BCE, Eratosthenes showed how you could filter out the prime numbers from the composites by a method we call the Sieve of Eratosthenes.

Mersenne primes Less remotely, in the 1500s and 1600s, Marin Mersenne, Pietro Cataldi and others studied the specific problem of trying to find numbers, n such that $2^n - 1$ is prime. These are known as a *Mersenne Primes* and no one knows whether there are infinitely many such primes. The first four Mersenne primes are:

$$
\begin{aligned}
M2 &= 2^2 - 1 = 3 \\
M3 &= 2^3 - 1 = 7 \\
M5 &= 2^5 - 1 = 31 \\
M7 &= 2^7 - 1 = 127
\end{aligned}
$$

and each time a new one is found by somebody using a computer algorithm, it is publicly announced in a geeky race to find the largest known Mersenne prime. The day after Christmas 2017, the 50th one called M77232917 was found after a fourteen year search by Jonathan Pace, an electrical engineer from Tennessee. He used the Great Internet Mersenne Prime Search (GIMPS) algorithm getting

$$2^{77,232,917} - 1$$

This prime number has $23,249,425$ digits. The GIMPS algorithm is being used to produce new larger Mersenne Primes at the rate of about one per year.

Meanwhile, during the past 2000 years, mathematicians have been raising other questions about primes.

Prime gaps

How long a gap can there be between two consecutive prime numbers? The answer is that gaps can be made larger than any size you specify. But gaps do not just constantly increase, sometimes they get longer and some times they get smaller. For the integers less than 130 there is a gap of 13 between 113 and 127; every integer between 113 and

127 is composite. Then it drops back down to 3 between 127 and 131. Is it possible for a gap of some size to repeat itself infinitely many times?   As we shall see, some major puzzles concerning the gaps in prime numbers have been resolved, as recently as 2014.

The prime number theorem

A big question in number theory is: Given any positive integer n, how many primes are less than n? The answer is given as $\pi(n)$. But, alas, this is not the π which stands for the ratio of the circumference of a circle to its diameter, it is just a use of the Greek letter π to stand for p in the word, *prime*.   The way you read $\pi(n)$ is simply, "the number of primes less than n." Despite what we said earlier about modern notation being the best there is for its purposes, we must confess that mathematics is riddled with enough notational ambiguities and double *entendres* to classify it as a branch of poetry.

It is easy to compute the number of primes less than any given positive integer, n, by *brute force*; that is listing all the primes and counting them.   Well, it is easy if the number n is not very big, like, say, $n < 100$.

For example, in the list of the primes $\{2, 3, ..., 83\}$ given at the beginning of this section you can see, by counting that there are 23 primes less than or equal to 83. That is, $\pi(83) = 23$. But what about a large number, like $1,000,000$?   Using a computer we can say:

$$\pi(1,000,000) = 78,498$$

Listing them and counting them is out of the question without extremely powerful computers.   Even then, we cannot, at this time, make a general statement such as, "If n is any positive integer, then there is a formula, f, written in terms of known functions of n which allows us to calculate the exact number of primes that are less than n."

In the 1700's, mathematicians were struggling to find some kind of formula that would even roughly approximate the number of primes less than n.   They were searching for a good *approximation* to the "pie in the sky," $\pi(n)$.   Nobody found one.   In 1798, the French mathematician Adrien Marie Legendre did come up with a pretty good one, which has been the one most used even up to today–it is this:

$$\pi(n) \approx n/\ln(n), \text{ for } n \geq 6$$

where $\ln(n)$ is the base e logarithm of n. (We will discuss logarithms in a later chapter.) If we apply this formula to the number 83, we expect to get something close to 23.

$$83/\ln(83) = 83/4.4188... \approx 19$$

Not great, but not too bad either. Since 1798, some refinements have been made to give us slightly better answers.[2]   But the important thing is that the Legendre formula approaches $\pi(n)$ as n gets larger and larger.   Two French mathematicians, Jacques Hadamard and Charles-Jean de la Vallee Poussin, proved this in the 1890's and it is known as the *prime number theorem*.

If $n > 2$ and $\pi(n)$ is the number of primes less than n, then

$$\lim_{n \to \infty} \frac{\pi(n)}{(n/\ln(n))} = 1$$

---

[2] One example is $\pi(n) \approx n/\ln(n) + n/(\ln(n))^2$

This is often stated as: "$n/\ln(n)$ is asymptotic to $\pi(n)$."

Examples: If $n = 100$, then by counting, the number of primes less than 100, we get 25 as the actual value of $\pi(100)$. But by the Legendre formula, $100/\ln(100) \approx 100/4.605 \approx 22$ So,

$$\frac{\pi(100)}{(100/\ln(100))} = \frac{25}{22} = 1.13636...$$

For $n = 10,000,000$, the Legendre formula gives us

$$10,000,000/\ln(10,000,000) = 620,420$$

instead of the actual value for $\pi(10,000,000)$ which is $664,579$. For this n the ratio is

$$\frac{\pi(n)}{n/\ln(n)} \approx 1.0711748...$$

Goldbach conjecture

We have already mentioned one unsolved problem in the prime numbers. In 1742, the German mathematician, Christian Goldbach wrote a letter to Leonard Euler, proposing two conjectures, which may be stated as follows:

GC1. *Every even integer ≥ 6 can be written as the sum of two primes.*

GC2. *Every odd integer ≥ 9 can be written as the sum of three odd primes.*

It is understood that the words *two* primes need not mean two *distinct* primes and, similarly, *three* primes also can mean that all three primes are the same. An example of GC1 for the even number $n = 6$ is $3 + 3$ and, in GC2, for the odd number $n = 9$, just use $3 + 3 + 3$.

He did not prove either conjecture in this letter, but Euler did observe that GC2 could be proved from GC1, but not the other way around. For this reason, GC1 has been called the *Strong* Goldbach Conjecture and GC2 has been called the *Weak* Goldbach conjecture.

Here is how we could prove the weak one by assuming the strong one.

Proof:

Assume GC1. Every even integer ≥ 6 can be written as the sum of two primes.

1. Let n be any odd number ≥ 9

2. So, n, can be written as $2k + 3$, for some integer k. Let $n = 2k + 3$

3. Since $n = 2k + 3$, then $n - 3 = 2k$, an even number

4. Being even, $n - 3$ can be written, by $GC1$, as the sum of two primes p and q, that is $n - 3 = p + q$

5. Therefore $n = p + q + 3$; the sum of three odd primes

Of course, this little exercise is *not* a stand alone proof of the weak, because it depends upon the strong, which has not been proven yet, as of the year 2020. What we need is a proof not dependent on the strong.

Actually, we have it, in 2013, the Peruvian mathematician Harald Helfgott proved GC2 for any odd number C greater than $n = 10^{27}$. How does that count as a proof for GC2?   Let us go back 90 years to 1923 when the two British mathematicians G. H. Hardy and J. E. Littlewood proved that every "sufficiently large" odd number, C, could be written as the sum of three odd primes, but only if they were allowed to assume a conjecture called the *Grand Riemann Hypothesis*, GRH (which we will not state here). There were two drawbacks with that: first, GRH had not been, and is still not yet, proven and second, the odd number C mentioned in the proof was (and still is) beyond the computational capacity of any computer. This means that we could not verify the result for all odd numbers less than that number C. This proof, however, was important because it showed that the weak was not unassailable as long as GRH is true.

In 1937, we had a breakthrough; the Russian mathematician I. M. Vinogradov proved, without using GRH, that every odd number, n greater than $C = 3 \times 10^{6846168}$ is the sum of 3 odd primes. The trouble with this is that computers today can verify that GC2 is true for only the odd numbers up to 8×10^{30}. This is where Helfgott's proof comes to the rescue; he showed that every odd number greater than $C = 10^{27}$ is the sum of three odd primes. Wow! This is well within the range for computer verification! Therefore, GC2 is true for all odds (greater than or equal to 9.)

What about the strong?

For the past 275 years, however, no one has ever been able to prove GC1, without assuming some other equivalent conjecture. So exactly what have these people been doing all these years, just wasting their time?   No, they have been adding new conjectures, proving theorems that are modified versions of GC1, such as having *extra* terms in the equation. In these cases, their goal is to find new proofs which reduce the extra terms, or better yet to eliminate them, in order to get down to just two primes.   These problems, GC1 and GC2 as well as other theorems and conjectures are called *additive prime number theory*, a branch of *analytic number theory*. Some examples are as follows.

In 1895, Emile Lemoine, a French mathematician, conjectured, but did not prove, that every positive odd integer greater than 9 is the sum of a prime and twice the product of two primes. *Mon Dieu!* A proof of this would have given us a special case of the GC2.

In 1919, Viggo Brun, a Norwegian mathematician proved that every even number is the sum of two numbers, each having at most nine prime factors. If we could reduce the number of prime factors to just *one* instead of nine, for each number, we would have it made!

In 1930, Lev Schnirelmann, a Russian mathematician, proved that every positive integer greater than 1 (including all even numbers) can be written as the sum of not more than C primes; his number for C was $800,000$. *Boje Moi!* Over a couple of years, this number was reduced to $C = 2,208$ by various mathematicians.   In 1936, Lev Landau reduced it to $C = 71$.   Then C was reduced to 67 in 1942 by the Italian Giovanni Ricci. The British mathematician, R. C. Vaughan brought it down, for even numbers, to $C = 6$ in 1995, and this stood as the record until Helfgott's proof in 2013. Now, for $n \geq 12$, it is $C = 4$. Because then $n - 3 \geq 9$, and thus by GC2, $n - 3$ can be written as the sum of three odd primes: p_1, p_2, p_3. So, $n - 3 = p_1 + p_2 + p_3$. In other words, $n = 3 + p_1 + p_2 + p$, the sum of 4 odd primes.

In 1953, the Russian mathematician,Yu V. Linnik, proved that every sufficiently large even number, $2n$, can written as the sum of two primes plus K powers of 2. It is called an "Almost Goldbach Theorem." This had been proved for numbers as large as $K = 54,000$, but, later K was reduced by several mathematicians. In 1999 Tianze Wang, from China, brought it down to $K = 2,250$. Finally, in 2002, two British mathematicians D. R. Heath-Brown and J. C. Puchta, proved it for $K = 13$. *Gorblimey!* To get GC2, we need both $K = 0$, and a number $2n$ small enough to be within computer range.

Probably the best result yet is one called the "$(1,2)$ theorem", proved by the Chinese mathematician, Chen Jing Run, stated in 1966, and proved in 1972.

Chen's (1,2) Theorem:

For a large enough n, every even number, 2n, can be written either as the sum of two primes or the sum of a prime and a semiprime.[3]

We will not attempt to write out this difficult proof here. Unfortunately, here "large enough" means a number $C \geq 10^{10^{16}}$. This might take a computer the size of our solar system ten million years to verify the theorem for all even numbers $< C$.

But there is a new theorem, using *coprime* instead of semiprime. It is almost as neat, as Chen's Theorem and can easily be proved. It happens to overlap Chen's

First, we need to see a definition for what it means to say two integers are coprime.

Definition: *Two numbers a and b are coprime means that they have no common factors other than 1.*

Examples: 9 and 35 are coprime; the prime factors of 9 are $(3,3)$ and the prime factors of 35 are $(5,7)$, so they have no common factors, other than 1. But 9 and 39 are not coprime because they both have 3 for a common factor, since $9 = 3 \times 3$ and $39 = 3 \times 13$.

I will present this theorem here without a proof. It is scheduled to appear in the March 2021 issue of the Mathematical Gazette.

Prime plus Coprime Theorem:

For all integers $m \geq 4$, the even numbers 2m can be written either as the sum of two primes or the sum of a prime and a coprime to m.

Notation: Let us call Chen's Theorem the PSP, for prime-semiprime theorem and the new theorem, just stated, the PCP, for prime-coprime theorem.

I say that PSP overlaps PCP because sometimes the semiprime in Chen's theorem *is* coprime to n, and sometimes it *is not*. Also the coprime number, r, in PCP can be, but is not always a semiprime.

Examples: If $2m = 44$, then, by *both* PSP and PCP

$$44 = 13 + 31, \text{ prime} + \text{prime}$$

By PSP

$$44 = 11 + 33, \text{ prime} + \text{semiprime. (not coprime)}$$

By PCP

$$44 = 17 + 27, \text{ prime} + \text{coprime (not semiprime.)}$$

Sometimes an attempt to prove Goldbach will depend upon some assertion that is either erroneous or turns out to be actually *equivalent* to Goldbach. *Equivalent* means

---

[3] A semiprime is a composite number that has exactly two (odd) prime factors, not necessarily distinct. For example, 9 is a semi-prime because $9 = 3 \times 3$, but 45 is not a semiprime, because $45 = 3 \times 3 \times 5$. It has three prime factors.

that the assertion can be proved by assuming Goldbach and Goldbach can be proved by assuming the assertion.   There is nothing wrong with this, and it could even be useful in giving us another way to look at Goldbach.   Actually, both of them could be false, or true.

Here is an equivalent conjecture, that we can use to help shed some light on Goldbach.
Conjecture (Prime Symmetry).

SYM. *If $n \geq 3$ there is always at least one non-negative integer $r < n$ such that each of $n - r$ and $n + r$ is a prime.*

Among other things, this says,

1. Every number $n \geq 2$ either is prime or half-way between two primes.
2. *Every pair* of primes whose sum is $2n$, must be symmetric to n.

By saying that a pair of primes, (p, q) is *symmetric* to any number $n \geq 3$, we mean that n is the average of the two primes.   In other words, we are saying that there is a number r, such that $0 \leq r \leq n$ and one of p or q must be $n - r$ and the other must be $n + r$.   If n is prime then $r = 0$ and $p = q$.

Theorem (GC1 is equivalent to SYM)
The SYM is true if and only if the GC1 is true.

Hint for a proof: If GC1, then $2n = p + q$ and either $p = q = n$, or $p > n > q$ or $p < n < q$.   Let's take the case where $p > n > q$,   Since, in all cases, $p = 2n - q$, therefore (subtracting n from both sides) $p - n = n - q$, so $p - n$ and $n - q$ are equal to the same number, r, etc.   And if SYM, then the sum of the two primes primes $n - r$ and $n + r$ is $2n$.

Neither GC1 nor SYM has been proved true nor false, for every integer yet.   Apparently, the prime symmetry conjecture, itself, is *not easy* to prove because if it were then Goldbach's would already be affirmed.   So far, since the Goldbach conjecture is known to be true for all integers up to $2n = 4 \times 10^{17}$, a four followed by seventeen zeros, then the prime symmetry conjecture is also true for equally large numbers.   So SYM gives us an easy way to find Goldbach pairs, (primes p and q that add up to $2n$) for any even number $2n$, such that

$$4 \leq 2n \leq 400,000,000,000,000,000$$

Here it is, we will call it the SYM Algorithm for finding Goldbach Pairs for any such small even number[4].

SYM ALGORITHM

1. Let n be any positive integer ≥ 3, then for any even number $2n$

2. If n is *odd,* find $n \mp r$ for every *even* number, r less than n.   Include $r = 0$, but not $r = n - 1$.

3. If n is *even,* find $n \mp r$ for every *odd* number, r less than n.   Include $r = 1$, but not $r = n - 1$.

4. Any number r for which both $n - r$ and $n + r$ are primes, $n \mp r$ is a Goldbach pair.

Example: Suppose $2n = 28$.

---

[4]The even number $n = 2$ is a special case, since 2 is the only even prime.   Therefore, $r = 0$ and $4 = (2 - 0) + (2 + 0)$

Let $n = \frac{1}{2}28 = 14$

Find $14 \mp r$, where r is one of the the 6 odds, $\{1, 3, 5, 7, 9, 11\}$, (not including 13)

$$\{14 \mp 1, 14 \mp 3, 14 \mp 5, 14 \mp 7, 14 \mp 9, 14 \mp 11\}$$

Of these, there are exactly two prime pairs

$$\{(11, 17), (5, 23)\}$$

If you had wanted to find only one Goldbach just stop when you get the first one $(11, 17)$.

In the graph, Figure 3.4 below, we will illustrate small even numbers having 2 or more Goldbach pairs.

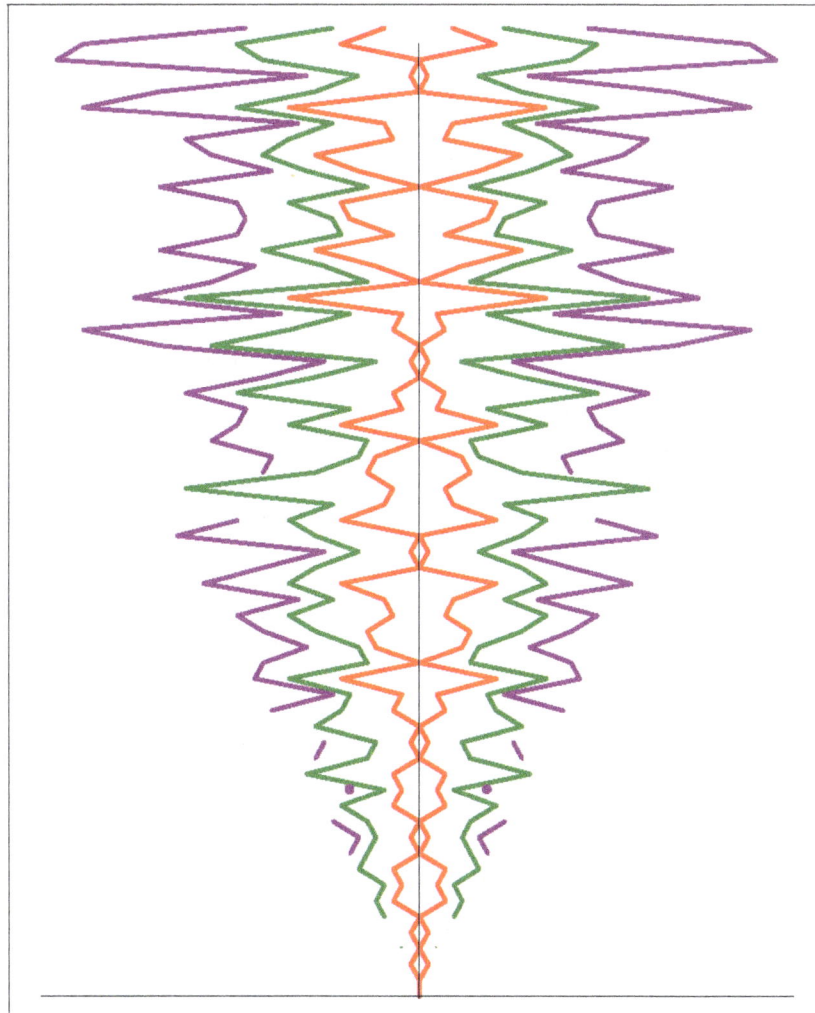

FIGURE 3.4 PRIMES OF THE FORM $n \mp r$

In Figure 3.4, we run n from 3 to 67 and compute up to three values of r, such that $n \mp r$ are primes. It is an interesting pattern and has a nice pleasant symmetrical graph. It was constructed as follows:

1) Red: for each integer, n, we select the *smallest* value of r that can be used to find primes of the form $n \mp r$, and we plot $n - r$ and $n + r$, in red.

2) Green: for each integer n, we find the second smallest value of r (if it exists) that makes $n \mp r$ prime and plot those in green.

3) Purple: For each n, the next smallest value (if it exists) of r that makes $n \mp r$ prime is plotted in purple.

Millions of Goldbach Pairs

In the above table and graphs, we have computed only the first few values of r for which $n \mp r$ are prime. It can be shown by examples, that for large enough numbers n there can be *hundreds*, *thousands*, even *millions* of Goldbach pairs that add up to $2n$. But in the midst of all these riches, we still (in 2020) have not proved that it is true that for any unspecified number n, there must be at least one single pair of primes whose sum is $2n$.

Other paths on the Prime number trail

But the Goldbach conjecture is not the only interesting shining gem hidden in the counting numbers. There are many theorems and conjectures about primes, prime gaps and prime formulas.

One major unsolved problem is: If n is any positive integer does there always exist a prime between the two consecutive squared integers n^2 and $(n + 1)^2$? For example, we know that there is a prime number between 49 and 64, that is between 7^2 and 8^2, and that there is a prime between 400 and 441, (20^2 and 21^2) But is this always true? The general form was conjectured by Legendre in 1800, and has not been proved yet. What about the question: If $n \geq 4$, is there always a prime number between n and $2n - 2$? This was claimed to be true, by a French mathematician Joseph Bertrand in 1845; it was finally proved by the Russian mathematician Pafnuty Chebyshev in 1919. It is known as the Bertrand-Chebyshev theorem.

I want to introduce you to one of the most obscure prime number conjectures that, in my opinion, needs to be more widely known because it tells us something about the ratio of two consecutive primes. Knowing how close two consecutive primes can be to each other might shed some light on some other prime number problems. It was posed in 1982 by the Iranian mathematician, Faridah Firoozbakht.

Firoozbakht conjecture:

For all $n \geq 1$, if p_n is the nth prime number, then the next prime p_{n+1} is less than $p_n \times$ the nth root of p_n.

She actually stated it as

$$p_{n+1} < p_n \times p_n^{1/n}$$

It has been verified for n up to 10^{19}.

Examples: Look at a few primes, $\{p_1 = 2, p_2 = 3, p_3 = 5, p_4 = 7, p_5 = 11\}$, then

$$3 < 2 \times 2, \; 5 < 3 \times 3^{1/2}, \; 7 < 5 \times 5^{1/3}, \; 11 < 7 \times 7^{1/4},$$

If you check out the computations you will see that

$$3 < 4, \ 5 < 5.196, \ 7 < 8.549, \ 11 < 11.386$$

Try it for any prime you want (but make sure you stay less than 10^{19}).

Dirichlet's Theorem

In 1837, German mathematician Peter Gustav Lejeune Dirichlet proved that given any two relatively prime integers a and b, there are infinitely many non-negative integers, n, such that $a + nb$ is a prime.

Examples: There are infinitely many primes of the form $4 + 7n$.

$$\{11, 53, 67, 109, ...\}$$

There are infinitely many primes of the form $5 + 7n$.

$$\{5, 19, 47, 61, 89, ...\}$$

Twin Primes

A conjecture about primes may have more to do with the gaps in the distribution of primes, rather than the primes themselves. One such is the following:

There are infinitely many pairs of twin primes.

This statement has never been proved. Nor has it been proved that there are only a finite number of twin primes.

What are twin primes, anyway? A pair of twin primes are two primes that are two apart, such as $\{(3, 5), (5, 7), (11, 13), (17, 19), (29, 31), ...\}$.

Consecutive primes

An important theorem related to twin primes has *just recently*, (in 2013) been proved, and we will tell you about it shortly. This new theorem is about *consecutive primes*, but what are they?

Two primes, p and q, are called *consecutive* primes if there are no primes between them. For example, 31 and 37 are consecutive primes because all of the numbers between them $\{32, 33, 34, 35, 36\}$ are composite, but the two primes 41 and 47 are *not consecutive* because one of the numbers between them, namely 43, is a prime. Twin primes are consecutive primes.

If p and q, with $p > q$, are two consecutive primes both greater than 3, then $p - q = m$, an even number, is called the *gap* between these two primes. It means that there are $m - 1$ composite numbers in a row between the two primes. As we have just noted, twin primes are consecutive primes and have a gap of 2. Now let some fixed even number $2k$ be the gap between a pair of consecutive primes. We have wanted to know, since the time of Euclid, whether or not there are infinitely many pairs of consecutive primes with that same size gap of $2k$.

Zhang's discovery

In April, 2013, a Chinese-American mathematician, Yitang Zhang, while teaching at the University of New Hampshire, wrote a paper giving a proof that there are infinitely many pairs of consecutive primes that have a gap of $70,000,000$. This is a long way from a gap of 2, and this seemingly small step toward proving the twin prime conjecture is actually a giant leap. Up until that time, no one had ever proved that there are infinitely many pairs of consecutive primes that are some given fixed number apart.

Many mathematical discoveries these days are the results of a team effort of several mathematicians working together at a large research oriented university or a well-financed institute; so, it is remarkable that Dr. Zhang developed this result while working alone, as a lower ranking faculty member in the mathematics department. (He had to supplement his salary by working in a fast food restaurant.)

Of course, he did have access to other papers related to this discovery, but he persevered on his own to ferret out this result. After his paper was published, other mathematicians started using Zhang's method. Within weeks, a team called the *Polymath 8 Project* narrowed the gap considerably, proving that there are infinitely many consecutive prime pairs with a gap of 5414. They later narrowed the gap to 246. They are marching toward a proof of the twin prime conjecture.

Dr. Zhang became an instant celebrity, winning 3 mathematical prizes, and becoming the "most famous mathematician in the world" in 2013. He was promoted to full professor at New Hampshire, gave invited lectures at several other universities, and, in 2015, was lured away to join the faculty at the University of California at Santa Barbara.

What attracts us to primes?

Other than purely academic reasons, why do we want to study these conjectures and theorems about prime numbers? The practical answers, in the 20th and 21st Centuries, are that we need them for computer security codes used in financial, military, and other operations. For example, a bank or a credit card company will assign large composite numbers to a system of accounts. The prime factors of these composite numbers are known only to the institution, itself. This is supposed to provide security against anyone who wants to pry into the records of clients.

Unfortunately, this can break down as computers become more capable of factoring larger and larger numbers and as algorithms and hackers become more sophisticated. In a sort of "arms race," the institutions fight back with even more cryptic codes. Even the matrix codes we discussed in Chapter 2 have their limitations against the really powerful super computers.

On the other hand, mathematicians want to prove theorems about prime numbers in order to better understand the number system. From this they hope to gain insights into many interesting and important truths which, for a mathematician is the Holy Grail.

How Mother Nature uses integers

The counting numbers and the prime numbers are always available for all human beings, and other creatures, to work with or play with or to otherwise make use of. Ravens, and other members of the *covidae* will count how many hunters entered a cabin and how many left, so they could know when it was safe to come out of hiding and do their ravenly things.

Nature even has her own agenda in the use of prime numbers. Certain species, such as the 13 year locust or the 17 year locust have evolved to hide their reproductive cycles in prime numbers. For example, if one of their predators has a population burst every 4 years, it will not be able to find the 17 year locust for 68 years, 68 being the least common multiple of 17 and 4. In other words, the predator would have gone through 17 four-year generations without a single locust to eat.

Of course, this does not mean the predators would die out, but their preferred food would be harder to get, and the prey has improved its survival rate. Otherwise, if against this same 4 year predator, the locus species re-emerged every 16 years, the predator would be there waiting for them every fourth generation, thereby reducing the 16 year locusts' survival rate. This is a simplified version of what really happens, but it does illustrate an interaction between the periodicity of multiples of prime numbers.

In April 2018, the journal, *Communications Biology* published an article, reporting that if the 13 year Locust and 17 year locust emerge, at the same time, they hybridize, thus improving their genetic vigor. In order for their emergence to occur simultaneously, the two cycles must coincide at a number that is a multiple of both 13 and 17. The least common multiple of these two prime numbers is 221. This means that every 221 years they get together and have a big party, and they have been doing this for about $100,000$ to $200,000$ years.

What is hidden between the integers?

This wraps up what we wanted to say about number theory. It feels like we have been travelling through a forest, the counting numbers. Our ride is down a special path between the giant sequoias, the prime numbers. But now it will be necessary to dismount and spend some time looking at ground between the trees: the squirrels, the small wild flowers, the Stellar Jays, the butterflies the spider mites, and the microbes. What is going on with the life of these fractions, both rational and irrational? In the next chapter, we propose to study the number system, with which you are most familiar—the real number line. Fortunately, the group axioms can provide us with steady support and guidance.

CHAPTER 4

THE REAL NUMBERS

Distance and time

Ask anyone in Santa Rosa CA, "How far is it from Petaluma, to San Francisco?" They might say, "50 miles" or they might say "one hour." Of course, these two concepts are quite different, but they are both correct if you know that you can travel at an average rate of 50 miles per hour.

One other interesting thing about these two concepts (space and time) is that both of these are interpretations of the *real number line*. Each is a *continuum* of numbers; one is the set of points between two geometric points A and B and the other is the set of times between two temporal points A and B. By a continuum, we mean a set of numbers that can be cut infinitely into smaller and smaller pieces. The distance between any two such points is a positive number going in one direction (to the right) and a negative number going in the other direction.

If a is to the left of b, and you take any interval $[a, b]$ consisting of all of the numbers (points or times) between a and b including both a and b, then that interval has positive length, $b - a$. Each point (either geometrical or temporal) itself has zero length.

Selecting a single point would be the like standing in one place forever, or freezing time, as in pushing "pause" button while playing a video. Albert Einstein and all modern physicists consider the space–time continuum to be the real world; consisting of four dimensions, three space and one time. Of course, this is a greatly over-simplified idea of what physical models are today. Most of them are formulated in multidimensional space with more than four number–lines of real and imaginary numbers, and other things called vectors and tensors. It is also true that the number of dimensions need not be finite.

But it all begins with the real number line. Fortunately, we can precisely describe the real numbers. They are defined by axioms, which we are about to reveal to you right now. Not that they are big secrets, but we want them all in one handy place so that you can refer to them, later on, as you read the proofs of theorems or as *you* prove the theorems for yourself. As we say in philosophy or law, "let us define our terms."

Definitions

Definition: (Field) *Let S be a set of elements and define two binary relations $+$, and $\times$ on S such that S is a commutative group with respect to both operators, with the condition that the two identity elements are different from each other, and that these two operators, $+$ and $\times$ satisfy the distributive law:*

$$x \times (y + z) = x \times y + x \times z$$

then the system $(S, +, \times)$ is a field.

Note that this gives us eleven axioms, because the group $(S, +)$ has five axioms and the group $(S, \times)$ has five axioms and the distributive law is the eleventh axiom.

   Definition: (Ordered Field) In addition to the eleven axioms defining a field, we will state the four *order axioms* which give us rules for handling inequalities, totaling 15 axioms defining what is called an *ordered field*.

   Definition: (Well-ordered field) Axiom 16, the *well-ordering principle* further restricts the properties of "greater than" and "less than." It tells us that every set of positive integers has to have a first element, which is, as we shall see, the basis of *mathematical induction*.

   Definition: (Real numbers) Finally, when we include Axiom 17, the completeness axiom, which basically tells us that the number line is continuous, not full of holes, then we will have defined the real number system. In other words, the real number line is a *complete well-ordered field*. It satisfies all 17 axioms.

If you are ready to embark on this mathematical adventure, here is your itinerary.

The itinerary

We will systematically introduce all seventeen axioms in several coherent sub-collections. We will state theorems whose proofs require only the first few axioms. Using the previously proved theorems and a few more axioms, we will state and prove more theorems. However, in this book we will not prove all of the theorems stated, but leave some of them for the reader to prove. We will continue this process until we find ourselves stating theorems using all seventeen axioms.

It is a fun trip, and you as a tourist will get to participate in the proofs as much as you want. Your reward will be that you will have glimpsed a major part of the wonderful world of mathematics. This could be compared to having been exposed to the world's greatest literature, music and art.

Here are the *mathematical world heritage* sites to look for:

- Axioms and theorems of the additive group[1]

- Axioms and theorems of the multiplicative group

- The distributive axiom and theorems

- The order axioms and theorems

- The induction axiom and theorems

- The completeness axiom and theorems

---

[1] The Group axioms stated here are the same ones we learned back in Chapter 3.

Packing for the trip

We will need two suitcases for the trip, *Logic* and *Equality*.

Some logic

Before we get started we need to take note of an assumption we will make about logic. We can and will confidently assert that *a properly expressed statement must be true or false*, and not both, of course. Also, we will assume that it is not possible to prove or disprove every conceivably statement. This type of logic is what everyone uses all the time and is called, by mathematicians, *two-valued logic*[2], and is essential for *indirect proofs*.

Many theorems may be stated in the "If A then B " format. For example: "If S is a square then S is a rectangle. This is saying that every square is a rectangle and its proof is based upon the definitions of squares and rectangles. The "If ___ " part is called the *Given* and the "then ___" part is called the *Conclusion*. Usually, you cannot reverse the *if* and *then* clauses and expect the new statement to also be true. For example, it is *not* true that if S is a rectangle then S is a square because there are some rectangles that are not squares.

Some theorems, however, are true "both ways." For example: the statement "If $3x > 1$, then $x > \frac{1}{3}$" is true and so is the statement "If $x > \frac{1}{3}$, then $3x > 1$." These are two theorems, both of which can be proved independently from the axioms and definitions. But we must not try to use one of them to prove the other one until we have actually proved at least one of them. When we have two theorems like this we can write them as one theorem by using the phrase, "if and only if". Thus, we may write these two theorems as the single statement:

$$3x > 1 \text{ if and only if } x > \frac{1}{3}$$

And, even though we are writing both theorems in *one* statement we must still prove them backward and forward from the axioms and definitions. A widespread notation for the clause "if and only if" is "iff", thus the above statement can be written:

$$3x > 1 \text{ iff } x > \frac{1}{3}$$

Properties of equals–Reprise

One more preparation we must make concerning this plan is that we assume that the operations of *addition* $+$, *multiplication* $\times$, and the relation *equals*, $=$, applies to all real numbers $x, y, z, \ldots$ as follows

1. **The reflexive law of equals**

   For any number x, $x = x$

2. **The symmetric law of equals**

   If $x = y$, then $y = x$

---

[2]Later, in Chapter 7, we will examine other, newer types of logic called *multi-valued* logic, developed in the 20th century.

3. **The transitive law of equals**

   If $x = y$ and $y = z$, then $x = z$

4. **The translation law of equals**

   If $x = y$ and z is any number, then $x + z = y + z$

5. **The magnification law of equals**

   If $x = y$ and z is any number then $x \times z = y \times z$

The reflexive law says that anything is equal to itself. The symmetric law says that you can switch the sides of any equation back and forth. The transitive law says that things equal to the same thing are equal to each other. The translation law says you can *add any number to both sides* of an equation and it will stay an equation. That is, you can translate the numbers on both sides along the x-axis by *adding* or *subtracting* the same amount and you will still have an equation. The magnification law says you can magnify (or shrink) numbers positively or negatively on both sides by *multiplication* or *division* by the same amount and you will still have an equation.

The axioms
The Group $(S, +)$

There is an operation called "addition" with an operator "+" called *plus* such that:

    Axiom 1. The + operator is closed

        If x is a number and y is a number then $x + y$ is a number

    Axiom 2. The + operator is commutative

        If each of x and y is a number then $x + y = y + x$

    Axiom 3. The + operator is associative

        If each of x, y and z is a number then $x + (y + z) = (x + y) + z$

    Axiom 4. The + operator has an identity element

        There exists a number 0 such that if x is any number then $x + 0 = x$

    Axiom 5.   Each element has an additive inverse

        If x is any number, there is a number $(-x)$ such that $x + (-x) = 0$

The Group $(S, \times)$

Note: For multiplication we will use the notation "$\cdot$" rather than "$\times$" to avoid confusion with the number x.

    There is an operation called "multiplication" with an operator "$\cdot$" called *times* such that:

    Axiom 6. The operator $\cdot$ is closed

        If x is a number and y is a number then $x \cdot y$ is a number

    Axiom 7. The operator $\cdot$ is commutative

        If each of x and y is a number then $x \cdot y = y \cdot x$

    Axiom 8. The operator $\cdot$ is associative

        If each of x, y and z is a number then $x \cdot (y \cdot z) = (x \cdot y) \cdot z$

    Axiom 9. The operator $\cdot$ has an identity element

There exists a number 1 not equal to 0, such that if x is any number then $x \cdot 1 = x$

Axiom 10. Each element has a multiplicative inverse

If x is any number, and $x \neq 0$, then there is a number $(1/x)$ such that $x \cdot (1/x) = 1$

The Distributive Law

There is the following relationship between the operators of addition and multiplication, called the *distributive law*:

Axiom 11. If each of x, y and z is a number then $x \cdot (y + z) = x \cdot y + x \cdot z$

The order Axioms

The symbol $<$ is read "is less than"; so $a < b$ reads "a is less than b", and it means that a is to the *left* of b on the number line. Given any number x and any number y, then

Axiom 12. *One and only one of the following statements is true:* $x = y$, $x < y$, or $y < x$.

Axiom 13. If $x < y$ and $y < z$, then $x < z$

Axiom 14. If $x < y$, and z is any number, then $x + z < y + z$

Axiom 15. If $x < y$ and $0 < z$, then $x \cdot z < y \cdot z$

The well-ordering principle

The following axiom is assumed to be true.

Axiom 16. If M is any set of positive integers, then there is a first number in M.

The Dedekind Cut Axiom, the "completeness axiom"

The first sixteen axioms allows us to solve any problem in the rational numbers, which would please Herr Kroneker, remember him, he was the mathematician who favored the rational numbers as the works of God, but in order to solve problems involving irrational numbers, we now need to turn to "all else"—the work of man. So here is the assumption we need to accomplish human mathematics.

The following axiom is assumed to be true.

Axiom 17.

Let S_1 and S_2 be any two non-empty sets on the X-axis such that every point of the X-axis belongs to either S_1 or S_2, and every point of S_2 is to the right of every point of S_1, then there is either a right-most point of S_1 or a left-most point of S_2.

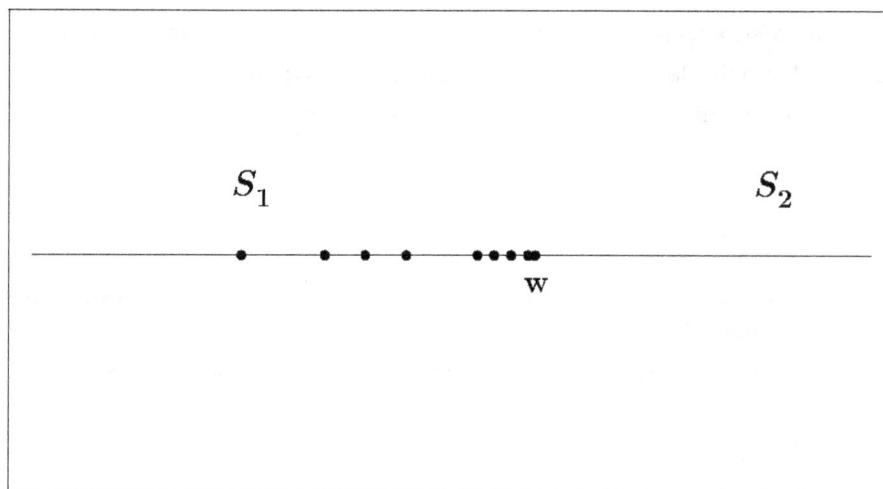

Figure 4.1 S_1 IS THE SET OF LOWER BOUNDS OF S_2.

In Figure 4.1, w is either the Greatest Lower Bound of S_2 or the Least Upper Bound of S_1.

The field theorems

Let's have some fun proving some of the most elementary theorems in the real number system.   With the 17 axioms that we have just stated, we could spend the next 200 pages proving 200 theorems that would carry you well into the topics of advanced college algebra, trigonometry and calculus.   Actually with these seventeen axioms and a whole bunch of definitions we can prove any theorems based on the real number system.

Don't worry, we are not going to do that.   But a few illustrative examples will give you a flavor of the nature of theorem and proof procedure for algebra very much like the theorem and proof structure in geometry presented by Euclid 2300 years ago.   Some of my students have told me that they loved geometry but hated algebra in high school.   I believe that this reveals a "mathematical spirit" in those students.

Because not enough time is devoted to mathematics in school, there is an attempt to rush through, covering a lot of material.   Consequently, algebra is sometimes taught as a set of rules without taking the time to leisurely allow students to develop the results on their own by proving these rules as theorems.

What I am proposing to do here is to state a few theorems with proofs, and a few others with hints but no proofs.   Then we will just discuss a few algebraic topics, such as binomials, exponents, graphs and logarithms, not formulating them into the theorem-proof format.   At that point we can show how they might be used in practical applications

Theorem 1: The number -0 is the same as the number 0

Proof:

$$0 + (-0) = 0 \quad \text{Axiom 5, additive inverse law}$$
$$-0 + 0 = 0 \quad \text{Axiom 2, Commutative law for addition}$$
$$-0 = 0 \quad \text{Axiom 4, additive identity law}$$

Note: Next, we want to prove that 0 is the only additive identity element.

Theorem 2: If there exists a number y such that for every number x, $x + y = x$, then $y = 0$.

Proof:

$x + y = x$	Given, for any number, x
$0 + y = 0$	Letting 0 be one of the numbers x
$y + 0 = 0$	Axiom 2 additive commutative law
$y = 0$	Axiom 4 additive identity law

Hence, there is only one zero.

Theorem 3: If x is any number and y is a number such that $y + x = 0$, then $y = -x$.

Proof:

$y + x = 0$	Given
$(y + x) + (-x) = 0 + (-x)$	Property of equals adding $(-x)$ to both sides
$y + (x + (-x)) = 0 + (-x)$	Axiom 3 additive associative law
$y + 0 = 0 + (-x)$	Axiom 5 additive inverse law
$y + 0 = (-x) + 0$	Axiom 2 additive commutative law
$y = -x$	Axiom 4 additive identity law

Note: $(-x) = -x$, because when there is no ambiguity, the use of parentheses is optional. This theorem proves that the additive inverse element of x is unique, for each x there is only one $-x$.

Definition: (negative)

If x is any number, the then additive inverse $-x$ defined in Axiom 5 is called the *negative* of x.

Definition: (subtraction)

If a is a number and b is a number then $a + (-b)$ is a number called the *difference, a minus b*. When this difference is written as $a - b$ and we say that b *is subtracted from a*.

Theorem 4: If x is any number, then $-(-x) = x$.

Proof:

$x + (-x) = 0$	Axiom 5 additive inverse law
$x = -(-x)$	Theorem 3

Problem: Show that

$x - (-y) = x + y$.

Solution:

By definition of subtraction and Theorem 4, $x - (-y) = x + -(-y) = x + y$.

Note: This says that if you subtract the *negative* of a number then you are really *adding* that number.

Theorem 5: If x, a, and b are numbers such that $x + a = b$, then $x = b - a$.

Hint Start with the hypothesis and subtract a from both sides..

Problem:

Show that the sum of two negatives is negative, that is $(-x) + (-y) = -(x + y)$.

Hint: Start with $(x + y) + (-y) = x$ and add $(-x) - (x + y)$ to both sides

The proofs of the preceding theorems needed only the axioms of the additive group $(S, +)$, the first five axioms. We could continue proving some other theorems in this group, but things get more interesting when we start including the axioms for multiplication, namely the group $(S, \times)$.

First, we note that the operation of multiplication is not the same as addition.   If they were the same, then the identity element, 0 in Axiom 4 would be the same as the identity element, 1 but Axiom 9 prohibits that.

Theorem 6: The number 1/1 is 1.

Proof:

$$
\begin{array}{ll}
1 \neq 0 & \text{Given in Axiom 9} \\
1 \cdot (1/1) = 1 & \text{Axiom 10, multiplicative inverse} \\
(1/1) = 1 & \text{Axiom 9, multiplicative identity}
\end{array}
$$

Theorem 7: The multiplicative identity element, 1, is *unique.*   That is, if y is a number such that $x \cdot y = x$, for all x, then $y = 1$.

Theorem 8: If x is any number and $x \neq 0$, then the multiplicative inverse of x is *unique*. That is, if $x \cdot y = 1$, then $y = 1/x$.

Definition: (reciprocal)

By Axiom 10, the number $1/x$, the multiplicative inverse of x is also called the *reciprocal of* x.   Another notation for the reciprocal is $\frac{1}{x}$, in which case $x \cdot \left(\frac{1}{x}\right) = 1$.

Definition: (division)

If x is any number and y is any number not zero, then $x \cdot \frac{1}{y}$ is a number called the *product,* x times $\frac{1}{y}$. This product is written as $\frac{x}{y}$, and when it is written this way, we say it is the *quotient x divided by y.* This is our definition of *division of* x by y. Other notations for quotient are x/y, and $x \div y$.

Definition: (fraction)

The number $\frac{x}{y}$ is also called a fraction and x (the top part) is the *numerator* and y (the bottom) is the *denominator.*   These names are fickle because a fraction can be frequently manipulated to change numerators and denominators around.

Example: If $b = \frac{1}{3}$, then 1 is the numerator and 3 is the denominator.   In the following strange looking fraction

$$
\frac{1}{b} = \frac{1}{\frac{1}{3}}
$$

1 is the numerator and $\frac{1}{3}$ is the denominator and this fraction is equal to 3, which is $\frac{3}{1}$ where 3 is the numerator and 1 is the denominator.

Problem:

How do we know that $\frac{1}{1/3} = 3$?

Solution:

$$
\frac{1}{3} \cdot \frac{1}{\frac{1}{3}} = 1 \quad \text{By Axiom 10, and}
$$

$$
\frac{1}{3} \cdot 3 = 1 \quad \text{By Axiom 10}
$$

We will now prove this, in general, as a theorem.

Theorem 9: If $x \neq 0$, and then $x = 1/(1/x)$.

Proof:

$$
\begin{array}{ll}
x \cdot (1/x) = 1 & \text{Axiom 10 multiplicative inverse law} \\
x = 1/(1/x) & \text{Theorem 8}
\end{array}
$$

Theorem 10: If a, b, and x are numbers and $a \neq 0$, and $a \cdot x = b$, then $x = \frac{b}{a}$.

Note: We could also solve the original equation, $a \cdot x = b$, for a (if we know that $x \neq 0$) then $a = \frac{b}{x}$.

Problem: If d is distance, r is rate and t is time, and neither t or r is zero, then the verbal equation, "distance is rate times time" can be solved for either rate, or time by Theorem 10.

$$
\begin{aligned}
d &= r \cdot t \\
r &= \frac{d}{t} \\
t &= \frac{d}{r}
\end{aligned}
$$

Theorem 11: If a, b, and c are numbers with $a \neq 0$, and x is a number such that $a \cdot x + b = c$, then $x = \frac{(c-b)}{a}$.

Hint: Start with $a \cdot x + b = c$, then subtract b from both sides then use Theorem 10.

What is zero times any number?

We can prove that $x \cdot 0 = 0$, for any number x, by using Axiom 11, the distributive law[3].

So, here is our first theorem needing the use of the distributive law, and some of the preceding axioms and theorems.

Theorem 12: If a is any number, then $a \cdot 0 = 0$.

Proof: Let a be any number then

$1 + 0 = 1$	Axiom 4 additive identity
$a \cdot (1 + 0) = a \cdot 1$	Property 5 of equals, multiply both sides by a
$a \cdot 1 + a \cdot 0 = a \cdot 1$	Because of Axiom 11, the distributive law
$a + a \cdot 0 = a$	Because of Axiom 9, multiplicative identity
$a \cdot 0 = 0$	By Theorem 2 , 0 is the only additive identity

Indirect arguments

The proof of Theorem 12 was a proof by *direct argument.* We started with an axiom and made step by step changes to come out with the result stated by the theorem. The following is a theorem we can prove by an *indirect argument.* We assume the theorem is *false* and then show that such an assumption leads to a *contradiction*, thus proving the theorem is *true.*

Theorem 13: $\frac{1}{0}$ is not a number.

Proof: Suppose that $\frac{1}{0}$ *is* a number

$\frac{1}{0} \cdot 0 = 1$	Axiom 10, multiplicative identity
$\frac{1}{0} \cdot 0 = 0$	Theorem 12, $x \cdot 0 = 0$, for any x
$1 = 0$	By the preceding two steps
$1 \neq 0$	Axiom 9
$\frac{1}{0}$ is not a number.	Assuming that it is leads to a contradiction.

---

[3]It seems to be unlikely that anyone could ever prove this from just the first ten axioms, that is without Axiom 11. But you are welcome to try.

Theorem 14: If $x \cdot y = 0$ then ether $x = 0$ or $y = 0$.
Corollary[4]: If a and b are numbers and both $a \neq 0$, and $b \neq 0$, then $a \cdot b \neq 0$.
Theorem 15: If b is a number, then its negative $-b = (-1) \cdot b$
Theorem 16: $(-1) \cdot (-1) = 1$
Corollary: $(-x) \cdot (-y) = x \cdot y$
Proof: Left to the reader. Hint: Use Axioms 7, 8, 9 and Theorems 15 and 16.

A poem

We have just discussed the subject of this little poem by W. H. Auden:

Minus times minus equals plus
The reason for this we need not discuss

Practical applications of the distributive law

Here are a few practical problems that can be solved by the distributive law.
   Problem:
Say you have a room that is 12 ft. wide and 22 feet long and you want to extend the width another 3 ft. Consider the following three questions:

1. What is the area of the original room?

2. How much area is added?

3. What are the final dimensions of the room?

See Figure 4.2.
When you answer these questions you will realize that

$$22 \cdot (12 + 3) = 22 \cdot 12 + 22 \cdot 3$$

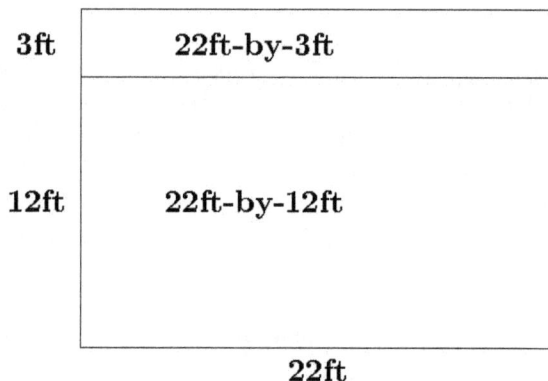

3ft | **22ft-by-3ft**

12ft | **22ft-by-12ft**

22ft

FIGURE 4.2 WHAT IS $22 \times (12 + 3)$?

Thus, the original area (22×12), 264 ft^2, plus the new area (22×3), 66 ft^2, equals the area of the new room (22×15), 330 ft^2. It says that any two numbers added together times a given number is the same as multiplying those original two numbers times the given number and then adding the two products together.

---

[4] A corollary is another statement closely related to and easily proved by the main theorem.

The common factor law

The distributive law written in reverse is the common factor law.

$$x \cdot y + x \cdot z = x \cdot (y + z)$$

How it it possible to reverse an equation this way? By the symmetric property of equals, any equation $A = B$ can be written as $B = A$.

Leaving out the multiplication notation, "$\cdot$", we simply write the distributive and the common factor laws as:

$$
\begin{aligned}
x(y + z) &= xy + xz, \text{ distributive law} \\
xy + xz &= x(y + z), \text{ common factor law}
\end{aligned}
$$

When we use the common factor law, we say we "factored out" the common factor x. Without the common factor law the number system would be very limited in what it can do. For example we could not simplify an expression such as $6\pi \cdot x + 7\pi \cdot x$, we would forever be writing larger and larger expressions with no ability to reduce them to simpler terms.

Example:

$$
\begin{aligned}
6\pi \cdot xy + 7\pi \cdot xy &= (6 + 7) \cdot \pi xy, \text{ or} \\
&= 13\pi xy
\end{aligned}
$$

A surprise in the common factor law!

A surprising appearance in the distributive law (common factor law) is

$$x + xy = x(1 + y)$$

Say what? Where did the 1 come from? Don't panic, by Axiom 9, any number x can be written as $1 \cdot x$ (or $x \cdot 1$), therefore

$$
\begin{aligned}
x + xy &= x \cdot 1 + x \cdot y, \text{ so } x \text{ is a common factor and} \\
x + xy &= x \cdot (1 + y)
\end{aligned}
$$

Problem: (Simple interest)

$$A = P + P \cdot rt$$

is the formula for the total amount A, after a principal amount P, earns interest at rate r, for a period of time t. It is the sum of the principal amount plus the interest earned. Using the common factor law, we write this is as:

$$A = P(1 + rt)$$

If you invest a principal $P = \$10,000$ at annual interest rate $r = 4.2\%$ for one year, $t = 1$, then what will your amount A, be?

Answer:

$$A = \$10,000(1 + 0.042)$$
$$A = \$10,420$$

Problem:

If this same investment had been done for a time of two years, $t = 2$, the amount would have been

$$A = 10,000(1 + 2 \cdot 0.042)$$
$$A = 10,840.$$

Problem:

(Compound interest)

If, after one year, we reinvest the first amount, $\$10,420$ as the new principal amount, call it P_2 and we will get: $A_2 = P_2(1 + rt) = 10,420(1 + rt)$, or

$$A_2 = 10,420 \cdot (1 + 0.042)$$
$$= 10,420 + 437.64$$
$$= 10,857.64$$

this means that the interest earns interest. Now the amount is $\$17.64$ more than the two-year simple interest. Here is how the repeated use of the distributive law gives us that extra 17.64

$$A_1 = 10,000(1 + 0.042)$$
$$A_2 = P_2(1 + 0.042)$$
$$= 10,000(1 + 0.042)(1 + 0.042)$$
$$= 10,000(1 + 0.042)^2$$

But what is $(1 + 0.042)^2$? It is $(1 + 0.042) \times (1 + 0.042)$, and upcoming Theorem 17 will let us write it as

$$(1 + 0.042)^2 = 1^2 + 2 \cdot (0.042) + (0.042)^2$$
$$= 1 + 0.084 + 0.001764$$
$$= 1.085764$$

which we can then multiply by $10,000$. In Chapter 6 when we study infinite series, we will see what happens when we invest in continuously compounded interest.

But where in the world would we get $1 + 2(0.042) + (0.042)^2$, from $(1 + 0.042)^2$? This is a special case of *squaring a binomial.*

Theorem 17: If x and y are numbers then $(x + y)^2 = x^2 + 2xy + y^2$.

Proof:

$$(x+y)^2 = (x+y) \cdot (x+y) \qquad \text{Definition, } A^2 = A \cdot A$$
$$(x+y)^2 = (x+y) \cdot x + (x+y) \cdot y \qquad \text{Axiom 11, distributive law}$$
$$(x+y)^2 = x \cdot x + y \cdot x + x \cdot y + y \cdot y \qquad \text{Axiom 11, again}$$
$$(x+y)^2 = x^2 + 2xy + y^2 \qquad \text{Distributive law again}$$

The sum of two numbers such as $x + y$ is called a binomial and Theorem 17 is called *squaring a binomial*. Notice that there is a middle term, $2xy$; so $(x+y)^2$ is not just x^2+y^2. This becomes more apparent if we show what happens geometrically. See Figure 4.3.

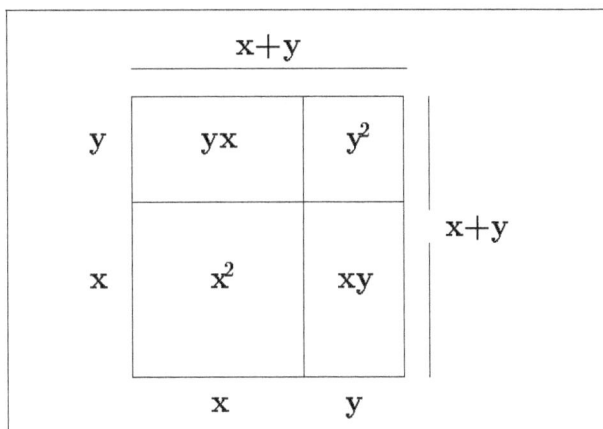

FIGURE 4.3 $(x + y)^2 = x^2 + 2xy + y^2$

Hardy-Weinberg Principle in genetics

The square of a binomial has an interesting application in biology. In a simple model in genetics, some trait such as the color of a rose is controlled by a gene which is made up of two forms, called *alleles*. One allele can be dominant and the other recessive. Say a gene controlling color is made up of two alleles, a dominant one, R, for red and a recessive one, w, for white. Since R is dominant, if any rose has a gene with an R allele it will be red. That is, its *phenotype* (what it looks like) is red, but its *genotype* (what its genetic make-up is) may be either RR, or Rw or wR. If the genotype of a rose is ww, then its phenotype will be white. You can tell by looking at a white rose that its genotype is ww, because white is recessive and white will only be manifested if there is no dominant red allele. But you cannot tell by looking at a red rose whether its genotype is RR or Rw or wR. When you cross pollinate two roses the allele split up and the two parent roses contribute one of their alleles to the offspring according to the square of a binomial.

This model is widely used and was formulated independently by a German physician Wilhelm Weinberg and a British mathematician G. H. Hardy each publishing a paper on this subject in 1908.

Problem:

Suppose you are cross breeding two red roses each with genotype Rw what are the possible outcomes? What proportion of the outcomes do you expect to be red?

Solution:

Square the binomial $(R + w)$; that is, $(R + w)^2 = R^2 + 2Rw + w^2$, three red roses and one white rose. In biology, we usually write this out as a table.

×	R	w
R	R^2	Rw
w	wR	w^2

Table 4.1

Problem: Show that if you cross pollinate a red rose with the RR genotype with a white rose, then the expected outcome is *four* red roses.

Solution: $(R + R)(w + w) = 2R \cdot 2w = 4Rw$.

In a table form it is:

×	w	w
R	Rw	Rw
R	Rw	Rw

Table 4.2

Fractions

Theorem 18: If neither a nor c is zero, then

$$\frac{1}{a} \cdot \frac{1}{c} = \frac{1}{a \cdot c}$$

Theorem 19:

$$\frac{x}{a} \cdot \frac{w}{c} = \frac{x \cdot w}{a \cdot c}.$$

Theorem 20: If neither a nor c is zero, then

$$\frac{x}{a} + \frac{w}{c} = \frac{c \cdot x + a \cdot w}{a \cdot c}$$

Definitions and theorems about inequalities

The expression $a > b$ reads "a is greater than b" and it means $b < a$. So, b is to the left of a or consequently, a is to the **right** of b on the number line. In the following definitions $\neq$ reads "not equal to", $\not<$ reads "not less than", and $\not>$ reads "not greater than."

By Axiom 12, we logically deduce the following true statements:

$$\text{If } x \ \neq \ y, \text{ then } x < y \text{ or } x > y$$
$$\text{If } x \ \not< \ y, \text{ then } x > y \text{ or } x = y$$
$$\text{If } x \ \not> \ y, \text{ then } x < y \text{ or } x = y$$

Shorthand notation: If $x > y$ or $x = y$, we can write $x \geq y$, read as "x is greater than or equal to y." The shorthand notation $x \leq y$ reads "x is less than or equal to y" and means $x \not> y$.

Positive and negative

The statement that z is *positive* means $z > 0$. That is, z is to the right of 0 on the number line. The statement that z is *negative* means $z < 0$. That is, z is to the left of 0 on the number line. The statement that z is *non-negative*, means that $z \geq 0$.

Theorem 21: If x is negative then $-x$ is positive. In symbols: If $x < 0$, then $-x > 0$

Theorem 22: If $x < y$ and $z < 0$, then $x \cdot z > y \cdot z$

Example: If $x < 0$, then multiplying both sides by x, you get

$$\begin{aligned} x \cdot x &> 0 \\ x^2 &> 0 \end{aligned}$$

Theorem 23: If x is any number then $x^2 \geq 0$.

In other words the square of a number cannot be negative.

Note: In a very subtle way the order axioms characterize the real number system. For example, Theorem 23 tells us that the square of any (real) number is non-negative. Thus, in this system, there can be no solution to any equation in which the square of a number is negative.

Example: The following equation cannot be solved in an ordered field.

$$x^2 = -4$$

Later, when we use graphs we will see that there is no point of the graph of the equation $y = x^2$ that is below the x-axis.

The next theorem has several parts, all of which are closely related to each other, therefore we will write it as one.

Theorem 24:

(a) $1 > 0$, that is, 1 is positive.

(b) $2 > 1$

(c) $0 < \dfrac{1}{2} < 1$

Theorem 25: If $0 < x$, then $0 < \frac{1}{x}$, and if $x < 0$, then $\frac{1}{x} < 0$

Theorem 26: If $a > b > 0$, then $\frac{1}{b} > \frac{1}{a} > 0$

Theorem 27: If x and y are any two numbers, and $x < y$, then the number $\frac{1}{2}(x+y)$ is neither x nor y and is between x and y, which we can write as: $x < \frac{1}{2}(x+y) < y$.

Note: This theorem says that no *two* numbers can be "next" to each other on the number line. There is always a number between any two numbers.

Theorem 28: An ordered field has infinitely many numbers between any two numbers.

Proof: left to the reader

Mathematical induction

Sometimes we will discover a statement, S, that seems to be true for several integers. For example: "The first odd number is 1^2, and the sum of the first 2 odd integers is 2^2, and the sum of the first 3 odd integers is 3^2, and the sum of the first 4 odd integers is

4^2." In other words,

$$
\begin{aligned}
1 &= 1^2 \\
1+3 &= 2^2 \\
1+3+5 &= 3^2 \\
1+3+5+7 &= 4^2
\end{aligned}
$$

This becomes a *rumor*, as in "Did you hear that the sum of the first n odd integers is n^2?" But is it true for *any* number of integers? You could go on computing these equations until you die and you still would not have proved this formula to be true for all positive integers. Even if it is, true for every integer, how do we prove it? Since there are infinitely many integers, it is obvious that we cannot possibly check out every integer.

Here is where Axiom 16, the well-ordering principle, comes to the rescue. We can use Axiom 16 in a negative way. For example, if there is any integer, n, such that this rumor is *not* true for it, then n must be an element of a set, C, of such contrary integers (it may be the only one in the set). By the well ordering axiom, Axiom 16, C must have a first element, k. Good! Now we got it pinned down. Since k is the first place the formula ceases to be true, then it must be true for $k-1$, the integer just below (less than) k.

$$
\begin{aligned}
1+3+...+(2(k-1)-1) &= (k-1)^2 \text{ or} \\
1+3+...+(2k-3) &= (k-1)^2
\end{aligned}
$$

Given this true equation we can add $2k-1$ to both sides.

$$
1+3+...+(2(k-3)+(2k-1) = (k-1)^2 + 2k - 1
$$

Therefore,

$$
\begin{aligned}
1+3+...+(2k-3)+(2k-1) &= k^2 - 2k + 1 + 2k - 1 \\
1+3+...+2k-1 &= k^2
\end{aligned}
$$

which means that the formula is true for k, which was supposed to be a number for which the formula was *not* true. This proves the formula for all positive integers, n.

Although we used the well ordering principle and an indirect argument to get this proof, the common practice for problems like this is to use *Mathematical Induction*. This is a type of reasoning derived from Axiom 16 that lets us verify whether or not the rumor (an ostensible formula) is true for all positive integers. An induction proof requires two steps.

1. Proof that the given statement S is true for the integer $n = 1$, (called the basis step) and

2. Proof that whenever S is true for some positive integer k it is also true for $k+1$, (called the induction step).

These two steps are tantamount to proving it for every integer.

The summation sign

At this point we would like to introduce some mathematical notation, $\sum$, that we will find useful here, and later in other settings.

Definition: In the notation,

$$\sum_{j=m}^{j=n} f(j) \tag{4.1}$$

the greek letter, sigma, $\sum$ is called the "summation sign". The number j is an integer called the "index" of the summation. The number $f(j)$ (read as " f of j") is a *function* of j, (a number that depends on j) is called the "summand". The small equation $j = m$ below the summation sign tells us that j starts at the positive integer m and the small equation $j = n$ on the top of the summation sign is the stopping place for j. When we compute this summation we let j run through all of the integers from m to n taking on every positive integer between m and n, including both m and n, then we compute the summand value $f(j)$ for every one of the values of j and then the $\sum$ tells us to *add* up all of the summands. Thus, the expression (4.1) is shorthand for the following one.

$$f(m) + f(m+1) + ... + f(n-1) + f(n) \tag{4.2}$$

We read the expression (4.2) as "The sum of f of j as j runs from m to n."

Example: If $f(j) = 3j(j+1)$ and $m = 2$ and $n = 4$, then

$$\sum_{j=2}^{j=4} 3j(j+1) = 114$$

is read as: "The sum of 3 j times the quantity j plus 1, as j runs from 2 to 4 equals 114."

Here is how we get it.

$$
\begin{aligned}
\sum_{j=2}^{j=4} 3j(j+1) &= 3 \times 2 \times 3 + 3 \times 3 \times 4 + 3 \times 4 \times 5 \\
&= 18 + 36 + 60 \\
&= 114
\end{aligned}
$$

We computed the summand $3j(j+1)$ for every j from 2 to 4, getting $18, 36, 60$, and added them up!

Problem: Find the sum,

$$\sum_{j=1}^{j=5} \left(\frac{1}{2}\right)^j$$

Solution: The function, $f(j)$, is $(1/2)^j$. The index, j, runs from 1 to 5, so the summands are:

$$\left(\frac{1}{2}\right)^1, \left(\frac{1}{2}\right)^2, \left(\frac{1}{2}\right)^3, \left(\frac{1}{2}\right)^4, \left(\frac{1}{2}\right)^5 \quad \text{or}$$

$$\frac{1}{2}, \frac{1}{4}, \frac{1}{8}, \frac{1}{16}, \frac{1}{32}$$

Therefore the summation is

$$\sum_{j=1}^{\frac{1}{2}} \left(\frac{1}{2}\right)^j = \frac{1}{2} + \frac{1}{4} + \frac{1}{8} + \frac{1}{16} + \frac{1}{32}$$
$$= 31/32$$

One more convention we need is the following, if the index starts and ends at the *same* number, there is just one single term in the sum. Thus,

$$\sum_{j=m}^{j=m} f(j) = f(m)$$

Now we are ready to state the following theorem:

Theorem 29: If n is any positive integer then

$$\sum_{j=1}^{j=n} (2j - 1) = n^2$$

Proof: We proved this above by Axiom 16.

Now, here is a theorem so that you can try your hand at using an induction proof.

Theorem 30: If n is any positive integer then the sum of the first n integers is $n(n+1)/2$. That is, if n is any positive integer, then

$$\sum_{j=1}^{j=n} j = \frac{n \cdot (n+1)}{2}$$

Proof: Left to the reader.

The existence of irrational numbers

Two lemmas about the Dedekind cut axiom   Here are two small theorems immediately deducible from the Dedekind cut axiom.   In general, an immediate consequence of an axiom is called a *lemma* or it is sometimes called a corollary of that axiom.

Let us restate the Dedekind cut axiom here:

Axiom 17 (Reprise)

Let S_1 and S_2 be any two non-empty sets on the X-axis such that every point of the X-axis belongs to either S_1 or S_2, and every point of S_2 is to the right of every point of S_1, then there is either a right-most point of S_1 or a left-most point of S_2.

Lemma 1: It is not possible for a point to belong to both S_1 and S_2.

Proof:

Suppose a point of S_1 belonged to S_2, then it would be to the right of every point of S_1, including itself.

Lemma 2: It is not possible for both S_1 to have a right-most point and S_2 to have a left-most point.

Proof:

Suppose S_1 has a right-most point z_1 and S_2 has a left-most point, z_2. They can not be the same point as just proved in Lemma 1, and neither one of them can be to the right of the other. For example, if $z_1 < z_2$, then there is a real number x between them that is to the left of the left-most point of S_2 and to the right of the right-most point of S_1, contradicting Axiom 17 big time. Why can't $z_2 < z_1$?

Rational vs irrational

Rational numbers are those that can be written as the ratio of whole numbers, such as, for example, $\frac{3}{400}$, or $\frac{7}{33}$. The decimal equivalent of any rational number either terminates in all zeros, or has an infinitely repeating "block" of digits..

Examples:

$$\frac{3}{400} = 0.00750000..., \text{ ends in all zeros}$$

$$\frac{7}{33} = 0.21212121..., \text{ repeats 21 forever}$$

Quick Trick: If you are given a decimal that repeats some block of digits, you can easily find two whole numbers whose ratio is equal to that decimal. For example suppose you are given the decimal : 0.783783783...where the block of 3 digits 783 repeats for ever. Just write down those three digits and divide by the integer 999. That is, use as many 9's as the number of digits in the repeated block.

$$0.783783783783... = \frac{783}{999}$$

This method will always give you two integers that you can use in order to write the repeating decimal as a ratio of two whole numbers. In some cases the two integers will not be relatively prime (that is they may have a common factor, other than 1). In the above example, the fraction $\frac{783}{999}$ can be reduced to $\frac{29}{37}$ because 27 is a common factor to both 783 and 999.

$$\frac{783}{999} = \frac{27 \times 29}{27 \times 37} = \frac{29}{37} = 0.783783783...$$

This can be done for any repeating decimal. In the example, $\frac{7}{33}$, above, the repeating block is 21, and if we use this quick trick we will write $\frac{21}{99}$, which is actually $\frac{7}{33}$ because we can cancel out the common factor 3 from the numerator and denominator getting::

$$\frac{21}{99} = \frac{7}{33} = 0.21212121...$$

If the decimal equivalent of a number *never repeats* and *does not end in zeros*, then that number cannot be written as the ratio of two whole numbers. It is *irrational*, that is not rational. The number $\sqrt{3}$, is such a number

$$\sqrt{3} = 1.732050808...\text{not ending in all zeros and never repeating}$$

We shall show in Theorem 31(b), below, that $\sqrt{3}$ cannot be written as the ratio of two whole numbers.

When we want to solve a practical problem involving irrational numbers we simply cut off the decimal at some number of terms accurate enough to satisfy the amount of precision we want in the applied problem. For example, π is irrational,

$$\pi = 3.1415926535897...\text{not ending in all zeros and never repeating}$$

and we often use the decimal 3.14 as an approximation for π in problems that do not require a high degree of accuracy.

How do we even know that there are irrational numbers? Axiom 17 can be used to prove that there are. As an example, using Axiom 17, we will establish the existence of at least one irrational number, $\sqrt{3}$. In order to do this we will prove two things: (a) $\sqrt{3}$ exists and (b) it is impossible to find two whole numbers m and n such that $\frac{m}{n} = \sqrt{3}$.

Theorem 31 (a): $\sqrt{3}$ exists.

Proof: First, let us say what we mean by $\sqrt{3}$. It is any positive number x, such that $x^2 = 3$. We know that if x is a number then x^2 exists because by Axiom 6, the product $x \cdot x$, (that is, x^2) is a number.

By Axiom 12, either $x^2 < 3$, $x^2 > 3$, or $x^2 = 3$, which says $x < \sqrt{3}$ or $x \geq \sqrt{3}$. Start by supposing that $x < \sqrt{3}$, and let S_1 be the set of all positive numbers x such that $x^2 < 3$, (that is, $x < \sqrt{3}$). And let S_2 be the set of all numbers such that $x^2 \geq 3$.

S_1 does **not** have a right most point because, suppose otherwise, there *is* a right-most number t in S_1, $t < \sqrt{3}$, but by Theorem 28, there is a number s such that $t < s < \sqrt{3}$ so $s^2 < 3$ and this makes s a number in S_1 to the right of the right-most point of S_1. Impossible! Thus, by Axiom 17, S_2 has a left-most point and it is $\sqrt{3}$.

Theorem 31 (b): $\sqrt{3}$ is not rational, in other words, $\frac{m}{n} = \sqrt{3}$ is not possible for any two whole numbers m and n.

Proof: Suppose there is a pair of integers m and n with no common factors such that $\frac{n}{n} = \sqrt{3}$.

$\left(\frac{m}{n}\right)^2 = 3$	Assumption
$\frac{m^2}{n^2} = 3$	Theorem 19, $\frac{m}{n} \cdot \frac{m}{n} = \frac{m \cdot m}{n \cdot n}$
$m^2 = 3n^2$	Multiplying by n^2
$m = 3k$	Because m^2 is a multiple of 3
$m^2 = 9k^2$	The square of $m = 3k$
$3n^2 = 9k^2$	Both $3n^2$ and $9k^2 = m^2$
$n = 3h$	Because $n^2 = 3k^2$, so n is a multiple of 3

Therefore, m and n have a common factor, 3, when they weren't supposed to have any common factors. This contradicts the assumption that $\sqrt{3}$ could be a rational number.

But why can't there be a rational number $\frac{a}{b} = \sqrt{3}$ where a and b are integers that have *do* have common factors? If there were, we could cancel the common factors and reduce the fraction to $\frac{m}{n}$, one with no common factors. One other question you might ask is this: If n^2 is a multiple of 3 how do we know that n must be a multiple of 3? Just suppose n is *not* a multiple of 3 and see what happens to n^2.

The arithmetic of irrational numbers

What can we say about a mathematical system that satisfies the well ordered field axioms, and the Dedekind cut axiom? An answer is that it consists of *rational and irrational* numbers and these two sets together make up what we call the *real* number system. It is also clear from the axioms that we can add subtract, multiply and divide, except by zero, these rational and irrational numbers all we want to get other real numbers. For example we can add $\sqrt{5}$ to $\sqrt{7}$ and get a number and that number is exactly $\sqrt{5} + \sqrt{7}$. Which is fine for a mathematics class or a textbook, but virtually *useless* for a carpenter who may want to find the length of wood needed for two cross beams. He or she might like to have a decimal approximation such as the following calculator approximations.

$$\sqrt{5} + \sqrt{7} \approx 2.23 + 2.65 = 4.88$$

Or even better, say if the measurements were in feet; approximately 4 *feet* 11 *inches*.

When we designate *two* numbers, x and y, rational or irrational, on the real number line, we know that there is *some* number between them. Without attempting a proof, we will take as a fact that between any two numbers there is a rational number and between any two numbers there is an irrational number. One example we will use in particular is that for any positive number x, (rational *or* irrational) there is a positive integer n such that $0 < \frac{1}{n} < x$. This is called the Archimedes principle and it is derivable from the Dedekind cut axiom.

Archimedes Principle:

If $x > 0$, there exists a positive integer n such that $n \cdot x > 1$.

In other words, no matter how small a positive number x is you can always find an integer n big enough to multiply by x to get a number greater than 1.

Graphs of some theorems

From the last few theorems we can conclude that the real numbers can be represented continuously by points on the real number line. We can also conclude that each point in the (two-dimensional) plane can be represented by a pair of numbers (x, y) corresponding to points on two mutually perpendicular real number lines which we will call the *x-axis* and the *y-axis*. Thus, we will treat the two dimensional plane as a kind of road map and treat graphs in the plane somewhat like paths on the map, where you can locate a point on the path by referring to the letters (or numbers) along the bottom edge of the map and the numbers along the left edge.

As we mentioned Chapter1, René Descartes introduced the concept that every point in a plane could be assigned a pair of numbers, referring to two coordinate axes.

Example: (Cartesian coordinates)

The Figure 4.4, shows three points $A = (1.15, 2.07)$, $B = (1.9, 0.9)$, and $C = (-2.8, -2.5)$

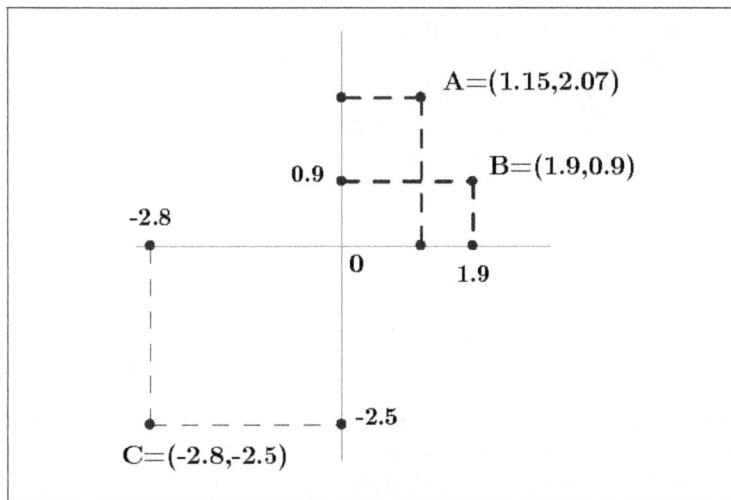

FIGURE 4.4 THREE POINTS A, B AND C IN THE PLANE.

Now, what about a collection of such points in which the variables x and y satisfy some kind of *equation* relating them to each other? The collection of such points is called a *graph of that equation*.

Example: The graph of the equation $y = x^2$, is the set of all points (x, y) in the coordinate plane for which it is true that y is the square of x.

In other words, all of the points on this graph have coordinates of the form (x, x^2).

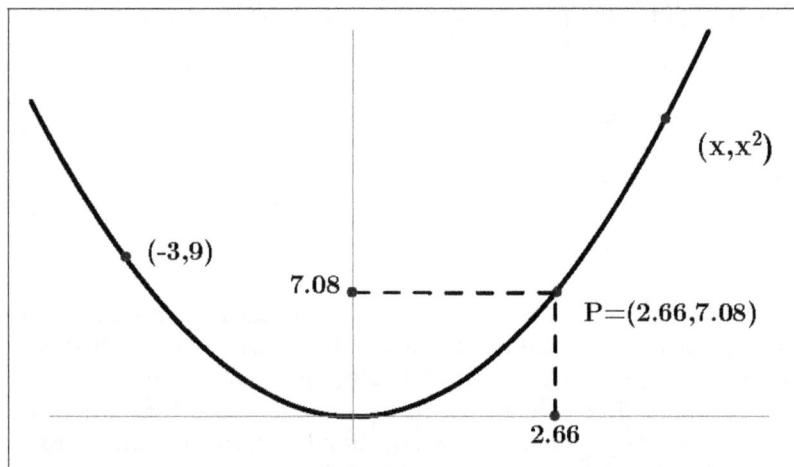

FIGURE 4.5 GRAPH OF $y = x^2$, $y \geq 0$ FOR ALL x.

Some equations will express a restriction in the relationship between the variables x and y. For example, a graph of the equation:

$$y = \sqrt{x}$$

will not have any points in which x is negative. That is, the numbers, $x < 0$ are not suitable for the function $\sqrt{x}$.

Another example of an equation in which there is a value of x that is *not* suitable is the following:

$$y = \frac{1}{x+1} \tag{4.3}$$

The number $x = -1$ is not suitable; there is no point whose coordinates are $(-1, y)$, because if x were to be -1, we would have

$$y = \frac{1}{-1+1} = \frac{1}{0}$$

which we know is not a number by Theorem 13. The graph of the Equation 4.3 shown here in Figure 4.6 depicts the fact that there is no point (x, y) on the graph for which $x = -1$. The dashed vertical line, $x = -1$, contains no point on the graph.

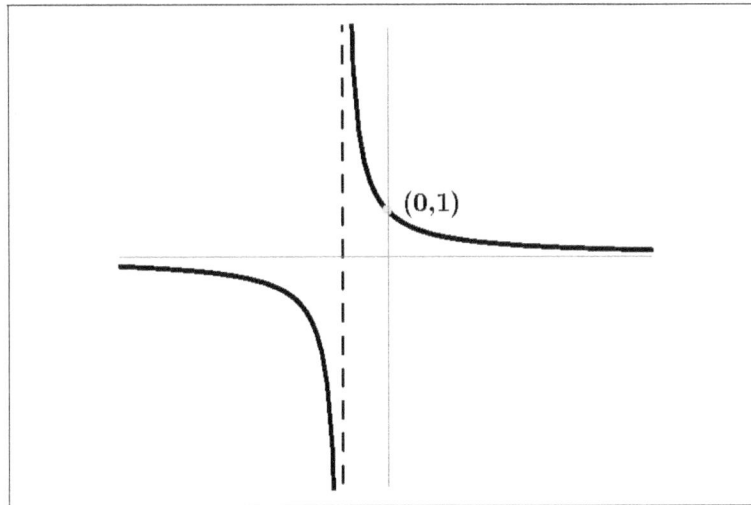

FIGURE 4.6 GRAPH OF $y = \frac{1}{x+1}$.

Theorem 32: (Rule of cubes) If x is positive, then so is x^3 and if x is negative, then so is x^3.

Proof: We will first assume x is positive and prove that x^3 is positive. Then assume x is negative, then prove that x^3 is negative. We can use Theorem 23 to get a head-start on this proof.

By Axiom 12 any number that is not zero is either positive or negative. Let $x \neq 0$, then by Theorem 17 $x^2 > 0$.

$$
\begin{array}{ll}
x^2 > 0 & \text{Theorem 23, when } x \neq 0 \\
x^2 \cdot x > 0 \cdot x & \text{If } x > 0, \text{ Axiom 15} \\
x^3 > 0 & \text{Definition of } x^3 \\
x^2 \cdot x < x^2 \cdot 0 & \text{If } x < 0, \text{ Axiom 15} \\
x^3 < 0 & \text{Definition of } x^3
\end{array}
$$

This proves that if $x > 0$, then $x^3 > 0$ and if $x < 0$, then $x^3 < 0$.

The graph in Figure 4.7, below, shows that for all positive numbers, x, we see that

x^3 is above the x-axis and for all negative numbers, x, the graph for $y = x^3$ is below the x-axis.

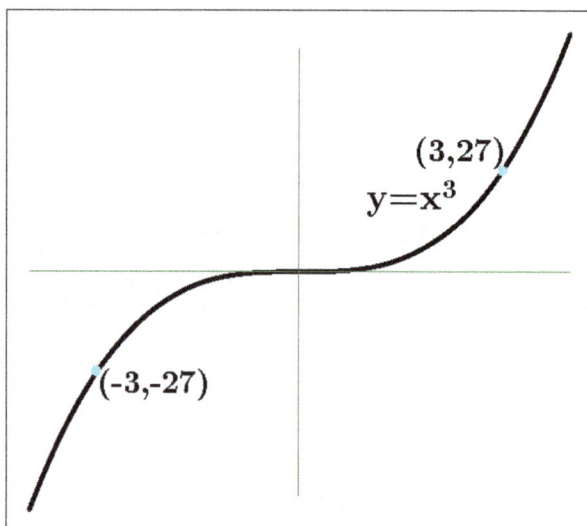

FIGURE 4.7 GRAPH OF Y=X^3

Powers

No, this section is not about super heros with special powers, but it is about b^n and $b^{1/n}$, b^x and other neat things.

Whole number exponents

What is b^n? If $n = 1$, then b^1 is just b, itself. The number $b^2 = b \times b$, the area of a square whose sides are b, and $b^3 = b \times b \times b$, the volume of a cube whose sides are b. In the 3rd century BCE, the Greek mathematicians easily dealt with powers up to three because they could represent them geometrically as figures in one, two and three dimensions. Geometry does not provide an easy way to depict a four dimensional figure whose sides are b. But if we think of the algebraic applications and not just the geometric ones we can conceive of b^4 or b^5, or any other b^n for any number n. Here is the formal definition for b^n.

Definition: (nth power, b^n)

If b is any number, then $b^1 = b$, and if n is any integer, positive, or negative or 0, then $b^n = b^{n-1} \cdot b$.

This expression, b^n, is called an *exponential expression*, and is read as, "b to the nth power". The number b is called the *base* of this expression and the number n is called the *exponent*.

Note: You may wonder: Is this really a definition? Does it tell us what b^2 is for example? Yes, laboriously so.

$$b^2 = b^{2-1} \cdot b \qquad \text{Definition of } b^n, \text{ when } n = 2$$
$$b^2 = b^1 \cdot b \qquad \text{Because } 2 - 1 = 1$$
$$b^2 = b \cdot b \qquad \text{Definition of } b^1$$

Similarly b^3 is defined since, $b^3 = b^{3-1} \cdot b = b^2 \cdot b = b \cdot b \cdot b$. That is, b^3 is b times itself 3 times.

What is b^0?

If we are allowing n to be any integer, then even b^0 is defined. Look!

$$
\begin{array}{ll}
b^1 = b^{1-1} \cdot b & \text{Definition of } b^n, \text{ when } n = 1 \\
b = b^0 \cdot b & \text{The exponent is } 1 - 1 = 0 \\
b^0 = 1 & \text{Theorem 7 unique multiplicative identity}
\end{array}
$$

One other interesting fact we get from this definition is, the meaning of b^{-1}, when $b \neq 0$.

What is b^{-1}?

Answer: It is $\frac{1}{b}$. Why?

$$
\begin{array}{ll}
b^0 = b^{0-1} \cdot b & \text{Definition of } b^n, \text{ when } n = 0 \\
b^0 = b^{-1} \cdot b & \text{The exponent } 0 - 1 = -1 \\
1 = b^{-1} \cdot b & \text{Because } b^0 = 1 \\
b^{-1} = \frac{1}{b} & \text{Theorem 8, unique multiplicative inverse}
\end{array}
$$

Sample Questions:

1. What is $\left(\frac{2}{3}\right)^{-1}$? Answer: $\left(\frac{2}{3}\right)^{-1} = \frac{1}{(2/3)} = \frac{3}{2}$

2. What is $(0.75)^0$? Answer: $(0.75)^0 = 1$

3. What is $\frac{1}{(3/4)^{-1}}$? Answer: $\frac{1}{(3/4)^{-1}} = \frac{1}{(4/3)} = \frac{3}{4}$

Theorem 33: For any integer m, $b^m \cdot b = b^{m+1}$.

Proof:

$$
\begin{array}{ll}
b^{m+1} = b^{(m+1)-1} \cdot b & \text{Definition of } b^n, \text{ when } n = m + 1 \\
b^{m+1} = b^m \cdot b & \text{Because } (m + 1) - 1 = m
\end{array}
$$

We really want to prove that, for any n, not just for 1, that $b^m \cdot b^n = b^{m+n}$.

The proofs of the next five theorems are left to the reader; use mathematical induction and the proof of Theorem 33 as a hint.

Theorem 34: Given any integer, m, then for any integer n,

$$b^m \cdot b^n = b^{m+n}$$

In words, Theorem 34 says, "When you multiply two powers of the same number you *add* the exponents."

Example:
$$x^5 \cdot x^3 = x^8$$

Theorem 35: Given any integers m and n, and $b \neq 0$, then

$$\frac{b^m}{b^n} = b^{m-n}$$

Theorem 36: If a and b are any numbers, not zero, then $a^n \cdot b^n = (a \cdot b)^n$

Theorem 37: If $a \neq 0$ and $b \neq 0$, then for any integer n ,

$$\left(\frac{a}{b}\right)^n = \frac{a^n}{b^n}$$

Theorem 38: If $b > 0$, and m and n are integers, then,

$$(b^m)^n = b^{m \cdot n}$$

Theorem 38 says: "When *raising* the power of a number to another power you *multiply* the exponents."

Example: When raising the fifth power of x to the third power you multiply 5 times 3.

$$\left(x^5\right)^3 = x^{15}$$

When we stated Theorems 33 through Theorem 38, we required the bases b , or a to be not equal to zero; this is necessary because if any of the exponents, m , or n etc. where to be negative, such as $n = -1$, and b were to be zero, then b^n would be dividing by zero. However, with integer exponents, the numbers a , b , etc. could be negative, because a negative number b raised to the -1 does not present the same problem as 0 to the -1 .

There is one precaution you have take, and that is, what happens when your raise a negative number to an even *vs* an odd power?

Now, $(-1)^n$ is either $+1$ or -1 as n is even or odd. If n is even $(-b)^n = b^n$, and if n is odd $(-b)^n = -b^n$, that is $-1 \times b^n$.

An even power of -1 is $+1$, but an odd power of -1 is -1 .

Rational exponents

Must we restrict the theorems on exponents (powers) to only integers such as a^3, a^{-55} , etc.? No, we can make a case for fractional exponents such as $n = \frac{1}{2}$ or $n = -3/4$, etc. That is, we will be able to give meanings to such things as $a^{\frac{1}{2}}$ or $a^{-3/4}$. But in these cases, we will have to impose some restrictions upon whether or not the bases, a , b , etc. are negative.

Guessing Problem: What would you guess $9^{\frac{1}{2}}$ to be?

Answer: Guess 3, because...Why? See footnote[5].

The number 3 is called a *square root* of 9 because 3 squared is 9.

Definition: (square root)

If x and y are numbers such that $y = x^2$, then we say "y is x *squared*" or "x is a number whose square is y ." We denote this by saying $x = \sqrt{y}$, which reads, "x is the square root of y ." But it turns out that another number $-\sqrt{y}$, is also one whose square is y , because:

$$\begin{aligned}
(-\sqrt{y})^2 &= (-1 \cdot \sqrt{y})^2 \text{ and then} \\
(-\sqrt{y})^2 &= (-1)^2 \cdot (\sqrt{y})^2 \text{ by Theorem 25} \\
(-\sqrt{y})^2 &= +1 \cdot (\sqrt{y})^2 = y
\end{aligned}$$

---

[5]From Theorem 38, $(9^{\frac{1}{2}})^2 = 9^{\frac{1}{2} \cdot 2} = 9^1 = 9$ But , $3^2 = 9$. So, we guess that $9^{\frac{1}{2}} = 3$.

Therefore, if we are given the equation

$$y = x^2$$

then we know that either:

$$x = \sqrt{y} \text{ or}$$
$$x = -\sqrt{y}$$

Example If we know that x is a number such that $9 = x^2$, then $x = 3$ or $x = -3$, since both $(3)^2 = 9$ and $(-3)^2 = 9$.

Question: Let us try another example, what is $10^{\frac{1}{2}}$?

Answer If $x = 10^{\frac{1}{2}}$ then $x = \sqrt{10}$. Here is why?

$$\text{If } x = \sqrt{10}, \text{ then } x^2 = 10, \text{ by definition of square root}$$
$$\text{But}$$
$$\text{If } y = 10^{\frac{1}{2}} \text{ has } any \text{ meaning, then by Theorem 38}$$
$$y^2 = (10^{\frac{1}{2}})^2 = 10^{\frac{1}{2} \cdot 2} = 10$$
$$\text{So it reasonable to define } 10^{\frac{1}{2}} \text{ as } \sqrt{10}$$
$$\text{and } -10^{\frac{1}{2}} \text{ as } -\sqrt{10}$$

We just defined $\pm 10^{\frac{1}{2}}$ as $\pm\sqrt{10}$. This reasoning leads us to the following definition.

Definition: Fractional exponents.

If $n \geq 2$ is any even integer and $b \geq 0$, we will define $b^{\frac{1}{n}}$ to be $\sqrt[n]{b}$, the nth root of b. Here we are requiring b to be non-negative because a negative number does not have a square root or a fourth root, or a sixth root, or any *nth* root where n is even. See Figure 4.8 in which the graph of $y = \sqrt{x}$ is not defined for $x < 0$.

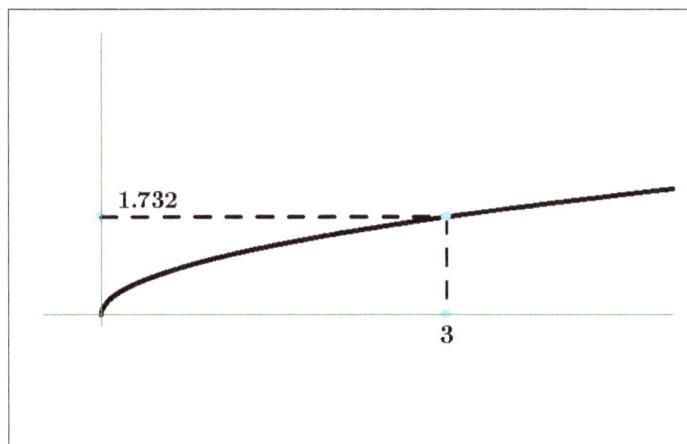

FIGURE 4.8 PLOT OF $y = \sqrt{x}$ FOR $x \geq 0$.

If $n \geq 3$ is any odd integer and b is any number, we will define $b^{\frac{1}{n}}$ to be $\sqrt[n]{b}$, the nth root of b. In this case it is OK for b to be negative because negative numbers do have cube roots, and fifth roots, etc. See Figure 4.9, in which the graph of $y = \sqrt[3]{x}$ is defined for $x < 0$.

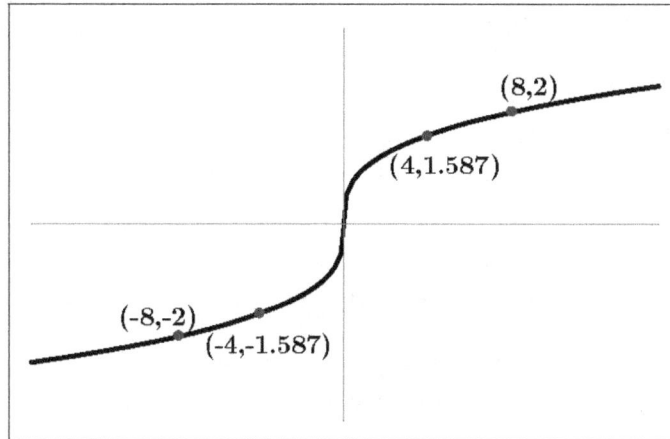

FIGURE 4.9 PLOT OF $y = \sqrt[3]{x}$, FOR ALL x.

Using Figure 4.9, what is $(-8)^{(1/3)}$? It is -2, because $\sqrt[3]{-8} = -2$, that is $(-2)^3 = -8$. What is $(-4)^{(1/3)}$, approximately?

$$(-4)^{\frac{1}{3}} \cong -1.587...$$

What is $(4)^{(1/3)}$?

$$(4)^{\frac{1}{3}} \cong 1.587...$$

Calculator warning: If you want to use your calculator to raise a number b such as 4, to a fractional power such as $(1/3)$, you need to put the fractional power in parentheses.

Example: Using the up carat key $\boxed{\char`\^\phantom{x}}$ if you INCORRECTLY enter $\boxed{4}\boxed{\char`\^}\boxed{1}\boxed{\div}\boxed{3}$ you will get $1.33333...$because the calculator interpreted your key strokes as: $\frac{4^1}{3} = \frac{4}{3}$. It did the $\boxed{4}\boxed{\char`\^}\boxed{1}$ first, getting 4, then divided that result by 3.  You need to key this in as $\boxed{4}\boxed{\char`\^}\boxed{(}\boxed{1}\boxed{\div}\boxed{3}$), thus, capturing the $1 \div 3$ in parentheses, so you now have $4\char`\^(1 \div 3)$, or $4^{(1/3)}$.

Problem: Using the calculator method just described, find the seventh power of the fifth root of ten. That is, find the seventh power of $10^{1/5}$.

Solution: It is

$$
\begin{aligned}
(10^{\frac{1}{5}})^7 &= 10^{\frac{7}{5}} \\
&= 10\char`\^(7 \div 5) \\
&= 25.1188...
\end{aligned}
$$

Real exponents

We know how to use *integral* exponents, such as a^3, and *rational* exponents, such as $a^{7/2}$, but what about real (not rational) exponents such as a^π ? This is the next step we need to consider for exponents.

What if we wanted to raise 10 to the $\sqrt{10}$?

$$10^{\sqrt{10}} = ?$$

Some calculators have built-in programs, executed by a key that looks like this: $\boxed{a^b}$ or $\boxed{x^y}$, that give you very accurate approximations, so if you entered $10^{\sqrt{10}}$, you would get:

$$10^{\sqrt{10}} \cong 1453.040302...$$

This is because we can use Axiom 17 to get a rational approximation of $\sqrt{10}$, then use it as a power on 10. If you ever needed to know what π^{π} is (approximately), here it is

$$\pi^{\pi} \approx 36.462...$$

Rules for real exponents

But what are the theorems that we can prove for formulas using real exponents? To begin with we need the following important fact.

$$\text{If } b > 1 \text{ and } x > 0, \text{ then } b^x > 1 \tag{4.4}$$

While this result may seem to be intuitively obvious, (after all, you are raising a number greater than one to a positive power, so why wouldn't the result be greater than 1?) It is actually *difficult* to prove. If x is irrational the definition of b^x requires a sophisticated proof using limits. In this book we will simply assume (4.4) without proving it.

Theorem 39: If $b > 1$, then $b^x > 1$ if and only if $x > 0$.

(This means: if $x > 0$, then $b^x > 1$, and if $b^x > 1$, then $x > 0$.)

Proof: omitted.

Theorem 40, below, has several parts; their proofs are similar to each other and they constitute what are known as the "Rules for real exponents." We will assume them without proofs. Diligent readers who are interested in doing so, can derive their own proofs by applying Axioms 1 through 17.

We are going to call these E-1, E-2, etc., so we can refer to them as the exponent rules. Similar rules will be stated later for another concept, called *logarithms*. They will be labeled L-1, L-2, etc.

Theorem 40: If $a > 0$, $b > 0$ and each of x and y is any number, then the following rules are true.

Rule E-1(a)	$b^0 = 1$
Rule E-1(b)	$b^1 = b$
Rule E-2	$b^x \times b^y = b^{x+y}$
Rule E-3	$b^x / b^y = b^{x-y}$
Rule E-4	$(a \times b)^x = a^x \times b^x$
Rule E-5	$(a/b)^x = a^x / b^x$

Going back to Theorem 39, we can get the following important theorem.

Theorem 41: If x and y are positive and $b > 1$ then $b^x > b^y$ if and only if $x > y$.

Proof:

We need to prove this theorem "back and forth." That is: if $x > y$, then $b^x > b^y$ and, if $b^x > b^y$, then $x > y$.

$\Longrightarrow$

$$
\begin{aligned}
x &> y \quad \text{Given} \\
x - y &> 0, \text{ Axioms 5 and 14} \\
b^{x-y} &> 1, \text{ Theorem 39} \\
b^{x-y}b^{y} &> 1b^{y}, \text{ Multiply by } b^{y} \\
b^{x} &> b^{y}, \text{ Theorem 40}
\end{aligned}
$$

$\Longleftarrow$

$$
\begin{aligned}
b^{x} &> b^{y}, \text{ Given} \\
b^{x}b^{-y} &> b^{y}b^{-y}, \text{ Multiply by } b^{-y}, \text{ Theorem 25} \\
b^{x-y} &> 1, \text{ Theorem 40} \\
x - y &> 0, \text{ Theorem 39} \\
x &> y, \text{ Axioms 5 and 14}
\end{aligned}
$$

Corollary:

If $0 < b < 1$, then $b^{x} < b^{y}$ if and only if $x > y > 0$.

The corollary follows from Theorem 41 by using $\frac{1}{b}$ for b, because if $0 < b < 1$, then $\frac{1}{b} > 1$.

One conclusion we can make from Theorems 39, 40, and 41 is that if $b > 1$, then $b^{x} = b^{y}$ if and only if $x = y$.

A second conclusion is: if b is positive and x is any number then b^{x} is a real number. And, oh, by the way, if $b \neq 1$, and, y is any positive number, then there exists a number x such that $y = b^{x}$. Thus, a continuous graph of the equation $y = b^{x}$ can be drawn. This means we can plot a few points (x, b^{x}) when the values of x are ones for which it is easy to compute b^{x}. Then we can connect these points into a continuous smooth curve.

For example, if $b = 10$, what does the graph of $y = 10^{x}$ look like? Let us compute 10^{x} for a few convenient values of x such as $x = -2$, $x = -1$, $x = 0$, etc. See the following table:

x	10^{x}
-2.0	0.010
-1.0	0.100
0.0	1.000
0.5	3.162
1.0	10.000
2.0	100.000

Where did we get these numbers? Well, $10^{-2} = \frac{1}{100}$, $10^{-1} = \frac{1}{10}$, $10^{0} = 1$, $10^{0.5} = \sqrt{10} \approx 3.162$, $10^{1} = 10$, and $10^{2} = 100$. We will use these values to plot points on the graph of $y = 10^{x}$.

Note: The negative values of x, (to the left of the y-axis) give you negative powers of 10, which are small *positive* numbers, so the graph stays above the x-axis and y just gets closer and closer to zero as you move farther to the left. After, plotting these few points and assuming that this can be done for all numbers, x (because of the completeness axiom, Axiom 17) we get Figure 4.10.

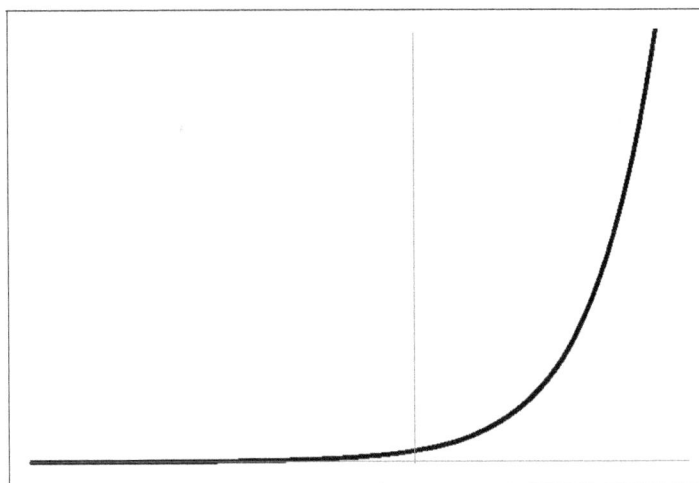

FIGURE 4.10 PLOT OF $y = 10^x$

This is called the exponential graph of $y = 10^x$. Every horizontal line above the x-axis will intersect this graph as some point. We can plot similar exponential graphs using numbers other than 10 for our base. Thus, graphs of equations like $y = 2^x$, or $y = e^x$, will be similar to Figure 4.10, and represent what is called *exponential growth*.

Exponential growth

A practical application of exponential growth is found when we study *populations* in nature. Under certain favorable conditions, animal populations grow exponentially. This is because the rate of growth, over time, is proportional to the size of the population. This is true even for humans, urban deer, bacteria, or money in an interest paying saving account.

Example *Compound interest*

Let $A(n)$ be the amount in your bank account at time n and suppose the annual interest rate per year is 8 percent. The formula for compound interest over n years is:

$$A(n) = A(0)(1 + .08)^n$$

Where $A(0)$ is the amount at some beginning time, $t = 0$. Say that $A(0) = 100$, then amount, with the interest being compounded just one time a year is

$$A(n) = 100(1.08)^n$$

The graph of this equation is:

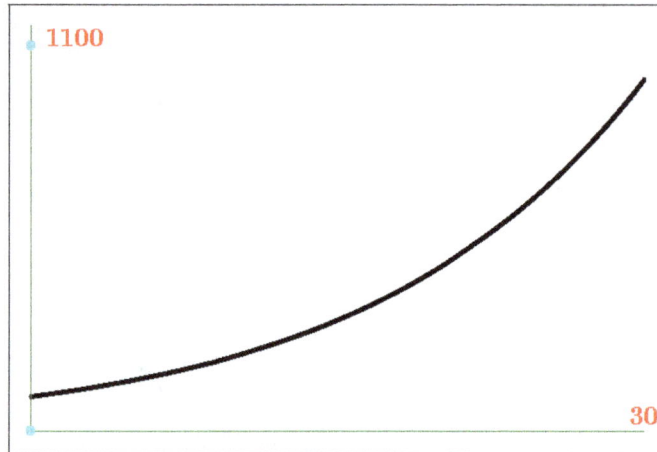

FIGURE 4.11 GRAPH OF $A = 100(1.08)^n$

The exponential graph for the equation $y = 10^x$ increases very rapidly and soon runs out the top of the page. Look back at Figure 4.10 and try to imagine how you would find a value of y for some given large value of x. Furthermore, suppose you reversed the problem and were given some value of y and you wanted to find the value x needed to produce that y. One attempt to solve this problem is to change the scale on the y-axis, so that instead of assigning equally spaced marks–as they are–being 10 apart, we would assign values to unequally spaced marks as being *powers* of 10 apart, such as 10, between the first two marks, then 100 between the second two, etc. Such varying marks on the y-axis of a graph are called *logarithmic scales*, the reason for which you will soon see.

Logarithm base 10

The graph in Figure 4.10 is the set of points satisfying the equation $y = 10^x$. This is exactly what you need if you were given a number x and wanted to find out what the answer is when you raise 10 to the x power.

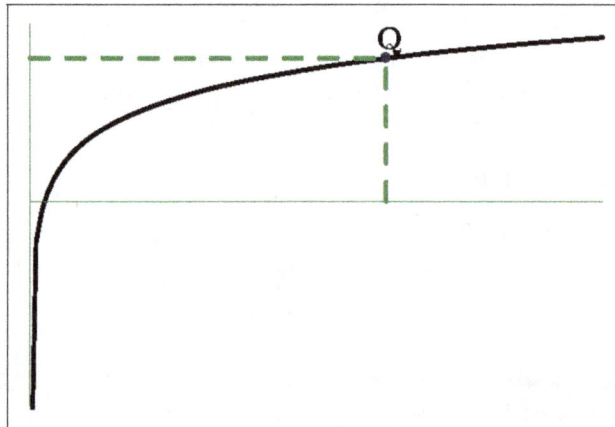

FIGURE 4.12 GRAPH OF $y = \log_{10}(x)$

But suppose you were confronted with the problem of finding out what power you needed to raise 10 to in order to get some specific number, say, for example, 25? That is if $25 = 10^x$, what is x? Fortunately, we can solve this kind of problem by examining a graph of the equation Unfortunately, however, it would have to be a very precise detailed graph.

It turns out that there are several applications, in which scientists, engineers and mathematicians need to solve problems where the values of y are known quantities and the values of x need to be computed

Before the invention of computers and other calculating machines such problems were quite tedious. An early attempt to compute a value of x that would yield a given a value of y in an exponential equation was originally invented by a Scottish mathematician, John Napier around 1600. By treating multiplication of two numbers such as 10^3 and 10^4 as addition (of the two exponents) he could read the answer from a type of calculator made from sticks (called rods) that had numbers marked on them. He also compiled tables of numbers that corresponded to various powers of 10. This motivated him to abstractly define a function, $y = f(x)$, that had the following property:

$$f(x \times t) = f(x) + f(t)$$

Napier called this function a *logarithm,* abbreviated as *log.*

$$f(x) = \log_{10}(x)$$

This equation reads, "$f(x)$ is the logarithm to the base 10 of x." It is called a *logarithmic equation* and it is the solution to the problem: "What is the number, $f(x)$, that you have to raise 10 to in order to get a given number x?" In the xy coordinate system, the definition is stated as follows.

$$y = \log_{10}(x) \text{ means } 10^y = x.$$

The graph of this logarithm is exactly the same as the graph of the exponential graph with the variables x and y *interchanged.*   In the graph above, Figure 4.12, the point Q has coordinates $(25, 1.39794001)$. Therefore,

$$\log_{10}(25) \approx 1.39794001$$

Which means
$$10^{1.39794001} \approx 25$$

Some applications of logarithms

Over the past three centuries logarithms became quite useful in arithmetical calculations and in measuring natural and scientific phenomena such as earthquakes, and chemical reactions. In practical problems, base 10 logarithms are used to reduce certain formulas to manageable numbers.

Arithmetic problems

Example: (Product of large numbers)

Suppose we want to multiply an awkwardly large number, A and by an awkwardly small number B. $A = 123,456$, $B = 0.0038817$.

If we had an absurdly large graph of Figure 4.12 on the side of a building several blocks long and several stories high, we could find a point where $A = 123,456$ and $\log(A) = 5.091512$, and another point where $B = 0.0038817$, and $\log(B) = -2.410978$. Then we could use these logarithms to make the job easier.

$$
\begin{aligned}
\log_{10}(A \times B) &= \log_{10}(A) + \log_{10}(B) \\
\log_{10}(123,456 \times 0.0038817) &= \log_{10}(123,456) + \log_{10}(0.0038817) \\
&\approx 5.091512 + (-2.410978) \\
&\approx 2.680534
\end{aligned}
$$

So, then using this same large graph (or a similarly large graph of Figure 4.10) to get:

$$
\begin{aligned}
A \times B &\approx 10^{2.680534} \\
&\approx 479.219
\end{aligned}
$$

I suggest that you use your calculator to verify this approximate answer. And I would like to point out how this very act of using your calculator obviates the use of logarithms and giant graphs to solve problems like this, today.

But logarithms are important in another way, even in this calculator age, and that is *mathematical modeling*. They can be used to model certain real phenomena such as measuring earthquakes, predicting weather, testing chemical compositions, describing astronomical distances, and modelling population growth. Logarithms may be regarded as being the handiest way to record data and express scientific results.

ρH *scale*

In 1909, a Danish chemist, Soren Peder Lantz, developed the ρH scale for measuring acidity by the logarithm formula:

$$
\rho H = -\log_{10}(a_{H+})
$$

Where a_{H+} is the hydrogen ion activity in solution. From Rule L-3 in Theorem 42, below, this formula is the same as

$$
\rho H = \log_{10}(1/a_{H+})
$$

Richter scale

In 1935, Charles Richter, a seismologist from the California Institute of Technology developed a formula that measures the strength of an earthquake as a number M_L according to the formula:

$$
M_L = \log_{10}(A) - \log_{10}(A_0(\delta))
$$

Where A is the seismograph reading at a given station, and $A_0(\delta)$ is a function of the distance δ from that station to the epicenter of the earthquake. Again by Rule L-3, Theorem 42, this formula may also be written as:

$$
M_L = \log_{10}\left(\frac{A}{A_0(\delta)}\right)
$$

Logarithms were used because the seismograph measurements produced a range of magnitudes that were inconveniently large   The use of logarithms helped seismologists get understandable numbers to study and report.

Slide rules

One of the interesting mathematical tools, built upon logarithmic scales is the *slide rule*. This was a handy computational device, about 12 inches long and 2 inches wide, invented in the 17th century by William Oughtred and was based on Napier's Rods.   Problems requiring multiplication, division, powers, and calculation of trigonometric functions were solved by adding, and subtracting sliding logarithm scales.

The instrument also served as a portable table of functions. It was the tool most commonly used by mathematicians, physicists, and engineers for over 200 years. But when electronic calculators came into widespread use in the 1970's, slide-rules disappeared. You can find ghostly pictures of departed slide rules on the Internet.

Logarithm base b

Not every logarithmic model, however, is in base ten. Depending upon the phenomenon being modeled, it may be much more convenient to pick some base other than ten. Any positive number that is not equal to 1 can be used as a base for logarithms. We could use $b = 5$ or $b = 7$, or $b = \pi$, and so forth. So what is the logarithm to the base b for just any old base b?

Definition (*Logarithm to the base b*)

If $b > 0$ and $b \neq 1$, then $y = \log_b(x)$ means that $b^y = x$

That is "The logarithm to the base b of x is the number y that you have to raise b to in order to get x." One way to learn this is to repeat it to yourself a couple of times, using actual numbers for b, x, and y.

Example:

Repeat to yourself, "The logarithm to the base 7 of 12 is the number, y, you have to raise 7 to in order to get 12." That is,

$$y = \log_7(12) \text{ means } 7^y = 12$$

Using your calculator, you could experiment with a few trial numbers by raising 7 to various powers like $y = 1.25$, or $y = 1.3$, etc., and finally, by trying various values of y notice that, for $y = 1.28$

$$7^{1.28} \approx 12.0705...$$

and conclude that

$$\log_7(12) \approx 1.28$$

In Rule L-5 of the following theorem will give you this answer without using trial and error.

This theorem has multiple parts and it constitutes what are known as the "Rules for logarithms (to any base, b)".

Theorem 42: If $a > 0$, $b > 0$ and $a \neq 1$ and $b \neq 1$, and each of x and y is a positive number, then the following rules are true.

Rule L-1(a) $\log_b(1) = 0$

Rule L-1(b) $\log_b(b) = 1$

Rule L-2 $\log_b(x \times y) = \log_b(x) + \log_b(y)$

Rule L-3 $\log_b(\frac{x}{y}) = \log_b(x) - \log_b(y)$

Rule L-4 $\log_b(a^x) = x \cdot \log_b(a)$

Rule L-5 $\log_b(x) = \log_b(a) \cdot \log_a(x)$

These correspond, roughly, to the rules for real exponents, (see Theorem 40).

Notes:

Rules L-1 (a) and (b): $0 = \log_b(1)$ and $1 = \log_b(b)$, are proved for any base b, by observing that $b^0 = 1$ and $b^1 = b$.

Rule L-5 is handy if you are trying to solve a problem in some base, but your calculator does not have logarithm keys in that base.

Example: Going back to the example we solved by trial and error in trying to find a number y that would make $7^y = 12$, we can use your calculator, even if it has only a $\log_{10}$ key, to find $\log_7(12)$. Start with the equation from Rule L-5.

$$\begin{aligned} \log_7(12)\log_{10}(7) &= \log_{10}(12) \\ \log_7(12) &= \log_{10}(12)/\log_{10}(7) \\ &\approx 1.0792/.8451 \\ &\approx 1.277 \end{aligned}$$

This answer is better than $y = 1.28$, because $7^{1.277} \approx 12.00025$.

This Chapter has been about the real numbers, that is a set of elements for which we defined two binary relations, addition and multiplications.   Then, with Axioms 11 through 17, we have the real numbers, which we have known and loved all our lives.

In Chapter 3 we were introduced to other kinds of things that we can and do call numbers, things like *symmetry, congruences,* and *matrices.* Each of these had their own limiting relationships and confining axioms. We saw semigroups, and monoids and groups and rings; so, now we seem to have all of the definitions we need for just about any kind of thing we want to call numbers. What more could there be other than applications to science, business, art and philosophy? Well, let me tell you. We have not exhausted the kind of things we call numbers.

Let us look into numbers that have 2 or 3 or more dimensions, not just confined to the one-dimensional number line.   Also, there are new meanings assigned to *multiplication* and *addition*.   Right now, I am thinking of a new kind of multiplication that says that the two dimensional number $(0,1)$ times itself is $(-1,0)$. That is, $(0,1)^2 = -1$. Read on, my friend.

CHAPTER 5

MULTIDIMENSIONAL NUMBERS

For the past few centuries, engineers, physicists and mathematicians have been working with things, called *vectors, quaternions, tensors, matrices,* and other entities. They are manipulated by their own special definitions of binary operators, which we still call multiplication and addition. But, these new operators are not just the usual multiplication and addition.

The elements, themselves, are not confined to a single number line, but are rather, have more than one "part". For example, we will see that the complex numbers have a *real part* and an *imaginary part.* We say that such numbers are in a higher dimension, say two dimensions, or three or four or even infinitely many dimensions. If I had wanted to avoid the word "multidimensional numbers" in the title of this chapter, I would have been required to name it COMPLEX NUMBERS, REAL AND COMPLEX VECTORS, AND QUATERNIONS. Whew!

We will start with the invention of imaginary numbers, which, by themselves, are still confined to a number *line*–the y-axis. But when we combine an imaginary number with a real number we get a *complex number* in a two dimensional *plane*, a sort of "fat" number line. Since the complex numbers have two parts, they can be interpreted as points (x, y) in the two dimensional complex plane, where we give you specific definitions for the addition and multiplication of such points. This puts a burden on the reader to understand when (x, y) is a point in the *real* two dimensional plane, or when it is a complex number, or when it is the end point of a vector. Fortunately, this book is not a "who-dunnit" mystery novel and the author, or the context will provide the necessary clues.

Imaginary numbers

We know from the theorems and axioms of the preceding chapter that there is no real number solution to the equation,

$$x^2 = -1600$$

Here was another example of a roadblock problem. So, we're told there is no real solution to this equation? When faced with such a situation, we humans make a tool! A 17th century British mathematician, John Wallis, was active in geometry and algebra and a contributor toward the discovery of calculus. His work lead to the foundation of the Royal Society of London in 1663, one of the outstanding scientific organizations still going today. In 1673, Wallis wrote an algebra book in which he stated a problem related to the loss of 10 acres of land to the sea, as in the failure of a dike, for example. He used an old unit of land measure, called a *perch*, where 160 square perches equal 1 acre; thus, the land loss was 1600 square perches. He asked the question, "What would be the sides of a square whose area equals -1600 square perches?"

A problem such as this could only arise in some practical case in which a real-life person wants to resolve a real-life conundrum that he or she is facing at some real-life moment. Anyone well-versed in the field axioms and theorems might simply reply, there is no answer. But, fortunately, a mathematician or any other normal human being might try to think of some new approach to solving such a problem

Wallis introduced the idea of using some new quantity, $\sqrt{-1}$, called an *imaginary number*, that did not fit the real number axioms. He incorporated this new quantity into his answer *without* being fully able to justify it. The flexibility of trying new things, that are often initially rejected, then later justified and accepted into the mathematical system is one of the ways that mathematics adds to its utility in solving problems.

About 100 years after Wallis, in 1797, a Danish surveyor, Casper Wessel, invented a symbol, ϵ, for $\sqrt{-1}$ which later was changed to i for "imaginary". With this, he anticipated the idea of a "two-part" (or two dimensional) number using both 1 and i (and their negatives) as <u>units</u> in the Cartesian coordinates. But in this the new system Mr. Wessel had to find a new way to define the addition and multiplication of these units.

Wessel units

In his paper, Wessel designated four different units, $\{+1, -1, +i, -i\}$ in an x-y coordinate plane and he gave the rules for multiplying them. (See Figure 5.1.)

> Let $+1$ designate the positive rectilinear unit, and $+i$ be a certain other unit perpendicular to the positive unit and having the same origin; then the direction angle of $+1$ will be equal to $0°$, that of -1 to $180°$, that of $+i$ to $90°$ and that of $-i$ to $-90°$ or $270°$.

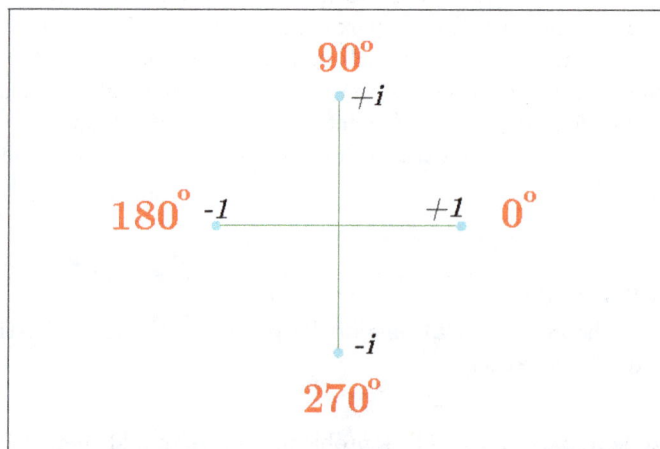

Figure 5.1 WESSEL UNITS $+1$, $+i$, -1, $-i$

He then defined what he meant by the *product* (multiplication) of any two of these units, which we paraphrase as follows:

When you multiply any two (of these four) units, you add their direction angles. The resulting sum of these angles will give you the direction angle of the product. In case the sum of the angles is greater than 360° we reduce it by mod(360) to get one of the four direction angles. Recall from "clock arithmetic" in Chapter 3 that *a is congruent to b mod n*, $a \cong b \bmod(n)$ means that $a - b$ is n or a multiple of n; for instance, $450 \cong 90 \bmod(360)$.

Examples:

1. When you want to multiply $(-i)$ by (-1), you notice that their direction angles are 270° and 180°, respectively; so you add 270° to 180° getting 450°. Using mod 360 arithmetic we get 90°. This is the direction angle for the unit $+i$. Thus, $(-i) \cdot (-1) = +i$.

2. Say you want to multiply $(-i)$ by $(+i)$, then add the degrees $90° + 270°$ getting 360°which is congruent to 0° mod(360), the direction angle of $(+1)$. Therefore, $(-i) \times (+i) = (+1)$.

3. Or how about i^2? Well, to get $i \times i$, you add the direction angles 90° and 90°, getting 180°, the direction angle of -1. Thus $i^2 = -1$.

Complex numbers
Argand diagram

In 1806, a few years after Wessel published his paper on the real and imaginary units, $+1, +i, -1, -i$, a Swiss mathematician by the name of Jean Robert Argand, noticed that Wessel's work, was restricted to just the four points $(1,0)$, $(0,1)$, $(-1,0)$, and $(0,-1)$. He introduced the idea of representing any point (x,y) in the plane as a *complex* number, not restricted to just the angles $0°, 90°, 180°$, and $270°$. He could then define the product of two points (x_1, y_1) and (x_2, y_2) by adding the corresponding angles made by the x- axis and the line segment from the origin $(0,0)$ to the points (x_1, y_1), and (x_2, y_2), respectively, in the same way that Wessel defined multiplication.

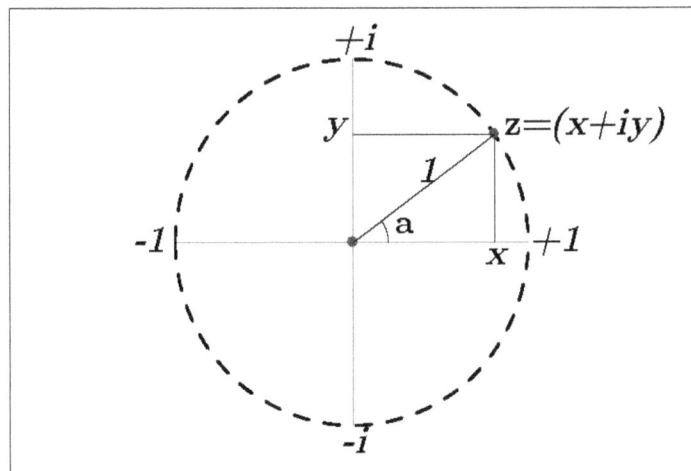

FIGURE 5.2 THE ARGAND DIAGRAM

Argand drew a circle whose center was at the origin and which passed through the four units $\{+1, +i, -1, -i\}$. Every point, (x, y) on the circle is designated as a *two part number* $x + iy$ in the xy-coordinate plane. This is a combination of the *real-number* unit 1 with the *imaginary number* unit i resulting in the number, z, written as

$$z = x \cdot 1 + i \cdot y = x + iy$$

called a *complex number*. Although both x and y are real numbers, x, the one multiplied by 1, is called the "real part" and y, the one multiplied by i is called the "imaginary part" of the complex number. So, for example, if the complex number $z = 0.8 + 0.6i$, then 0.8 is the real part of z and 0.6 is the imaginary part of z.

The line segment (the radius of the circle) drawn from the origin to the point (x, y) makes the angle α with the x-axis. In Figure 5.2, above, the measurement, α of the angle is *not* confined to being *just* 0°, or 90°, or 180°, but can be of *any* size, from 0° to 360°.

Pythagorean Theorem

An important theorem related to complex numbers is the famous *Pythagorean Theorem*, which was in use long before Pythagoras (about 500 BCE). In any given right triangle, OXP, the square of the hypotenuse is equal to the sum of the squares of the two legs. The *hypotenuse* is the side opposite the right angle; here it is $\overline{OP}$ and the legs are the two sides that are perpendicular to each other[1]. Here they are the sides $\overline{OX}$ and $\overline{XP}$.

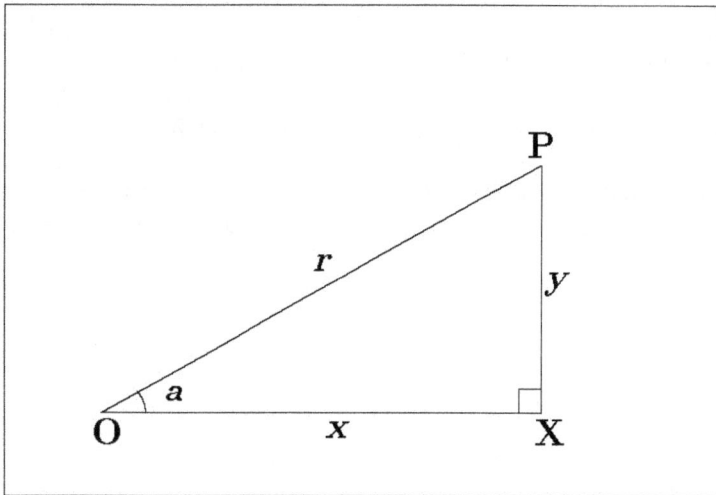

Figure 5.3 THE RIGHT TRIANGLE OXP

Thus, if we designate the length of $\overline{OX}$ as x and the length of $\overline{XP}$ as y and the length of $\overline{OP}$ as r, then the Pythagorean theorem says:

$$x^2 + y^2 = r^2$$

This theorem has been proved hundreds of times since it was first discovered in 2500 BCE. So many proofs were known through-out the world, that in the early 1900's a math-

---

[1]The notation $\overline{OP}$ represents the line segment between the points O and P, as distinguished from the entire line OP.

ematician by the name of Elisha Scott Loomis published a book called the *Pythagorean Proposition* in which he lists 367 different published proofs of the Pythagorean Theorem. One such proof was created by James Garfield who was a mathematics teacher before he became President of the United States of America. We will state it here as the next theorem, but the proof will be left to you, the reader. You may even come up with the 368th proof.

Theorem 1: *If* $\triangle ABC$ *is a right triangle, with perpendicular sides* $\overline{AB}$ *and* $\overline{BC}$ *and hypotenuse* $\overline{AC}$, *then,*

$$\overline{AC}^2 = \overline{AB}^2 + \overline{BC^2}$$

Proof: Left to the reader; you can look up President Garfield's proof on the Internet.

Complex numbers as a field

Riddle:

A real number is a complex number, but is a complex number really a number?

It has some interesting possibilities. The answer is, "Yes every real number, x, (on the x-axis) is a complex number whose imaginary part is zero, that is, $x = (x, 0)$, or expressed in terms of i, $x = x + 0i$." But the thing that might be a puzzle is this—do all of the complex numbers, written either as $z = x + iy$, or $z = (x, y)$, especially the ones in which $y \neq 0$, satisfy the axioms and theorems of the real numbers? The answer here is "No." The (non-real) complex numbers do not satisfy the *order axioms* in Chapter 4. If x is any number in a system satisfying the order axioms, then $x^2 \geq 0$. But we know that if $z = i$, (one of the Wessel units), then $z^2 < 0$.

On the other hand, if we can prove that the complex numbers *do* constitute a *field*, (satisfying the first eleven axioms in Chapter 4) then we can add and subtract, multiply, find inverses, and solve problems in the set of complex numbers, $z = (x, y)$; that is, $z = x + iy$.

Example:

Find z, if

$$z^2 + 1600 = 0$$

Solution:

$$z = \pm 40i$$

To understand how the complex numbers can be a field, we need to know a few things, like when is one complex number equal to another complex number? What are the definitions of *addition,* and *multiplication* of complex numbers? Can we find the *identity* elements and the *inverse* elements for these operators? And, this is a big one, do the operations of multiplication and addition of complex numbers satisfy the *distributive law?* It would be fortunate if the set of complex numbers were to be a field because then all of the field theorems would be true and available for use in solving algebra, trigonometric, engineering and calculus problems.

Here are the definitions we need. In this discussion we will opt to write the complex numbers either as $z = (x, y)$ or as $z = x + iy$, which ever is the more convenient form. In either case, we are emphasizing that complex numbers are two dimensional and can be represented as either a point in the plane *or* a two part number with a real and an imaginary part.

Equality of complex numbers

Definition: (Equality) Let z_1 be the complex number (x_1, y_1) and z_2 be the complex number (x_2, y_2), then z_1 *equals* z_2 if and only if $x_1 = x_2$ and $y_1 = y_2$.

That is, a complex number is equal to another complex number if the real part of the one equals to the real part of the other and the imaginary part of the one is equal to the imaginary part of the other. And they are *not* equal to each other if either the two real parts aren't equal or the two imaginary parts aren't equal.

Example 1:

$$
\begin{aligned}
(x^2 - y^2, 2xy) &= (1,1), \text{ if and only if} \\
x^2 - y^2 &= 1, \text{ and} \\
2xy &= 1
\end{aligned}
$$

Example 2:

$$
\begin{aligned}
x^3 + i(y - 9) &= 2 + 3i, \text{ if and only if} \\
x^3 &= 2, \text{ and} \\
y - 9 &= 3
\end{aligned}
$$

Adding complex numbers

Definition: (Sum) Let $z = x + iy$, and $w = p + iq$ then their *sum* $z + w$ is defined to be the complex number $x + p + i(y + q)$.

Example: If $z = 3 - 4i$, and $w = -7 + 8i$, then

$$
z + w = -4 + 4i
$$

The additive identity for all complex numbers

The complex number $\mathbf{0} = (0,0)$ is the additive identity. That is if, $z = (x, y)$ is any complex number, then

$$
\begin{aligned}
z + \mathbf{0} &= (x, y) + (0, 0) \\
&= (x + 0, y + 0) \\
&= (x, y) \\
&= z
\end{aligned}
$$

Of course, this could have been written as $(x + iy) + (0 + i0) = (x + iy)$.

The additive inverse of a complex number

Given $z = x + iy$ is any complex number then its additive inverse is $-z = -x - iy$ because $z + -z = \mathbf{0}$.

Multiplying two complex numbers

Multiplication is a much more involved. Its definition is motivated by the distributive law applied to the product of two binomials in the real numbers.

$$(a + b)(c + d) = ac + bd + ad + bc$$

Just use the imaginary unit, i, in the appropriate places and you will get the product that suggests the definition of the product of complex numbers.

$$(a + ib)(c + id) = ac + i^2bd + iad + ibc$$

For our "times" notation, we will use the symbol $\cdot$, but it is permissible to use the $\times$ notation any time you want.

Definition: (Product) If $z = (x, y)$ is a complex number and $w = (p, q)$ is a complex number, then

$$z \cdot w = (xp - yq, \, yp + xq)$$

In other words,

$$(x + iy) \cdot (p + iq) = xp - yq + i(yp + xq)$$

Example: If $z = (3, 1)$, and $w = (-5, 6)$, then

$$
\begin{aligned}
z \cdot w &= (3, 1) \times (-5, 6) = (3 \times -5 - 1 \times 6, \, 3 \times 6 + 1 \times -5) \\
&= (-21, 13)
\end{aligned}
$$

Multiplying by (1,0) vs multiplying by (0,1)

For any complex number (u, v), what is $(1, 0) \times (3, 4)$? It is still $(3, 4)$. Try it out for yourself. But look at the Figure 5.4 below to see what happens when you multiply by $(0, 1)$.

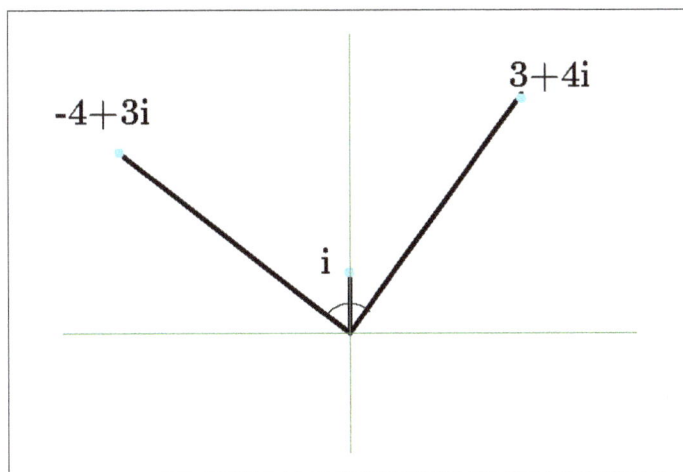

FIGURE 5.4 $(3, 4) \times (0, 1) = (-4, 3)$

This shows that, geometrically, when you multiply any (non zero) complex number by just i, that is by $(0, 1)$, you "rotate" the complex number through 90 degrees. For example, by the definition of multiplication:

$$(3,4) \times (0,1) = (3 \times 0 - 4 \times 1, 3 \times 1 + 4 \times 0) = (-4, 3)$$

What is remarkable about this is that the vector from $(0,0)$ to $(-4,3)$ makes a 90 degree angle with the original vector from $(0,0)$ to $(3,4)$..

The distributive law

From the definition of multiplication and addition of two complex numbers, we can prove the following theorem.

Theorem 2: *If each of* $z_1 = (x_1 + iy_1)$, $z_2 = (x_2 + iy_2)$, *and* $z_3 = (x_3 + iy_3)$, *is a complex number, then*

$$z_1 \cdot (z_2 + z_3) = z_1 \cdot z_2 + z_1 \cdot z_3$$

Proof: Left to the reader.   It is a straight-forward application of the definitions of addition and multiplication, but quite messy.   You must keep close track of the subscripts of x and y.

The complex conjugate

Another interesting thing happens when we multiply a complex number (x, y) by one in which we simply change the sign on the imaginary part $(x, -y)$; the result is a real number. We make this Theorem 3.

Theorem 3: *The product of the complex numbers* (x, y) *and* $(x - y)$ *is the real number* $x^2 + y^2$.

Proof:

By the definition of the product of two complex numbers, we have:

$$
\begin{aligned}
(x, y) \cdot (x, -y) &= (x \cdot x - y \cdot (-y),\ x \cdot y + (-y) \cdot x) \\
&= (x^2 - (-y^2),\ x \cdot y + -(x \cdot y)) \\
&= (x^2 + y^2,\ 0)
\end{aligned}
$$

It will be much easier to see this written out as:

$$
\begin{aligned}
(x + iy)(x - iy) &= x^2 + iyx - ixy - (i^2 y^2) \\
&= x^2 + 0 - (-y^2) = x^2 + y^2
\end{aligned}
$$

Amazing, because this product gives us an easy way to find the multiplicative inverse of a complex number, as we shall soon see.

There is a special name for this type of complex number, it is called the *conjugate* of the original number.   If $z = (x, y)$ then we denote the complex number $(x, -y)$ by $\overline{z}$, which you may read as "z bar".   It is called the *complex conjugate* of z.   In this notation we can say that if $z = (x, y)$ and $\overline{z} = (x - y)$, then $z \cdot \overline{z} = x^2 + y^2$, and this is the square of the distance of the complex number (x, y) as a point in the plane, from the origin $(0, 0)$. Thus, for any complex number z, the distance from the origin to z is $\sqrt{z \cdot \overline{z}}$.   Given any complex number, z, we can use the conjugate to define what we call the absolute value of z.

Definition: (*Absolute value*) If $P = (x, y)$ is a point defining the complex number $x = x + iy$, then the distance r from P to the origin O is called the *absolute value* of z, and is denoted by $\|z\|$, so

$$\|z\| = \sqrt{z \cdot \overline{z}}$$

The radian measure, α, of $\angle XOP$ for this point P is called the *argument* of z. We will see, later, that using the absolute value and the argument (the angle with the positive x-axis) is a way to represent the complex number z in what we call polar coordinates.

The multiplicative inverse of a complex number

Is it true that every complex number, except $(0, 0)$ has a reciprocal? In other words, if $z = (x, y)$ is any complex number not zero, does there exist a complex number $w = (p, q)$ such that $z \cdot w = (1, 0)$?

Let's try.

Problem: If $z = x + iy$ is any complex number and $z \neq 0$, find a complex number $w = p + iq$ such that $z \cdot w = 1$.

Solution: One approach is to suppose we can find a number w such that the equation $w \cdot z = 1$ is true. Then if we find a solution to this equation, we will test that solution to see whether or not it really is a multiplicative inverse of z. Let $z = x + iy$ and let $w = p + iq$ be a complex number, if any exists, such that

$$w \cdot z = (1, 0) = 1$$

We can use all of the field axioms and we can use the properties of equals because they can be proved for the definitions we gave of addition and multiplication; so, now, multiplying both sides by the conjugate $\overline{z}$, we get:

$$
\begin{array}{ll}
(w \cdot z) \cdot \overline{z} = 1 \cdot \overline{z} & \text{Property of equals} \\
w \cdot (z \cdot \overline{z}) = \overline{z} & \text{Associative law and Identity law} \\
w \cdot (x^2 + y^2) = \overline{z} & \text{Theorem 3 in this chapter} \\
w = \frac{\overline{z}}{x^2 + y^2} & \text{Solving for } w
\end{array}
$$

This says that *if there is a multiplicative inverse of z*, it must be

$$\frac{\overline{z}}{x^2 + y^2}, \text{ or } \left(\frac{x}{x^2 + y^2} - i \frac{y}{x^2 + y^2} \right)$$

But how do we actually know that this is an inverse? Check it out.

$$
\begin{aligned}
\left(\frac{x}{x^2 + y^2} - i \frac{y}{x^2 + y^2} \right) \cdot (x + iy) &= \left(\frac{x^2}{x^2 + y^2} + \frac{y^2}{x^2 + y^2}, \frac{-xy}{x^2 + y^2} + \frac{xy}{x^2 + y^2} \right) \\
&= \left(\frac{x^2 + y^2}{x^2 + y^2}, \frac{0}{x^2 + y^2} \right) = (1, 0)
\end{aligned}
$$

Problem: Show that the complex number

$$\frac{7 - 3i}{58}$$

is the multiplicative inverse of the complex number $7 + 3i$.

Solution: Multiply $(7 + 3i)$ times $\frac{7-3i}{58}$.

$$
\begin{aligned}
(7 + 3i) \cdot \frac{7 - 3i}{58} &= \frac{(7 + 3i)(7 - 3i)}{58} \\
&= \frac{58}{58} \\
&= 1
\end{aligned}
$$

Problem: If z is any point on the circle in the Argand diagram, Figure 5.2, show that $z \cdot \overline{z} = 1$.

Solution: The circle in Figure 5.2 has a radius of 1, therefore all of the points (x, y) on that circle satisfy the circle equation $x^2 + y^2 = 1$.

In Figure 5.5, below, we show two complex numbers $a = (x, y)$ and $b = (p, q)$ with absolute values r and s, respectively. Let α and β be their arguments. The product, $a \times b$, of these two complex numbers is a new complex number $c = (m, n)$, whose absolute value is $r \cdot s$ and whose argument is $\alpha + \beta$. Thus,

$$
\begin{aligned}
a = x + iy \quad & r = \|a\| = \sqrt{x^2 + y^2} \\
& \text{Argument of } a \text{ is } \alpha = \angle XOA \\
b = p + iq \quad & s = \|b\| = \sqrt{p^2 + q^2} \\
& \text{Argument of } b \text{ is } b = \angle XOB \\
c = a \times b \quad & \|c\| = \|a\| \cdot \|b\| = \sqrt{m^2 + n^2} \\
& \text{Argument of } c \text{ is } \alpha + \beta = \angle XOC
\end{aligned}
$$

In this figure we labeled the complex number a as the point A, b as the point B and c as the point C. Thus, an alternative way to find the product of two complex numbers, is to multiply their absolute values and add their arguments. We will come back to this idea when we discuss polar coordinates later in this chapter.

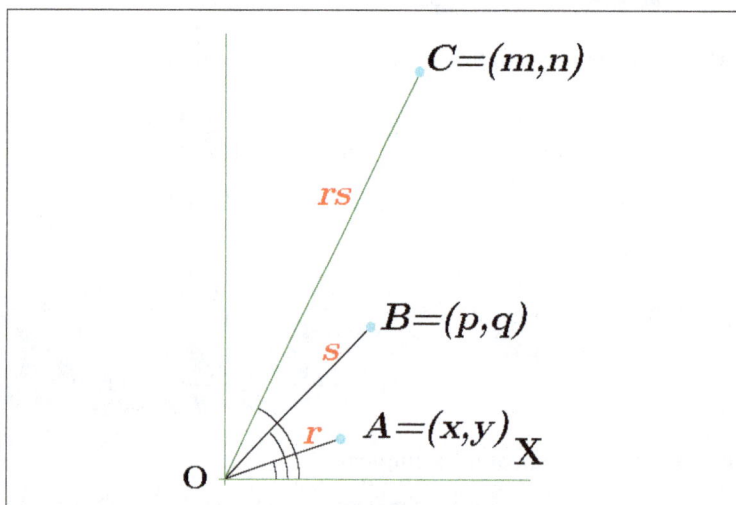

FIGURE 5.5 $\angle XOA = \alpha$, $\angle XOB = \beta$, $\angle XOC = \alpha + \beta$

A little bit of right angle trigonometry

The study of right triangles by the Egyptians and Babylonians goes back to around 2000 BCE. The word *trigonometry* itself means measurements of triangles.

At this point we need only a few definitions and theorems, the first is an important fact about the ratio of two sides of similar right triangles. Any two right triangles which have one common acute angle are called *similar*.

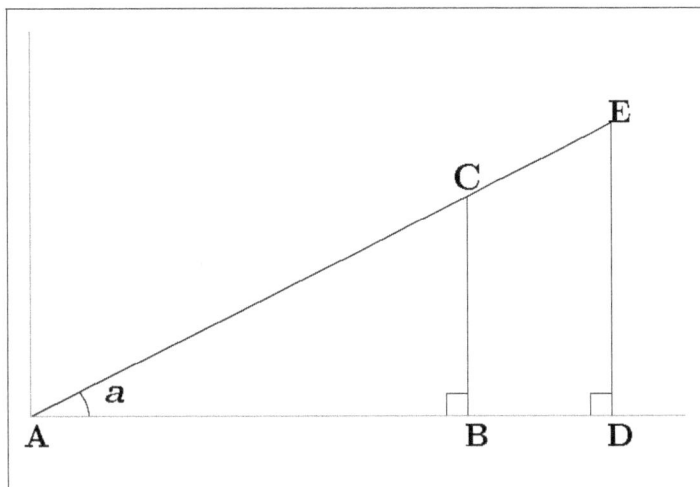

FIGURE 5.6 SIMILAR RIGHT TRIANGLES ABC, ADE

In Figure 5.6, α is any angle between 0 and 90 degrees and the triangles $\triangle ABC$ and $\triangle ADE$ are similar because they are right angle triangles and they have the angle α in common. This definition holds even if one of the triangles is the size of a sheet of paper, measured in inches and the other is the size of a mountain, measured in miles or an astronomical constellation measured in light-years.

The two similar triangles need not be nested one inside the other as shown in Figure 5.6. They can be far apart and of different sizes, and oriented in any direction, so long as they have one of their acute angles whose degree (or radian) measurement is exactly the same in both triangles. Why is this useful? Because, for example, we can use a small paper-triangle to measure the height of the mountain depicted in a large mountain-high triangle, just by comparing the ratios of two corresponding sides in the small paper triangle and the imagined mountain triangle.

Example: In Figure 5.6 the ratio of the sides $\overline{BC}$ and $\overline{BA}$ in triangle $\triangle ABC$ correspond to the ratio of the sides $\overline{DE}$ and $\overline{DA}$ in triangle $\triangle ADE$, as follows:

$$\overline{BC}/\overline{BA} = \overline{DE}/\overline{DA}$$

Theorem 4: *Corresponding sides of similar right triangles are proportional.*

Proof: Left to the reader.

Hint: This theorem can be derived from Proposition 2, Book VI in Euclid's *Elements*, in case you want to look it up. Its proof depends upon constructing *parallel* lines that contain corresponding sides of the similar triangles. Such constructions show that *transversals* (lines that cross parallel lines) cut off proportional segments.

For example, in Figure 5.6, the line segments $\overline{BC}$ and $\overline{DE}$ are on parallel lines, denoted as BC and DE, (without the bar notation). The line AE is a transversal crossing these two parallel lines, as is the line AD. These cut off proportional segments $\overline{CE}$ and $\overline{BD}$ respectively. This means that

$$\overline{AE}/\overline{AD} = \overline{CE}/\overline{BD}$$

Which eventually leads to the proportionality of the corresponding sides of the two triangles $\triangle ABC$ and $\triangle ADE$. Don't worry about understanding the details of the proof. If you are passionate about them you can succeed at completing a proof for yourself.

Example: Suppose we are given two similar triangles $\triangle ABC$, and $\triangle ADE$, and we know that:

$$\overline{AC}/\overline{AB} = m$$

then we can conclude that the corresponding ratio in triangle $\triangle ADE$ is also m; that is,

$$\overline{AE}/\overline{AD} = m$$

We can make up a list of acute angles measuring from $\alpha = 0°$ to $\alpha = 90°$, say for example, α is an angle in the set $\{0°, 1°, 2°, ...89°, 90°\}$ and compute ratios for all of the right triangles having one of those listed values of α. This lets us construct a *table of approximate ratios* that we can use for any right triangle that has at least one acute angle that is α, approximately. Thus, for the triangle in Figure 5.6, if we tabulate the ratios $\overline{AB}/\overline{AC}$ for angles α ranging in steps of 15° from 0° to 90°, we can get the table:

α	0°	15°	30°	45°	60°	75°	90°
$\overline{AB}/\overline{AC}$	1.00	0.97	0.87	0.71	0.50	0.26	0.00

TABLE 5.1 SOME ANGLES α IN FIGURE 5.6

The ratios entered in the table are *functions* of α, meaning that they depend on α. Fortunately, these tables are readily available to more precise values than two decimal places and for finer steps than 15° in your computer or even in your hand calculator. Actually, the values available in you calculators are not listed as a table; rather they are computed by an algorithm (recipe) based on infinite series, a topic that we will discuss in the next chapter.

Note: When $\alpha = 0$ in Table 5.1, the triangle $\triangle ABC$ collapses to just the line segment, $\overline{AB}$. That is, $\overline{AC}$ just becomes $\overline{AB}$. The ratio is then $\overline{AB}/\overline{AC} = \overline{AC}/\overline{AC} = 1$. That is why the ratio is 1.00 when $\alpha = 0$ in the table. And when $\alpha = 90°$, the triangle collapses to the vertical line segment $\overline{AC}$, therefore, $\overline{AB}$ just becomes a single point 0. So, $\overline{AB}/\overline{AC} = 0/\overline{AC} = 0$, when $\alpha = 90°$.

Names for the ratios: cosine, sine, tangent

Given any right triangle with a known acute angle α, the side "adjacent to" α, namely $\overline{AB}$ divided by the hypotenuse, $\overline{AC}$ is called the "cosine of α," denoted by $\cos(\alpha)$. The side "opposite to" α, namely, $\overline{BC}$, divided by the hypotenuse, $\overline{AC}$, is called the "sine of α", denoted by $\sin(\alpha)$ and the ratio of the side opposite α, $\overline{BC}$, divided by the side adjacent, $\overline{AB}$ is called the "tangent of α," denoted by $\tan(\alpha)$. Go back and look at Figure

5.6; these three functions are:

$$\cos(\alpha) = \overline{AB}/\overline{AC}, \frac{\text{Adjacent}}{\text{Hypotenuse}}$$

$$\sin(\alpha) = \overline{BC}/\overline{AC}, \frac{\text{Opposite}}{\text{Hypotenuse}}$$

$$\tan(\alpha) = \overline{BC}/\overline{AB}, \frac{\text{Opposite}}{\text{Adjacent}}$$

Lemma 1:

Using the notation back in Figure 5.3, we can write

$$\cos(\alpha) = \frac{x}{r}, \ \sin(\alpha) = \frac{y}{r}, \ \text{and} \ \tan(\alpha) = \frac{y}{x}$$

Proof:

In this figure the side opposite of the angle α has length y and the hypotenuse has length r, so by definition, the ratio $\frac{y}{r} = \sin(\alpha)$. The other ratios are similarly defined. You may notice that the fraction

$$\frac{y/r}{x/r} = \frac{y}{x}, \text{so}$$

$$\frac{\sin(\alpha)}{\cos(\alpha)} = \tan(\alpha) \qquad (5.1)$$

Equation 5.1 is not *just* an equation, true for just some specific value of α. It is a statement that is true for *all* values of α. A trigonometric equation that is true for *all* angles is called an *identity*, not to be confused with the identity elements we learned about in the group axioms. A "trigonometric identity" is one of the monsters of mathematics that strikes fear into the hearts of many students, especially when someone tells you that it is "obvious." A most unfortunate, and totally useless, word in any teaching situation.

Note: Since $r = \sqrt{x^2 + y^2}$, then Lemma 1 also says

$$\cos(\alpha) = \frac{x}{\sqrt{x^2 + y^2}}, \ \sin(\alpha) = \frac{y}{\sqrt{x^2 + y^2}}, \ \tan(\alpha) = \frac{y}{x}$$

Radians and degrees

When Euclid first talked about angles, he discussed *equal angles, straight angles,* and *right angles.* Only later did geometers start measuring angles in terms of some defined unit. For example, 1/360 of a full circle is called one *degree.* But such a measurement is arbitrary, they could have just as well divided the circle into 100 parts and taken a unit to be 1/100 of a circle and called it one *cent,* or something else. Of course, many other arbitrary angle measurements could be defined. Today, in mathematics and science, we use what is considered to be the most natural unit for measuring angles, the *radian,* and it is $1/(2\pi)$ of a full circle. We can easily relate radians to the other units of angle measurements.

The two most common angle measurements are degrees and radians. How do we get back and forth between angles that are measured in these units? Start by knowing that there are 360 degrees in a full circle and 2π radians in a full circle.

$$360 \text{ degrees} = 2\pi \text{ radians} \qquad (5.2)$$

So, dividing both sides of Equation 5.2 by 360, you get an equation for 1 degree.

$$1 \text{ degree} = \frac{2\pi}{360} \text{ or } \frac{\pi}{180} \text{ radians}$$

If you have an angle that is measured to be x degrees then it is $\frac{\pi}{180} \cdot x$ radians.

Go back to Equation 5.2 and divide both sides by 2π, to find what 1 radian is in terms of degrees, you get:

$$1 \text{ radian} = \frac{360}{2\pi} \text{ or} \frac{180}{\pi} \text{ degrees, about } 57.296 \text{ degrees}$$

If you have an angle that is measured to be x radians, then it is $\frac{180}{\pi} \cdot x$ degrees.

Example:

$$5.09 \text{ radians} = \frac{180}{\pi} \times 5.09 = \frac{916.2}{\pi} \approx 291.6 \text{ degrees}$$

Polar coordinate version of complex numbers

Every complex number $x + iy$ can be written in terms of polar coordinates. What does that mean? In rectangular coordinates, every point in the plane is located by the coordinates (x, y) where x is a number that we read off of the x-axis and y is a number we read off the y-axis (perpendicular to the x-axis). In polar coordinates, we locate a point by the following method. We start with a point in the plane called the origin, $O = (0, 0)$ and draw a line OX (usually a horizontal line) from O to a point X which is to the right of O. This is our set of positive numbers. Then we assign to each point, P, in the plane, a pair of numbers r and α where r is the distance from O to P and α is the radian measure of the angle $\angle XOP$. That is $P = (r, \alpha)$.

Rectangular coordinates are along city streets; to get from point A to point B, you travel from A along a street going East or West, then turn a corner and go on a North-South street to B. But, if you are out on a lake at point A, you use polar coordinates by first rotating your boat through some angle α degrees, say $60°$ North-by-East to line up on the point B, then zoom a straight line distance r directly to the point B. In either case you need a pair of numbers, (x, y) or (r, α) to locate the points in the plane, a two dimensional space.

Polar coordinates first came into use between 1630 and 1690 during a flurry of mathematical activity that saw the development of calculus by Descarte, Fermat, Cavalieri, and Newton. It was a natural out-growth of the idea of changing geometry into analytic geometry.

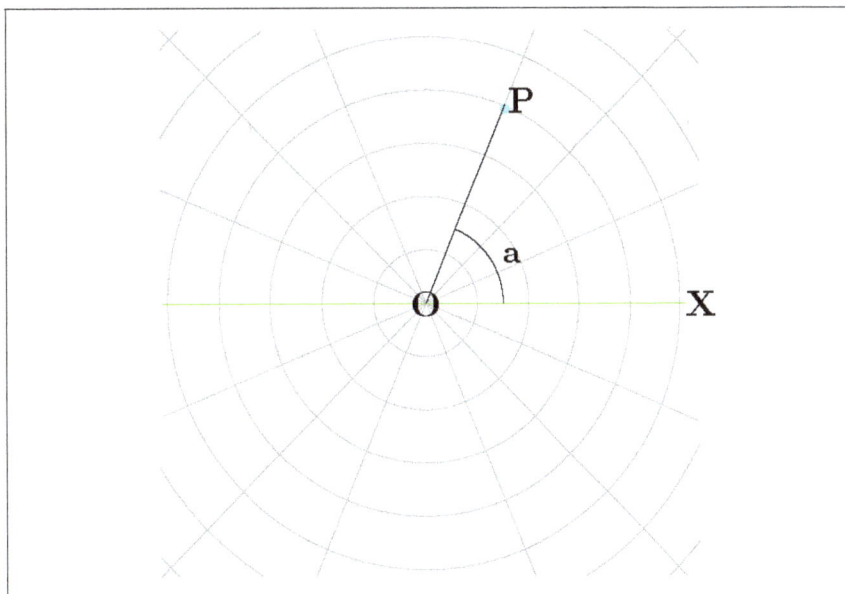

FIGURE 5.7 THE POINT $P = (\sqrt{10}, 3\pi/8)$

In Figure 5.7, above, the point P in polar coordinates (r, α) is $(\sqrt{10}, 3\pi/8)$ because it is a distance of $\sqrt{10}$ from the origin O and the angle $\angle XOP$ is $3\pi/8$ radians (about 67.5°).

In the rectangular coordinate system, this same point is $P \approx (1.21, 2.92)$, because

$$x = r\cos(\alpha) = \sqrt{10}\cos(\frac{3\pi}{8}) \approx 1.21 \text{ and}$$

$$y = r\sin(a) = \sqrt{10}\sin(\frac{3\pi}{8}) \approx 2.92$$

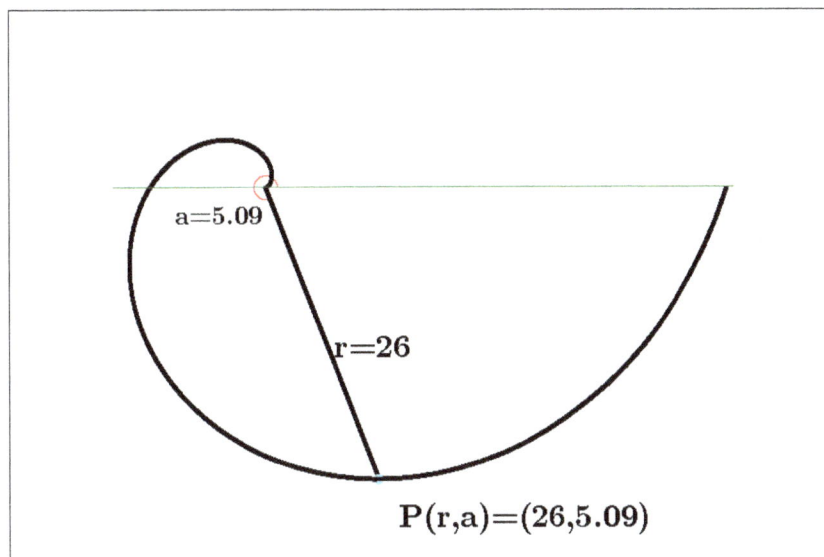

FIGURE 5.8 POLAR PLOT OF $r = \alpha^2$

If you are accustomed to the rectangular plot of an equation, you can often be startled by the polar plot of a similar looking equation.

For example, in the (x, y) plane (rectangular coordinates) the graph of the equation $y = x^2$ is a parabola. But in the (r, α) plane (polar coordinates) the graph of the equation $r = \alpha^2$ is a spiral as we see in Figure 5.8.

We can explain this by noting that the angle for the horizontal axis is $\alpha = 0$ radians. Then as we let the angle measure, α increase from zero, the equation $r = \alpha^2$ tells us that the radius, r increases as the square of α. Thus, for any point in which the angle has some value α, the distance, r of that point to the origin, is equal to α^2. This creates a set of points in which r reaches out further into the plane while the angle α increases in a counter-clockwise rotation. Think of the angle as *opening up* and the radius *growing outward* at the same time.

This is one example of how a *function of* α defines a set of points (r, α). If we use another function of α, other than α^2, we can generate a different beautiful graph in polar coordinates.

One cautionary note is that in polar coordinates, both r and α are real numbers; that is, the angle α is measured in *radians* rather than *degrees*.

The graphs of trigonometric functions such as cosine and sine are *periodic*; that is, they repeat the same values periodically. Every time an angle takes on a new value, α that is 2π more (that is $360°$ more) than a previous one, then the cosine of the new angle is the same as the cosine of the previous one. Thus, $\cos(\alpha + 2\pi)$ and $\cos(\alpha)$ are the same. Here is a graph in which $\alpha = 3t$, so every time t increases by $\frac{2\pi}{3}$, α increases by 2π, and

$$\cos(3t) = \cos(3(t + \frac{2\pi}{3}))$$

The rectangular graph of $y = \cos(3t)$

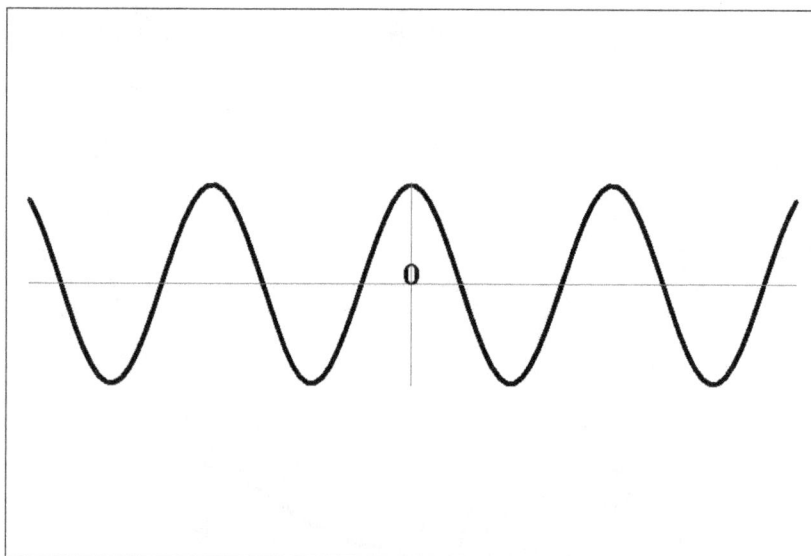

FIGURE 5.9 RECTANGULAR GRAPH OF $y = cos(3x)$

This graph continues to oscillate like this forever both to the right as $t \to \infty$ and to the left as $t \to -\infty$.

But guess what you get when you try to plot the polar coordinate graph of the same equation.

The polar graph of $r = \cos(3t)$

Question: What does the plot of the polar equation $r = \cos(3\alpha)$ look like?

Answer: See Figure 5.10. It is called a three-leaf rose. When $t = 0$, r starts out at 1, then as t increases, approaching $\frac{\pi}{2}$, $(90°)$, r gets smaller approaching 0. Then r becomes negative as t goes from $\frac{\pi}{2}$ to π, and so forth. When t gets all the way to 2π, where r becomes 1 again, and it retraces the entire figure at all angles t, whose measurements are now $t + 2\pi$. In this one graph we get to see the whole cosine graph with all of its infinitely many oscillations

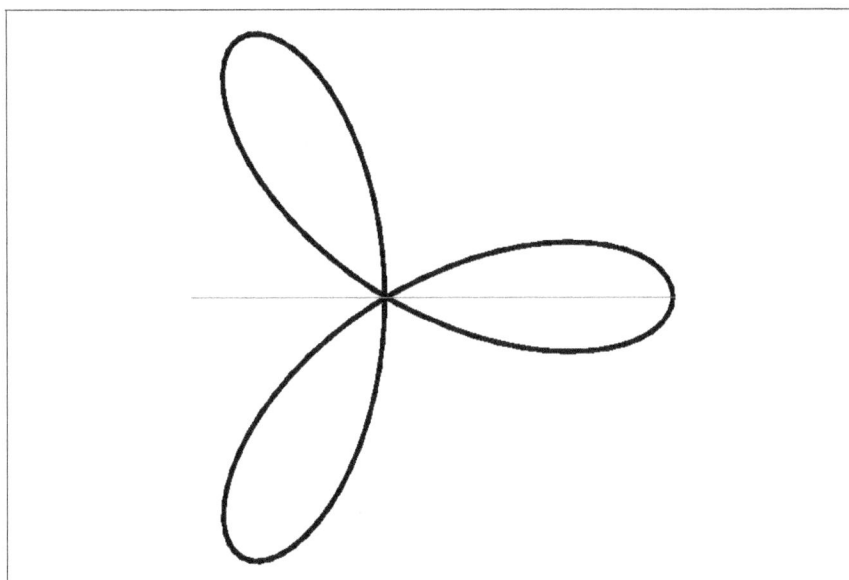

FIGURE 5.10 POLAR GRAPH OF $r = \cos(3\alpha)$

The next theorem illustrates the connection between the rectangular version of complex numbers and the polar version.

Theorem 4:

Any complex number $z = x + iy$ can be written as $z = r\cos(\alpha) + ir\sin(\alpha)$.

Proof:

If both $x = 0$ and $y = 0$, let $r = 0$ and α be any number, then both $0 + i0 = 0$ and $0(\cos(\alpha) + i\sin(\alpha)) = 0$.

If either $x \neq 0$ or $y \neq 0$, let $r = \sqrt{x^2 + y^2}$, which is not zero. Thus, we can write

$$x + iy = \sqrt{x^2 + y^2}\left(\frac{x}{\sqrt{x^2+y^2}} + i\frac{y}{\sqrt{x^2+y^2}}\right) \qquad \text{Distributive law}$$

$$x + iy = r\left(\frac{x}{\sqrt{x^2+y^2}} + i\frac{y}{\sqrt{x^2+y^2}}\right) \qquad \text{Definition of } r$$

$$\frac{x}{r} = \cos(\alpha) \text{ and } \frac{y}{r} = \sin(\alpha), \text{ when } \frac{y}{x} = \tan(\alpha) \qquad \text{Lemma 1}$$

$$x + iy = r\cos(\alpha) + ir\sin(\alpha) \qquad\qquad x = r\cos(\alpha) \text{ and } y = r\sin(\alpha)$$

In the next section we will get a spectacular equation as a result of having $x + iy = r\cos(\alpha) + i\sin(\alpha)$.

The exponential function, e^x

The number e is the base of the natural logarithm. An approximation is $e \approx 2.71828$, and it arises when we are studying the graph of a logarithm with a base e. It turns out to be the best number to use as the base of a logarithm because it simplifies all calculus formulas. Therefore, it is widely used in higher mathematics. The formula for e^x, as we shall see in Chapter 6 when we study a topic called *infinite series* is

$$e^x = 1 + x + \frac{x^2}{2} + \frac{x^3}{3!} + \frac{x^4}{4!} + \frac{x^5}{5!} + \dots$$

Where $3! = 3 \times 2 \times 1$, and $4! = 4 \times 3 \times 2 \times 1$, and so forth.

The Euler equation

A remarkable thing discovered by a Swiss mathematician Leonard Euler in the mid 1700's was that the infinite series for e^{ix} resulted in the sum of two infinite series one for $\cos(x)$ and one for $i\sin(x)$.

$$\begin{aligned}
e^{ix} &= 1 + ix + \frac{(ix)^2}{2} + \frac{(ix)^3}{3!} + \frac{(ix)^4}{4!} + \frac{(ix)^5}{5!} + \dots \\
&= 1 + ix - \frac{x^2}{2} - i\frac{x^3}{3!} + \frac{x^4}{4!} + i\frac{x^5}{5!} - \frac{x^6}{6!} + \dots \\
&= 1 - \frac{x^2}{2!} + \frac{x^4}{4!} - \frac{x^6}{6!} + \dots \\
&\quad + ix - i\frac{x^3}{3} + i\frac{x^5}{5!} - i\frac{x^7}{7!} + \dots
\end{aligned}$$

It turned out that the "even" infinite series,

$$1 - \frac{x^2}{2!} + \frac{x^4}{4!} - \frac{x^6}{6!} + .. -$$

is equal to the $\cos(x)$, and the "odd" series

$$x - \frac{x^3}{3} + \frac{x^5}{5!} - \frac{x^7}{7!} + \dots$$

is equal to the $\sin(x)$. Therefore

$$e^{ix} = \cos(x) + i\sin(x) \tag{5.3}$$

There are several different equations which we call, "Euler's Equation", this is one of them.

Euler is considered to be one of the greatest, if not the greatest mathematician of all time. There are so many mathematical formulas and procedures discovered by him, that we sometimes name such discoveries for subsequent mathematicians who were associated with these same ideas. But, Euler, himself, gets full credit as being the person to first prove Equation (5.3), which makes it much easier to calculate formulas in the complex variables. For instance, the products, quotients and conjugates of complex numbers can easily be computed from the laws of exponents. For example, if z is the complex number $re^{i\alpha}$ and w is the complex number $se^{i\beta}$, then

$$
\begin{aligned}
z \cdot w &= re^{i\alpha} \cdot se^{i\beta} = rs \cdot e^{i(\alpha+\beta)} \\
&= rs(\cos(\alpha + \beta) + i\sin(\alpha + \beta))
\end{aligned}
$$

and

$$
\begin{aligned}
\frac{z}{w} &= \frac{re^{i\alpha}}{se^{i\beta}} = \frac{r}{s}e^{i(\alpha-\beta)} \\
&= \frac{r}{s}(\cos(\alpha - \beta) + i\sin(\alpha - \beta))
\end{aligned}
$$

and

$$
\bar{z} = re^{-i\alpha} = r(\cos(\alpha) - i\sin(\alpha))
$$

Euler's equation, Equation 5.3, is probably the most elegant formula in mathematics. Here are some examples. Suppose you make $\alpha = \frac{\pi}{2}$, that is 90°, then $x + iy = \cos(\frac{\pi}{2}) + i\sin(\frac{\pi}{2}) = 0 + i \cdot 1$, because $\cos(\frac{\pi}{2}) = \cos(90°) = 0$ and $\sin(90°) = 1$. Thus,

$$
i = e^{i\pi/2}
$$

Another interesting case is when we let $\alpha = \pi$. Now, since $\cos(\pi) = -1$ and $\sin(\pi) = 0$, we get an even more spectacular result

$$
e^{i\pi} + 1 = 0
$$

This is an equation connecting five important constants in mathematics, $e, i, \pi, 1$, and 0. You didn't know that 1 was such a complex number, did you?

Problem:

One of the strangest looking numbers can be solved by Euler's equation:

What is $\sqrt{i}$?

Solution:

By Euler's equation for $\alpha = \pi/2$

$$
\begin{aligned}
i &= e^{i\pi/2} \\
\sqrt{i} &= e^{i\pi/4} \quad \text{taking the square root of both sides} \\
&= \cos(\pi/4) + i\sin(\pi/4) \\
&= \frac{\sqrt{2}}{2} + i\frac{\sqrt{2}}{2} \\
&= \frac{\sqrt{2}}{2}(1 + i)
\end{aligned}
$$

Actually there are two square roots, of i. We can show that

$$\left(\pm\frac{\sqrt{2}}{2}(1+i)\right)^2 = i$$

Another astonishing problem is what happens when you raise an imaginary number to an imaginary power?

Problem:

What is i^i ?

Solution:

Start with Euler's equation for $\alpha = \pi/2$

$$
\begin{aligned}
i &= e^{i\pi/2} \\
i^i &= e^{-\pi/2} \text{ raising both sides to the power } i \\
&\approx 0.2078795...
\end{aligned}
$$

Fantastic! The ith power of i is the real number $0.207879....$

Using Euler's equation, we can find the nth root of any complex number as follows:

$$
\begin{aligned}
\sqrt[n]{\cos(\alpha) + i\sin(\alpha)} &= \sqrt[n]{e^{i\alpha}} \\
\sqrt[n]{\cos(\alpha) + i\sin(\alpha)} &= e^{i\alpha/n} \\
\sqrt[n]{\cos(\alpha) + i\sin(\alpha)} &= \cos(\alpha/n) + i\sin(\alpha/n)
\end{aligned}
\qquad (5.4)
$$

What is a real number raised to an imaginary power?

Problem:

Find 3^i.

Solution: Look back at Chapter 4 and you will see that for the base $b = e$, the definition of a logarithm is:

$$y = \log_e(x) \text{ iff } x = e^y$$

Therefore, if

$$
\begin{aligned}
y &= \log_e(3), \text{ then} \\
3 &= e^y
\end{aligned}
$$

The notation for log base e is $\ln(x)$, so

$$3 = e^{\ln(3)}$$

Now raise both sides to the power i.

$$
\begin{aligned}
3^i &= e^{i\ln(3)} \\
&= \cos(\ln(3)) + i\sin(\ln(3)) \\
&\approx 0.45483 + i0.89058
\end{aligned}
$$

We can also raise other real or complex numbers to real or complex powers. Another application of complex numbers is their use in solving polynomial and other algebraic equations. The fundamental theorem of algebra states that a polynomial equation of degree n has n solutions. For example, the following quadratic equation is a polynomial equation of degree two.

$$z^2 + pz + q = 0.$$

Here the numbers p and q are real. The problem is to find two solutions in terms of p and q, that satisfies the given equation. If $p^2 \geq 4q$, this equation is solved for real numbers, by the *quadratic formula*:

$$z = x = \frac{-p \pm \sqrt{p^2 - 4q}}{2}$$

But what if $p^2 < 4q$? Then the square root is an imaginary number. Now we rewrite $\sqrt{p^2 - 4q}$ as follows.

$$\begin{aligned} \sqrt{p^2 - 4q} &= \sqrt{-1(4q - p^2)}, \text{ which is} \\ &= i\sqrt{4q - p^2} \end{aligned}$$

Therefore

$$z = \frac{-p \pm i\sqrt{4q - p^2}}{2}$$

By-the-way, if $p^2 = 4q$, then $\sqrt{p^2 - 4q} = 0$, the imaginary part of the answer is zero. So, we say the solution is the real number,

$$x = \frac{-p}{2}$$

What's up with that? Well, we just say that, this is a "repeated root" because, in this case, the original quadratic equation can be written as

$$\begin{aligned} z^2 + pz + \frac{p^2}{4} &= 0 \\ (z + \frac{p}{2})^2 &= 0 \end{aligned}$$

So the "two" roots are $-\frac{p}{2}$ and $-\frac{p}{2}$.

Pairs of conjugate roots

One of the most interesting things about z and $\overline{z}$, a complex number and its conjugate, is that for any real number t, the real part of z^t is the same thing as the real part of $\overline{z}^t$ and the imaginary part of z^t is the same thing as the *negative* of the imaginary part of $\overline{z}^t$. In other words the power of the conjugate of z is the conjugate of the power of z.

   Example: Let $z = re^{i\theta}$ and $\overline{z} = re^{-i\theta}$, and find the nth roots of z and $\overline{z}$. By Equation (5.4), we have

$$\begin{aligned} z^{1/n} &= r^{1/n}(\cos(\theta/n) + i\sin(\theta/n)) \text{ and} \\ \overline{z}^{1/n} &= r^{1/n}(\cos(\theta/n) - i\sin(\theta/n)) \end{aligned}$$

So, the nth root of the conjugate of z is the conjugate of the nth root of z. What is important about this is, that if we are given a function that is the sum of powers of a variable, and if that function has a complex root, z, then $\overline{z}$ must also be a root.

Example: For any positive integer n, and real numbers, p_1, p_2,..., p_n, let $P(z)$ be the polynomial

$$P(z) = z^n + p_1 z^{n-1} + p_2 z^{n-2} + ... + p_n$$

If $P(z) = 0$ has a complex root, z, then its conjugate, $\overline{z}$, *must also* be a root. This is true because $P(z) = 0$ means that

$$P(z) = 0 + i0$$

But, from our discussion about the powers of z and of $\overline{z}$, it is clear that the real part of $P(z)$ is zero and the real part of $P(\overline{z})$ is zero. The imaginary part of $P(z)$ is also zero. And the imaginary part of $P(\overline{z})$ is the negative of the imaginary part of $P(z)$. Therefore,

$$P(\overline{z}) = 0 + i0$$

Let us look at the roots of a simple fifth degree polynomial.

Problem: Find the five fifth roots of -1. That is, solve the following fifth degree polynomial.

$$z^5 + 1 = 0$$

Solution:

What we are looking for here are complex numbers z, that are fifth roots of -1. Our approach is to write $z = x + iy$ in polar form

$$z = \cos(t) + i\sin(t)$$

So we want to solve

$$(\cos(t) + i\sin(t)^5 = -1$$

which is the same thing as

$$\cos(5t) + i\sin(5t) = -1 + i \times 0$$

Not only is the cosine periodic as we have already seen, so is the sine.

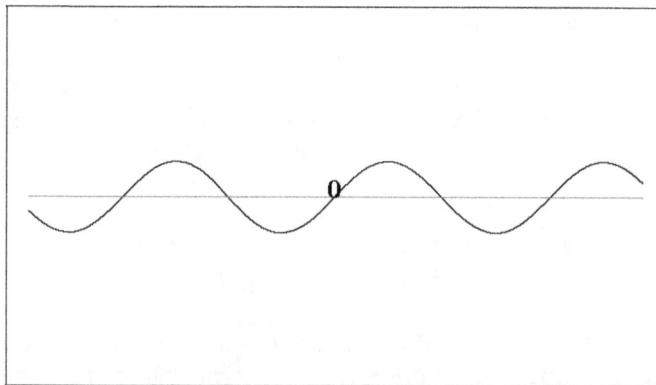

FIGURE 5.11 GRAPH OF $\sin(t)$

Since $\cos(\pi) = -1$, then, we are looking for t that makes

$$\cos(5t) = \cos(5t + 2\pi) = ... \cos(5t + 8\pi) = -1, \text{ and}$$
$$\sin(5t) = \sin(5t + 2\pi) = ... = \sin(5t + 8\pi) = 0$$

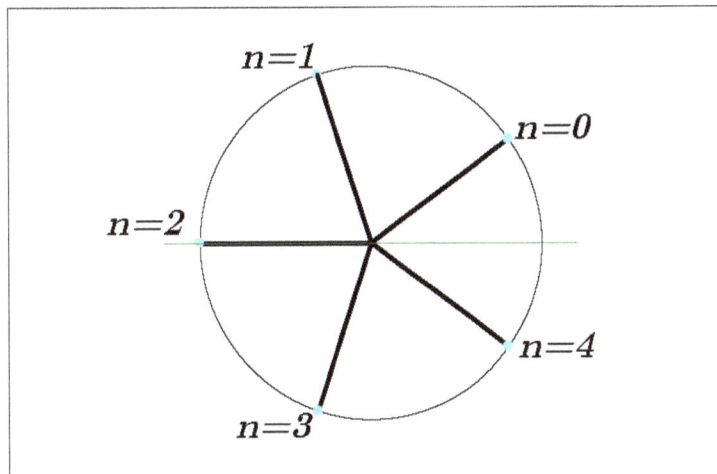

Figure 5.12 THE FIVE FIFTH ROOTS OF -1

Now we have 5 equations for t, where $5t = \pi + 2n\pi$. Letting $n = 0, 1, 2, 3, 4$, we get $5t$:

n	0	1	2	3	4
$\pi + 2n\pi$	π	3π	5π	7π	9π

So, $t = \pi/5, \, 3\pi/5, \, 5\pi/5, \, 7\pi/5$, and $9\pi/5$

$$\sin(5t) = \sin(5t + 2\pi) = \sin(5t + 4\pi) = \sin(5t + 6\pi) = \sin(5t + 8\pi) = 0$$

If we want to find all the fifth roots of -1, they are

When $n = 0$	$\cos(\pi/5) + i\sin(\pi/5)$	$0.8090 + i0.5878$
When $n = 1$	$\cos(3\pi/5) + i\sin(3\pi/5)$	$-0.3090 + i0.9511$
When $n = 2$	$\cos(5\pi/5) + i\sin(5\pi/5)$	$-1 + i0$
When $n = 3$	$\cos(7\pi/5) + i\sin(7\pi/5)$	$-0.3090 - i0.9511$
When $n = 4$	$\cos(9\pi/5) + i\sin(9\pi/5)$	$0.8090 - i0.5878$

These are known as the *five fifth roots of* -1. Notice that one of them, -1, is real and the other four are complex conjugate pairs. Figure 5.12 is a graphical representation of the five fifth roots of -1. For each integer $n = 0, 1, 2, 3, 4$, the complex number:

$$\cos\left[\frac{(2n+1)\pi}{5}\right] + i\sin\left[\frac{(2n+1)\pi}{5}\right]$$

is a fifth root of -1.

Problem:

Solve the fifth degree polynomial   That is, find the five fifth roots of $+1$.

$$z^5 - 1 = 0$$

Solution:

Let
$$+1 = \cos(0 + 2n\pi/5) + i\sin(0 + 2n\pi/5), \text{ where } n = 0, 1, 2, 3, 4$$

then compute
$$\cos(2n\pi/5) + i\sin(2n\pi/5)$$

for the five values, $0, 1, 2, 3, 4$ of n. This gives us

n	$\cos(2n\pi/5) + i\sin(2n\pi i/5)$
0	$1 + 0i$
1	$0.3090... + i0.9511...$
2	$-0.8090... + i0.5878...$
3	$-0.8090... - i0.5878...$
4	$0.3090... - i0.9511...$

These five points can be plotted on the unit circle as we did in Figure 5.12. There is one real root, $+1$, where $n = 0$, and the other four roots are two pairs of conjugate complex numbers.

Are there any applications using complex numbers other than these geometry and algebra problems? Yes, in physics, engineering and other sciences, we encounter the need for them. For example, in physics if we know that the acceleration of some object is negatively proportional to its displacement, then we will find its displacement in terms of e^{it}. This is a result that comes from calculus and differential equations. It roughly says, for any positive number a,

$$\text{If acceleration of } x(t) = -a^2 x(t), \text{ then } x(t) = C_1 e^{iat} + C_2 e^{-iat}$$

This means, if the acceleration of an object is negatively proportional to its location, then it is travelling in a linear combination of cosine and sine waves.

Vectors

Complex numbers are not the only denizens of the Euclidean x-y plane. We also find vectors there. What is a vector? In biology it is something, like a mosquito, that carries a disease or an infection from one organism to another; in medicine, a vector carries DNA information from one cell to another. But what is a vector in physics, or engineering, or mathematics? It is something that has length and direction. When we wrote a complex number $z = r\cos(\alpha) + ir\sin(\alpha)$, we depicted it as a vector in the plane (a *space* with two dimensions).

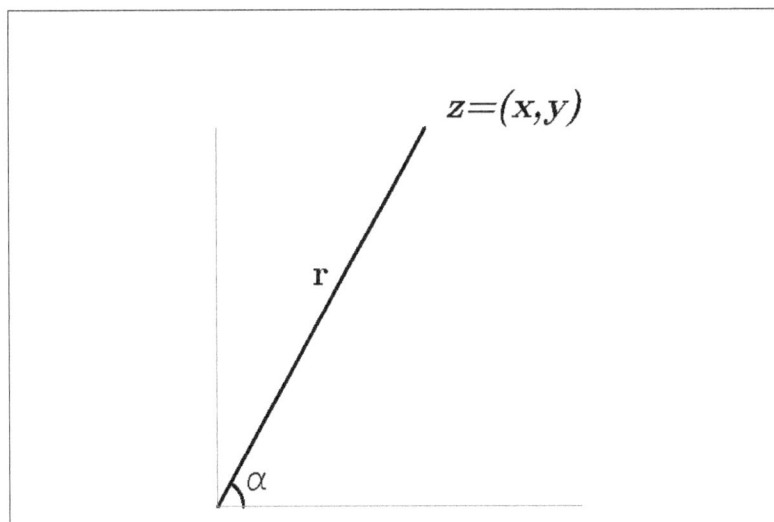

FIGURE 5.13 THE COMPLEX VECTOR (x, y)

In Figure 5.13 the complex number (x, y) is a two dimensional vector because it has length $r = \sqrt{x^2 + y^2}$, and a direction α, that is, the angle such that $\tan(\alpha) = \frac{y}{x}$. Recall that this means the same thing as $\cos(\alpha) = \frac{x}{r}$ and $\sin(\alpha) = \frac{y}{r}$ when we converted complex numbers to polar coordinates.

In physics, the length of a vector corresponds to some force that is applied to an object that propels it in a given direction. An example of such a force is gravity. On some planet, such as Earth, the force of gravity will pull a body straight down to the surface (actually toward the center of the Earth). But if some other force is acting on the body, such as a hurricane wind, acting from a different direction then the body could be pushed in another direction, not straight down.

Or, as another example, if a baseball is thrown by the pitcher with a horizontal force, then gravity acting vertically will make it arrive at the plate going in a downward angle where the batter will supply a force in the form of a swing of the bat hitting the ball sending it into an entirely new direction.

In electrical engineering, a vector equation relating *voltage*, *current* and *impedance* is written in complex variables. But since the notation for current is i, the electrical engineers use j for the complex number $\sqrt{-1}$.

Example:

$$\text{Voltage} = \text{current} \times \text{impedance}$$

Where *impedance*, also called *resistance*, is measured in ohms, which is defined as volts per amp. *Current* is measured in amps, so the "unit analysis equation" is:

$$\text{amps} \times \text{ohms} = \text{amps} \times \frac{\text{volts}}{\text{amps}} = \text{volts}$$

Problem:
Find the voltage if current is $3 + 2j$ amps, and impedance is $2 - j$ ohms.
Answer:

$$\text{Voltage} = (3 + 2j)(2 - j) = 8 + j \text{ volts}$$

It is instructive to look at the geometry of this electrical engineering equation.   In the complex plane with a real axis, x and a complex axis, jy let the current $c = (3 + 2j)$ and the impedance $r = (2 - j)$, and the voltage $v = 8 + j$.   Then geometrically, the three quantities, after changing over to polar form (r, t) are:

Rectangular (x, y)	Polar (r, t)
$(3, 2)$	$(\sqrt{13}, 0.5880)$
$(2, -1)$	$(\sqrt{5}, -0.4636$
$(8, 1)$	$(\sqrt{65}, 0.1244)$

as shown in Figure 5.14.

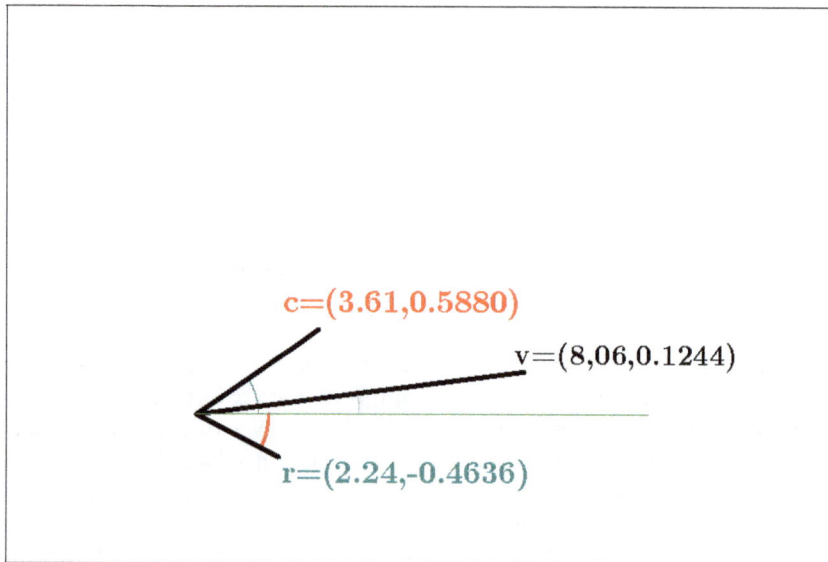

FIGURE 5.14 VOLTAGE = CURRENT × RESISTANCE

Vector spaces in three and higher dimensions

Anytime we study a *collections of vectors* and find out what happens when we add them or multiply them, we are studying a *vector space*.   For example the complex numbers are two-dimensional, with one real component and one imaginary component, so we call it a 2 space.   Other two, three, or more dimensional spaces might have *no* imaginary component at all, or some real and some imaginary components.

In most cases, *addition* will be defined simply as the addition of corresponding coordinates, as we just did in the complex plane or as was done in matrix addition (see Chapter 2).

The specialized properties of a given vector space usually come from the definition of *multiplication*.   It is what tells us the kind of practical applications that can be solved in that space.   In fact, some particular definition of multiplication may have been especially constructed to solve a specific problem, only to find out that the resulting space can be used in applications beyond that for which the multiplication was originally invented.   Any one of these spaces can be measured in rectangular, polar, spherical, cylindrical, or other *curvilinear coordinates*.

Just imagine that you have some bunch of points filling up some space, all around you, as if you were in an electrical substation somewhere. Then suddenly, you discover some rules of arithmetic that let you impose your own "grid" on your surroundings. Then you start locating certain key points and define vectors and distances suitable to your own practical application. This is what happened to William Rowan Hamilton in 1833 after spending ten years working on his inventions of a four dimensional space, called *quaternions*. Later, my friend!

A vector in real variables may have 3 or more dimensions; for example, an aircraft flying in the air can move in any one of three directions relative to some fixed coordinate system with an x-axis, y-axis, and z-axis. In the next figure, we depict a three dimensional vector. It has a length and directions. The distance from the origin to a point (x, y, z) is $\sqrt{x^2 + y^2 + z^2}$, and its directions are the three angles it makes with the positive branches of the three mutually perpendicular axes, x, y, and z. In this graph, (Figure 5.15) the length of the vector $\overrightarrow{OP}$ is $r = \sqrt{3}$.

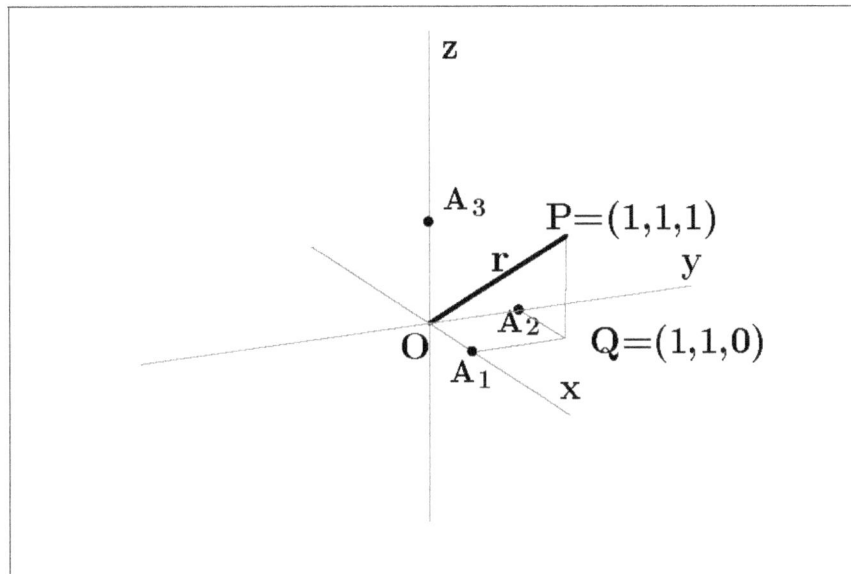

FIGURE 5.15 A VECTOR IN 3D RECTANGULAR COORDINATES

In this 3 space, coordinate axes all start at the single point O, the origin whose coordinates are all zeros; $O = (0, 0, 0)$. Let A_1 be a point on the x-axis and at a distance of 1 (1 *inch*, 1 *mile*, whatever unit of length you want to use) from the origin. A_1 has coordinates $(1, 0, 0)$. The point A_2 on the y-axis at a distance of 1 from the origin has coordinates $(0, 1, 0)$, and on the z-axis, $A_3 = (0, 0, 1)$.

We can use A_1, O, and P to create an angle $\angle POA_1$ and measure this angle in radians. Let us call the radian measure of this angle $t_1 = \angle POA_1$. Similarly we can find the radian measures, $t_2 = \angle POA_2$ and $t_3 = \angle POA_3$. These three angles give us the directions for the vector $\overrightarrow{OP}$.

The space-time continuum in physics uses the three space coordinates length, breadth, height and time to locate a point (x, y, z, t). Now we must think of four mutually perpendicular axes x, y, z, t

It is not possible to draw or even construct some kind of 4 space model made with straws or tinker toys, or pieces of wire. So we must *imagine* a four dimensional point. It helps to start with the idea that there is a single point, O, the origin that has four zeros for its coordinates.

$$O = (0, 0, 0, 0)$$

Then imagine four positive number lines, *axes*, all starting at O and going off to infinity in four different directions. You must be willing to call these the positive numbers, on one side of the origin and negative numbers on the other side. Then you must be willing to label four units on the positive branch of each axis and call each one a "unit" for that axis.

$$
\begin{aligned}
A_1 &= (1, 0, 0, 0) \\
A_2 &= (0, 1, 0, 0) \\
A_3 &= (0, 0, 1, 0) \\
A_4 &= (0, 0, 0, 1)
\end{aligned}
$$

All of the points in 4 space will have four coordinates that we can read off of the four axes. A 4 space is used to describe the real world–the place in which we live, breathe, work, and sleep, not only because we have to move back and fourth and from the left to the right, but also up and down and we have to take time to make these movements.

However, in the 1960's and 1970's, physicists introduced an idea that some concepts in physics, for example gravity at the sub-atomic level, need many more dimensions to be described properly. They have constructed a model with six or seven new dimensions, which wrap up fundamental particles. These, along with the usual 4 space, crates a ten or eleven dimensional space, *string theory*.

We don't even have to confine ourselves to only eleven dimensional space. For example for any positive integer n, we can simply define an n dimensional origin O and write out an ordered n-tple of numbers to stand for a point P in $n$space.

$$P = (x_1, x_2, x_3, x_4, ..., x_n)$$

An interesting example of a space in which $n = 13$, is the following.

Example: In an oceanographic model, in which you want to write equations for use in studying deep ocean currents, temperature, ambient light, dissolved minerals and gases at any location and any time, we can define a 13 space as follows.

Category	Variables	Coordinates representing:
Location and time	x_1, x_2, x_3, x_4	latitude, longitude, depth, time
Temperature	x_5	temperature
Solutes	x_6, x_7, x_8	Salt, CO_2, and Oxygen
Ocean currents	$x_9, x_{10}, x_{11}, x_{12}$	Magnitude and three directions
Ambient light	x_{13}	Sunlight not as a function of depth

One point, $\mathbf{x}$ in this model would have 13 different coordinates. Knowing the values for a sample of points in some region of the ocean we can define a weather condition for that region. Then if we know some theoretical, statistical, or even empirical equation that relates this condition to some event, like a cyclone or tornado, we could predict the intensity of a storm. Such an equation would be written as a function, $y = C(\mathbf{x})$

$$y = C(x_1, x_2, x_3, x_4, x_5, x_6, x_7, x_8, x_9, x_{10}, x_{11}, x_{12}, x_{13})$$

telling us how each variable would affect the outcome. Another function $y = F(\mathbf{x})$, might reveal its path, and so on.

How do we depict a 13 space? We can start with the origin as a point, O, all of whose coordinates are zero.

$$O = (0, 0, 0, 0, 0, 0, 0, 0, 0, 0, 0, 0, 0)$$

Imagine 13 axes starting at that point. On each axis we can imagine a point *one unit* away from the origin. Say A_1 is on the x_1 axis and one unit from O; let A_2 be a point on the x_2 axis one unit from O:

$$
\begin{aligned}
A_1 &= (1, 0, 0, 0, 0, 0, 0, 0, 0, 0, 0, 0, 0) \\
A_2 &= (0, 1, 0, 0, 0, 0, 0, 0, 0, 0, 0, 0, 0) \\
A_3 &= (0, 0, 1, 0, 0, 0, 0, 0, 0, 0, 0, 0, 0) \\
&\quad \dots \\
A_{13} &= (0, 0, 0, 0, 0, 0, 0, 0, 0, 0, 0, 0, 1)
\end{aligned}
$$

Then if P is any point, $(x_1, x_2, ..., x_{13})$ in 13 space it will make 13 angles $t_1 = \angle POA_1$, $t_2 = \angle POA_2, ..., t_{13} = \angle POA_{13}$ with the 13 coordinate axes. We can define the length of the vector $\overrightarrow{OP}$, as we did in two or three space, by:

$$\left\| \overrightarrow{OP} \right\| = \sqrt{p_1^2 + p_2^2 + p_3^2 + ... + p_{13}^2}$$

For the negative side of each of these thirteen axes, we would have the corresponding negatives of the unit vectors, $-A_1, -A_2$, etc. We can do the same thing for any other positive n, not just $n = 13$. We construct n unit vectors, A_k on the kth coordinate axis, and we construct n angles, $t_1 = \angle POA_1$, $t_2 = \angle POA_2$, ..., $t_n = \angle POA_n$.

Vector addition in n space

We need not have every vector in n space originate at the origin. We can start a vector at *any* point, and all we need to know is the direction and length, we can add it to another vector that does start at the origin (or anywhere else, for that matter). When we add two vectors in n space we add their coordinates. Thus, if V_1 and V_2 are two n dimensional vectors, their sum $V_1 + V_2$ is a vector whose coordinates are the sums of the corresponding coordinates of V_1 and V_2. If

$$
\begin{aligned}
V_1 &= (x_1, \ x_2, ..., \ x_n) \\
V_2 &= (y_1, \ y_2, ..., \ y_n)
\end{aligned}
$$

then

$$V_1 + V_2 = (x_1 + y_1, \ x_2 + y_2, ..., \ x_n + y_n)$$

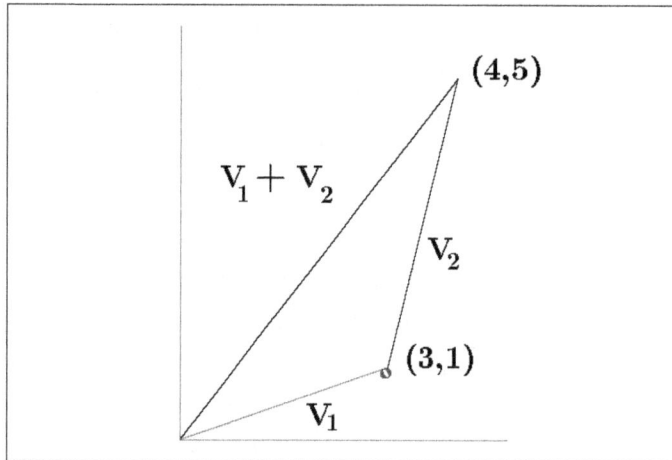

Figure 5.16 $(3,1) + (1,4) = (4,5)$

Example: Figure 5.16 illustrates the addition of vectors $V_1 = (3,1)$, and $V_2 = (1,4)$; their sum is $V_1 + V_2 = (4,5)$.

Lie Algebra

The difficulty in trying to establish a vector space in which we can "do the math", (add, multiply, and otherwise combine vectors) is inventing a suitable definition for these binary operators. For example, there is a definition for multiplication of real vectors that can be used to define a 3 space of vectors as an algebra called the *Lie Algebra*, named for the Norwegian mathematician Sophus Lie, (1875). His definition of multiplication is non-commutative and non-associative, (so it is not a multiplicative group). Never-the-less this space has a useful applications in a field known as *differential geometry*.

In the next section we will see a new system in a higher dimensional space (a 4 space to be exact) which has one real number axis and three *different* imaginary axes. The units (segments of length 1) in the system, are $1, i, j$ and k. Where $i^2 = -1, j^2 = -1, k^2 = -1$. Addition in this space is just adding coordinates, but an attempt to define *multiplication* that satisfied the group axioms (specifically the associative law) was extremely difficult.

Quaternions

In 1833, William Rowan Hamilton, an Irish mathematician, was trying to construct a three dimensional mathematical system that would satisfy the rules for algebra (the field axioms) in addition and multiplication. He had no problem showing how addition could be defined in order to satisfy the closure law and all the other laws for addition: commutative, associative, identity, and inverse to create an additive group. He simply defined addition to be the addition of the individual coordinates.

But multiplication seemed to be impossible. His goal was to create a definition of multiplication that would always result in having the product of two vectors result in a vector, and he wanted the *associative law* for multiplication $V \times (U \times W) = (V \times U) \times W$. Not only that, Hamilton wanted there to be an identity vector, I, that would have the property that for any vector V, $I \times V = V$. And he needed his multiplication to insure that if V was any nonzero vector, there would always be another vector W such that

$W \times V = I$. That is, he wanted every vector to have a multiplicative inverse. This was a big order to demand from abstract entities.

He finally decided to abandon making up a *real* space, unwittingly leaving it to Sophus Lie to invent his definition of a non-associative multiplication 40 years later. Instead, Hamilton turned to constructing a vector space that was a generalization of the complex plane because complex numbers were already a 2 space that satisfied all of the additive and multiplicative group axioms he needed. This approach lead to a 4 space in which one dimension was real and the other three were imaginary. He called vectors in this space *quaternions*. But how could multiplication be defined in that space?

This problem haunted Hamilton, day and night for ten years. He could not even devise a definition for multiplication that would make quaternions obey the *closure law*, so that the product of two quaternions comes out to *look like a quaternion*! He wanted to be able to write that two points, $\mathbf{u}_1$ and $\mathbf{u}_2$:

$$\mathbf{u}_1 = x_1 + \mathbf{i}y_1 + \mathbf{j}z_1 + \mathbf{k}w_1, \text{ and}$$
$$\mathbf{u}_2 = x_2 + \mathbf{i}y_2 + \mathbf{j}z_2 + \mathbf{k}w_2$$

could be *multiplied* by some method, yet to be defined, and result in a point:

$$\mathbf{u}_3 = x_3 + \mathbf{i}y_3 + \mathbf{j}z_2 + \mathbf{k}w_3$$

He invented the imaginary units whose squares are -1. $\mathbf{i}^2 = -1$, $\mathbf{j}^2 = -1$, $\mathbf{k}^2 = -1$.

Since quaternions are four dimensional points, we can *not* sketch a meaningful graph of a quaternion in all four dimensions, so we will look at a couple of three dimensional snapshots in which we let one of the variables be zero.

Example: In Figure 5.17, we look at a *pure imaginary point*, $(0, 1, 2, 2)$ in which we let the real part, $x = 0$, and the imaginary parts be $y = 1, z = 2, w = 2$.

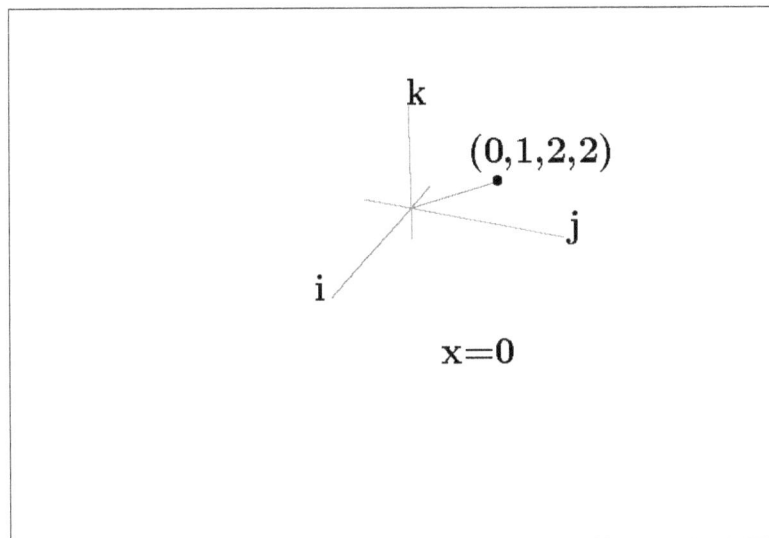

FIGURE 5.17 THE QUATERNION $(0, 1, 2, 2)$

Example: If the imaginary part, y, is zero then we get a quaternion that looks like Figure 5.18 below.

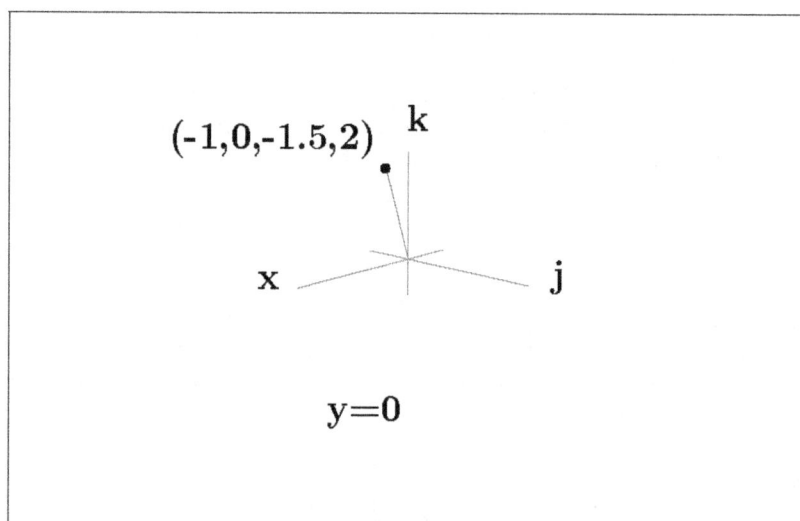

FIGURE 5.18 THE QUATERNION $(-1, 0, -1.5, 2)$

Where we set $y = 0$. This is a snapshot of the quaternion (x, y, z, w) when we let $x = -1$, $y = 0$, $z = -1.5$ and $w = 2$.

Here is the problem that Hamilton faced. When you multiply the two quaternions, $\mathbf{u}_1$ and $\mathbf{u}_2$, given above, together you get,

not only the $\mathbf{i}^2$, $\mathbf{j}^2$, and $\mathbf{k}^2$, terms, but you also get other terms like $\mathbf{ij}$, $\mathbf{ji}$, $\mathbf{ik}$, $\mathbf{ki}$, $\mathbf{kj}$ and $\mathbf{jk}$.

The $\mathbf{i}^2$, $\mathbf{j}^2$, $\mathbf{k}^2$ terms are OK.,.because you already know that each is -1. What do you do with other ones, the $\mathbf{ij}$, $\mathbf{ji}$, etc., terms? Hamilton did not have a way to relate the product of two *different* imaginary units. He needed some assumption that could reduce these disparate products to one or another of the imaginary units. Here is an example of what the difficulty is.

Example: Let $\mathbf{u}$ and $\mathbf{v}$ be two quaternions:

$$\begin{aligned} \mathbf{u} &= 3 + 1\mathbf{i} + 2\mathbf{j} + 2\mathbf{k} \\ \mathbf{v} &= -1 + 3\mathbf{i} - 1.5\mathbf{j} + 2\mathbf{k} \end{aligned}$$

Then to multiply these, we define the product, $\mathbf{u} \times \mathbf{v}$ as follows:

$$\begin{aligned} \mathbf{u} \times \mathbf{v} &= (3 + 1\mathbf{i} + 2\mathbf{j} + 2\mathbf{k})(-1 + 3\mathbf{i} - 1.5\mathbf{j} + 2\mathbf{k}) \\ &= -3 + 9\mathbf{i} - 4.5\mathbf{j} + 6\mathbf{k} + \\ &\quad -1\mathbf{i} + 3\mathbf{i}^2 - 1.5\mathbf{ij} + 2\mathbf{ik} + \\ &\quad -2\mathbf{j} + 6\mathbf{ji} - 3\mathbf{j}^2 + 4\mathbf{jk} + \\ &\quad -2\mathbf{k} + 6\mathbf{ki} - 3\mathbf{kj} + 4\mathbf{k}^2 \end{aligned} \qquad (5.5)$$

In Equations 5.5 we do not know how to find $\mathbf{ij}$, or $\mathbf{ik}$, or $\mathbf{ki}$, nor any other of the products of two different imaginary units. But we can simplify the expression a bit because we know the square of the three imaginary units, that is: $\mathbf{i}^2, \mathbf{j}^2, \mathbf{k}^2$, are each equal to -1.

$$\begin{aligned}
q \times r &= (-3 - 3 + 3 - 4) + (9 - 1)\mathbf{i} + \\
&\quad (-4.5 - 2)\mathbf{j} + (6 - 2)\mathbf{k} - 1.5\mathbf{ij} \\
&\quad + 2\mathbf{ik} + 6\mathbf{ji} + 4\mathbf{jk} + 6\mathbf{ki} - 3\mathbf{kj}
\end{aligned} \tag{5.6}$$

Re-write Equation 5.6 as:

$$\mathbf{u} \times \mathbf{v} = -7 + 8\mathbf{i} - 6.5\mathbf{j} + 4\mathbf{k} - 1.5\mathbf{ij} + 2\mathbf{ik} + 6\mathbf{ji} + 4\mathbf{jk} + 6\mathbf{ki} - 3\mathbf{kj} \tag{5.7}$$

To finish making the product actually look like a quaternion we need to know how to write each of $\mathbf{ij}$, $\mathbf{ji}$, $\mathbf{ik}$, $\mathbf{ki}$, $\mathbf{jk}$, and $\mathbf{kj}$ as one of the imaginary units: $\mathbf{i}, \mathbf{j}$, and $\mathbf{k}$. Can it be done? This is not yet the case in Equation 5.7.

A bit of graffiti

Oddly enough, the associative law for multiplication is easy to prove, and the multiplicative identity is known; it is: $1 + 0\mathbf{i} + 0\mathbf{j} + 0\mathbf{k}$. But the definition that would give him a closure law for multiplication eluded him. Then one day in 1843, as the story goes, while walking across Brougham Bridge in Dublin, like a bolt out of the blue, he hit upon the formula he needed. He realized that the product $\mathbf{i} \times \mathbf{j} \times \mathbf{k}$ had to be -1, to insure that he could reduce to equation to the proper form.

He took out his pen knife and carved the following formulas into a stone on the bridge.

$$\mathbf{i}^2 = -1, \ \mathbf{j}^2 = -1, \ \mathbf{k}^2 = -1, \text{ and } \mathbf{i} \times \mathbf{j} \times \mathbf{k} = -1$$

A replica of these can still be seen on the bridge today.

From these four equations we can derive every combination of products needed to get a proper definition of quaternion multiplication. For example, $\mathbf{j} \times \mathbf{k}$ equals $\mathbf{i}$. Here is a proof: we start by using $\mathbf{i} \times \mathbf{j} \times \mathbf{k} = -1$.

$\mathbf{i} \times \mathbf{j} \times \mathbf{k} = -1$	Given
$\mathbf{i} \times (\mathbf{i} \times \mathbf{j} \times \mathbf{k}) = \mathbf{i} \times (-1)$	Property of equals
$(\mathbf{i} \times \mathbf{i}) \times (\mathbf{j} \times \mathbf{k}) = \mathbf{i} \times (-1)$	Associative law
$(-1) \times (\mathbf{j} \times \mathbf{k}) = \mathbf{i} \times (-1)$	Because $\mathbf{i}^2 = -1$
$\mathbf{j} \times \mathbf{k} = \mathbf{i}$	Property of equals

Furthermore, we can use a similar argument to prove that $\mathbf{k} \times \mathbf{j} = -\mathbf{i}$.

Actually, starting with $\mathbf{i} \times \mathbf{j} \times \mathbf{k} = -1$, we can reduce each of the other undesirable products to one of the three imaginary units. Let us summarize all of the products of two imaginary units in the following table.

$\times$	1	$\mathbf{i}$	$\mathbf{j}$	$\mathbf{k}$
1	1	$\mathbf{i}$	$\mathbf{j}$	$\mathbf{k}$
$\mathbf{i}$	$\mathbf{i}$	-1	$\mathbf{k}$	$-\mathbf{j}$
$\mathbf{j}$	$\mathbf{j}$	$-\mathbf{k}$	-1	$\mathbf{i}$
$\mathbf{k}$	$\mathbf{k}$	$\mathbf{j}$	$-\mathbf{i}$	-1

TABLE 5.2

Here is how we finish off Equation 5.7

$$
\begin{aligned}
q \times r &= -7 + 8\mathbf{i} - 6.5\mathbf{j} + 4\mathbf{k} - 1.5\mathbf{ij} + 2\mathbf{ik} + 6\mathbf{ji} + 4\mathbf{jk} + 6\mathbf{ki} - 3\mathbf{kj} \\
&= -7 + 8\mathbf{i} - 6.5\mathbf{j} + 4\mathbf{k} - 1.5\mathbf{k} + 2(-\mathbf{j}) + 6(-\mathbf{k}) + 4\mathbf{i} + 6\mathbf{j} - 3(-\mathbf{i}) \\
&= -7 + (8 + 4 + 3)\mathbf{i} + (-6.5 - 2 + 6)\mathbf{j} + (4 - 1.5 - 6)\mathbf{k} \\
&= -7 + 15\mathbf{i} - 2.5\mathbf{j} - 3.5\mathbf{k}
\end{aligned}
$$

Whew! Now we can multiply any two quaternions with a definition of multiplication that satisfies the associate law! And, as we will show, the multiplicative inverse has a beautiful formation, reminiscent of the multiplicative inverse of a complex number. Look.

Dividing quaternions

To easily understand how to get the multiplicative inverse of a quaternion, we need to solve a small problem.

First let us find the square of a trinomial in the real number system.

$$
\begin{aligned}
(a + b + c)^2 &= (a + b + c)(a + b + c) \\
(a + b + c)^2 &= (a + b)^2 + (a + b)c + c(a + b) + c^2 \\
&= a^2 + b^2 + c^2 + \\
&\quad + ab + ba + ac + ca + bc + cb
\end{aligned}
\tag{5.8}
$$

Problem: If $\mathbf{h} = y\mathbf{i} + z\mathbf{j} + w\mathbf{k}$, find $\mathbf{h}^2$.
Solution:

$$
\begin{aligned}
(y\mathbf{i} + z\mathbf{j} + w\mathbf{k})^2 &= (y\mathbf{i} + z\mathbf{j} + w\mathbf{k})(y\mathbf{i} + z\mathbf{j} + w\mathbf{k}) \\
&\quad (y\mathbf{i})^2 + (z\mathbf{j})^2 + (w\mathbf{k})^2 + \\
&\quad + yz\mathbf{ij} + zy\mathbf{ji} + yw\mathbf{ik} + wy\mathbf{ki} + wz\mathbf{jk} + zw\mathbf{kj} \\
&= -y^2 - z^2 - w^2 + \\
&\quad + yz\mathbf{k} - zy\mathbf{k} - yw\mathbf{j} + wy\mathbf{j} + wz\mathbf{i} - zw\mathbf{i} \\
\mathbf{h}^2 &= -y^2 - z^2 - w^2 + 0 + 0 + 0
\end{aligned}
\tag{5.9}
$$

We will define the *conjugate* of a quaternion. Let x, y, z, w be real numbers not all zero and $\mathbf{v} = x + \mathbf{i}y + \mathbf{j}z + \mathbf{k}w$. Define

$$
\overline{\mathbf{v}} = x - \mathbf{i}y - \mathbf{j}z - \mathbf{k}w
$$

to be the conjugate of $\mathbf{v}$ just as we did in the complex numbers.
Theorem:

$$
\mathbf{v}^{-1} = \frac{\overline{\mathbf{v}}}{x^2 + y^2 + z^2 + w^2}
\tag{5.10}
$$

Proof: By definition of $\mathbf{v}$ and $\overline{\mathbf{v}}$, we have

$$
\mathbf{v} \times \overline{\mathbf{v}} = (x + (\mathbf{i}y + \mathbf{j}z + \mathbf{k}w)) \times (x - (\mathbf{i}y + \mathbf{j}z + \mathbf{k}w))
$$

which we can write as

$$\mathbf{v} \times \overline{\mathbf{v}} = (x + \mathbf{h})(x - \mathbf{h})$$

where $\mathbf{h}$ is the quaternion $y\mathbf{i} + z\mathbf{j} + w\mathbf{k}$. Now by the distributive law, we get that the *Right Hand Side* is the difference of two squares.

$$(x + \mathbf{h})(x - \mathbf{h}) = x^2 - \mathbf{h}^2$$

By Equation 5.9,

$$
\begin{aligned}
x^2 - \mathbf{h}^2 &= x^2 - (-y^2 - z^2 - w^2) \\
&= x^2 + y^2 + z^2 + w^2 \\
\mathbf{v} \times \overline{\mathbf{v}} &= x^2 + y^2 + z^2 + w^2
\end{aligned}
$$

$$\mathbf{v} \times \frac{(x - \mathbf{i}y - \mathbf{j}z - \mathbf{k}w)}{x^2 + y^2 + z^2 + w^2} = 1$$

Proving Equation 5.10.

The roll, pitch and yaw quaternions

An immediate application of quaternions is seen in computer programs that navigate ships, aircraft, space probes and even drones. The imaginary units, $\mathbf{i}$, $\mathbf{j}$, and $\mathbf{k}$ are the unit vectors and their products create rotations that are called *roll,* along the longitudinal axis, $\mathbf{i}$, *pitch* along the lateral axis, $\mathbf{j}$ and *yaw* along the vertical axis, $\mathbf{k}$. See Figures 5.19 *a, b* and *c*.

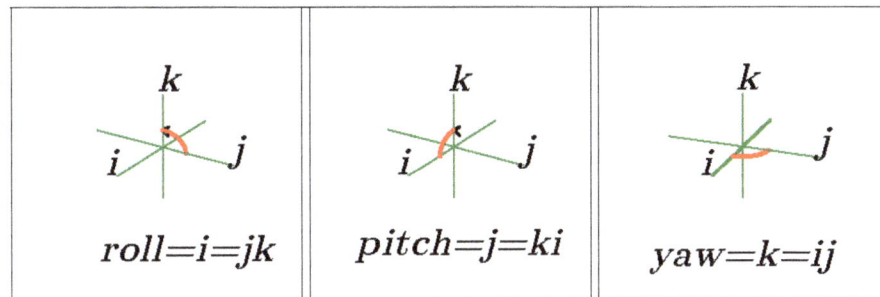

| FIGURE 5.19A | FIGURE 5.19B | FIGURE 5.19C |

See Figure 5.20 for a representational cartoon of the mathematical diagrams in Figures 19 *a, b* and *c*.

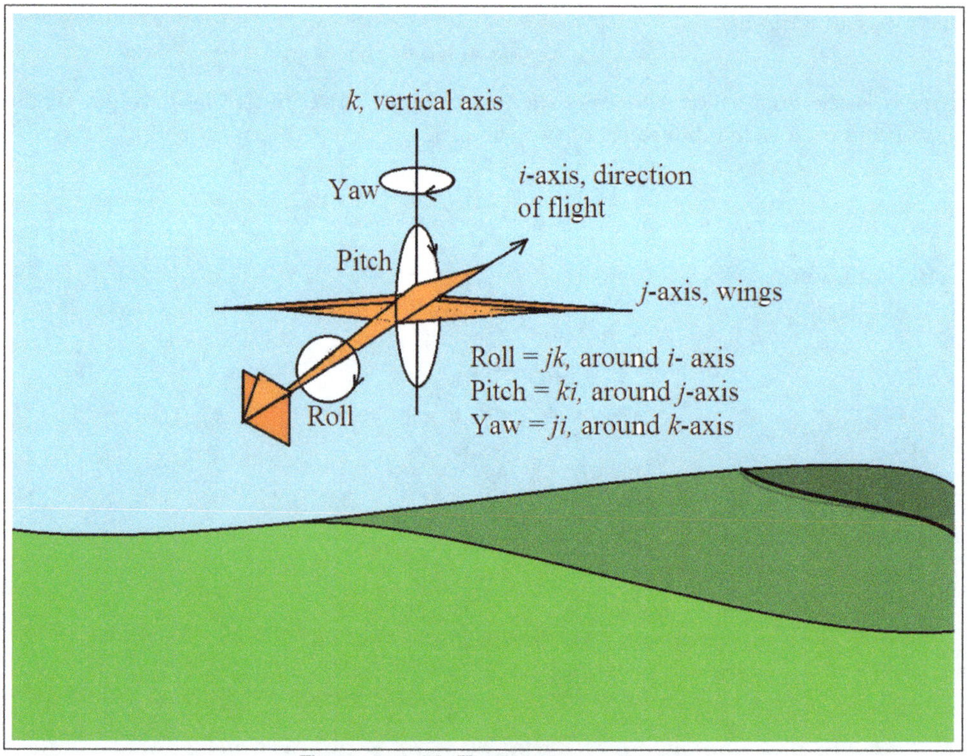

FIGURE 5.20 THE IJK AIRLINER

In the last half of the 19*th* century quaternions became an important tool in the study of rotations of objects in three dimensional space.   Eventually, newer concepts such as rotation vectors and matrices, and Euler angles have supplanted the explicit use of quaternions in many applications. Although quaternions faded into the background, they served as a basis for studying these rotational vectors.

With the advent of computers and more complex applications in the 20*th* century, quaternions have made a come-back. Pilot training modules, weather graphics, and other simulation models involving visual motion and rotation of objects on a screen employ computer programs that use quaternions.   The rotations, spins and tumbles that the character, Mario does in his video games are facilitated by quaternions.

When we derived the multiplicative inverse of a quaternions (above) we used some pretty fancy manipulation of binomials and trinomials. You may be wondering how we could get a result like Equation (5.8). Ha! You asked that question just in time, because in the next chapter we will show you how to compute powers of binomials and trinomials.

CHAPTER 6

BINOMIAL THEOREM

From the five group axioms for addition and the five group axioms for multiplication, in Chapter 4, we can solve a limited number of algebra problems. But any time we encounter a problem which needs *both* multiplication *and* addition in the same expression, these ten axioms are insufficient. For example when we proved Theorem 12 in Chapter 4 that $x \times 0 = 0$ we had to resort the distributive law, *aka* Axiom 11. This was because we were combining the operation of *multiplication* with the *additive* identity element, 0.

Separately the addition axioms tell us how to add numbers, $x + y$, and multiplication axioms tell us how to multiply numbers $p \times q$, and to raise numbers to various powers such as a^3 or $b^{\frac{1}{2}}$. But how do we *add* numbers and *then* raise them to powers? For example, what could be the result of $(x + y)^{\frac{1}{2}}$? This is not as easy as you may have thought. That is, supposing that you did think it was easy.

What is $(x + y)^n$?

It all comes down to the distributive law. We want to solve a problem like this:

Problem: For any number n, what is $(x + y)^n$?

Example: $n = 0, 1, 2, 3, ...$

Remember back in Chapter 4 when we defined what we mean by the power of a number? We stated that if b is any number, not zero, then $b^0 = 1$, $b^1 = b$, $b^2 = b \times b$, etc.? Well, the same thing is true for the expression $(x + y)$, which is called a *binomial*. That is, if $b = (x + y)$, then

$$
\begin{aligned}
b^0 &= (x + y)^0 = 1 \\
b^1 &= (x + y)^1 = (x + y) \\
b^2 &= (x + y)^2 = (x + y)(x + y) \\
b^3 &= (x + y)^3 = (x + y)(x + y)(x + y) \\
&\quad ...
\end{aligned}
$$

Now we want to take these results one step further: carry out the multiplication, and get

159

down to products of the individual terms x and y themselves. For example,

$$(x + y)(x + y) = x^2 + 2xy + y^2$$

This happens because of the distributive law[1]; thus writing $(x + y)$ as b, we get

$$
\begin{aligned}
(x + y)(x + y) &= b(x + y) \\
(x + y)(x + y) &= bx + by \\
(x + y)(x + y) &= (x + y)x + (x + y)y
\end{aligned}
$$

Then, again by the distributive law $(x + y)x = x^2 + yx$, and $(x + y)y = xy + y^2$

$$
\begin{aligned}
(x + y)(x + y) &= x^2 + yx + xy + y^2 \\
&= x^2 + 2xy + y^2
\end{aligned}
$$

But, what about the square of a *trinomial* such as $(x + y + z)$? Fortunately, we can treat this cases as if it were just a binomial.

$$(x + y + z) = (x + y) + z$$

So, for example

$$
\begin{aligned}
(x + y + z)^2 &= (x + y)^2 + 2(x + y)z + z^2 \\
&= x^2 + 2xy + y^2 + 2xz + 2yz + z^2 \\
&= x^2 + y^2 + z^2 + 2xy + 2xz + 2yz
\end{aligned}
$$

It may look like a magic trick pulled off by smoke and mirrors, but it is really just the distributive law used over and over again.

The same sort of thing works for the cube of a binomial $(x + y)^3$. It is $(x + y)(x^2 + 2xy + y^2)$, and if we use the distributive law;

$$
\begin{aligned}
(x + y)^3 &= (x + y)(x^2 + 2xy + y^2) \\
&= (x + y)x^2 + (x + y)2xy + (x + y)y^2 \\
&= x^3 + yx^2 + 2x^2y + 2xy^2 + xy^2 + y^3
\end{aligned}
$$

Gather up all the pieces that look like they belong to each other (like yx^2 and $2x^2y$), and after the smoke has cleared, we have

$$(x + y)^3 = x^3 + 3x^2y + 3xy^2 + y^3,$$

a nice peaceful Buddhist-like expression, symmetric with everything neatly in its place.

Question: Guess what $(x + y)^4$ ought to look like.

Answer:

$$(x + y)^4 = x^4 + 4x^3y + 6x^2y^2 + 4xy^3 + y^4$$

By the way, to find powers of $(x - y)$ instead of $(x + y)$, simply change y, to $-y$, so

---

[1]Recall that the distributive law is $b(x + y) = bx + by$.

that now, for example

$$\begin{aligned}(x-y)^4 &= x^4 + 4x^3(-y) + 6x^2(-y)^2 + 4x(-y)^3 + (-y)^4 \\ &= x^4 - 4x^3y + 6x^2y^2 - 4xy^3 + y^4\end{aligned}$$

An expression of terms in which the signs alternate between plus and minus is called an "alternating series."

In Chapter 4, we saw an application of the square of the binomial in the Hardy-Weinberg principal in genetics, but another application is in probability theory as we shall soon see.

There is definitely a pattern in the *coefficients* (the numbers multiplied by the powers of x and y) when we expand these powers of the binomial $(x + y)$.

Note: Pedagogically, it will be easier to see the pattern in the coefficients of $(x+y)^n$ if we just work on a simpler binomial, namely $(x+1)$ instead of $(x+y)$.

Let us look again at the expansions of $(x+1)^4$

$$(x+1)^4 = x^4 + 4x^3 + 6x^2 + 4x + 1$$

If you pay attention to the coefficients you may see the following pattern emerge from the zeroth, first, second, third, ... to the sixth powers of the binomial $(x + 1)$.

```
                        1

                   1         1

              1         2         1

         1         3         3         1

    1         4         6         4         1

1         5        10        10         5         1

1    6        15        20        15         6         1
```

FIGURE 6.0 PASCALS TRIANGLE

Each row begins and ends with a 1. After the second row, and between the 1's, the numbers are obtained by adding the two numbers above (left and right). The configuration, in Table 6.1, is a way to memorize the pattern of coefficients, and was known from early times.

Question. What are the coefficients of the seventh power?

Answer:

$$1 \quad 7 \quad 21 \quad 35 \quad 35 \quad 21 \quad 7 \quad 1$$

The pattern in this table was known to Chinese, and Persian mathematicians as early as 1040 *CE*, Jia Xian in China and Al-Karaji in Persia, both around the same time. Later, in 1100 *CE*, Omar Khayyam, also of Persia used it. There is some evidence that it was written in Sanskrit by Pingala in 200 *BCE*. It was published in 1520 by Petrus Apianus, a German humanist. Even later, in 1653, this same pattern was published by the French mathematician, Blaise Pascal, and has henceforth been called "Pascal's Triangle".

Pascal contributed to physics as well as to mathematics; as a teenager he constructed a calculating machine that could do all of the arithmetic functions including raising a number to some power. He became interested in predicting future outcomes in economics and he even developed the mathematics behind gambling. He did this by computing the probabilities for various outcomes in dice, coin tossing and card playing. The formulas he developed in these games became the foundation of probability theory as we know it today. For instance, the numbers in the Pascal' triangle gives you a way to predict the probabilities of various outcomes. This helps you know whether or not you should bet on that outcome and how much or little you should bet and what kind of odds you should demand on your bet.

Example Suppose there is a game in which a handful of coins, say one, two, three, or more coins, are to be tossed on the table and, prior to the toss, you want to bet on how many are going to be heads and how many tails. Start with just *one* coin where h stands for heads and t stands for tails. The toss of one coin looks like this.

$$(h + t)^1 = h + t$$

If you had bet tails (one tail) as the outcome for these two possibilities, your chance to win is 50%. The probability of tails is 0.5, so your chance of winning is the same as your chance of losing. You would then bet \$1 *even money*, meaning that the payoff to you would be −\$1 for each loss and +\$1 for each win.

As an example of a more complex bet, let's say you want to play the game with a toss of *six* coins, and you want to bet that at least *four* of them will come up *tails* (or, what is the same thing no more than 2 of them will come up heads). Assume that they are fair coins, meaning they are not weighted towards heads or tails and they are not two-headed or two-tailed coins. So you are betting that there are going to be either six tails and no heads or five tails and one head or four tails and two heads. Let us make a binomial out of heads and tails for six coins.

$$(h + t)^6 = 1h^6 + 6h^5t^1 + 15h^4t^2 + 20h^3t^3 + 15h^2t^4 + 6h^1t^5 + 1t^6$$

The coefficients, $1, 6, 15, 20, 15, 6, 1$, tell you how many ways there can be head and tail combinations among all the outcomes of a six coin toss. $1h^6$ says that there is only one way all the coins can be heads. $6h^5t^1$ tells you there are 6 ways that there can be 5 heads and 1 tail. This is because the single tail could be the *first* coin and the other five heads, or the tail could be the *second* coin and all the other five heads, or the tail could be the *third* coin, or the *fourth* or the *fifth* or the *sixth* with the other five being heads

The $15h^2t^4$, tells you there are fifteen different ways that there could be two heads and four tails. If you add up all of the coefficients

$$1 + 6 + 15 + 20 + 15 + 6 + 1 = 64$$

You can see that there are 64 ways the 6 coins can occur in some head-tail combination. The number $64 = 2^6$. How many of them will correspond to your bet that at *least four* coins come up tails? The answer is $15 + 6 + 1 = 22$ because there are 15 ways to get 4 tails, 6 ways to get 5 tails and 1 way to get 6 tails. So, there are 22 ways out of 64 that at least 4 tails would show up.

$$\frac{22}{64} = 0.34375$$

In other words, the probability of 4 or more tails is slightly greater than 34% This roughly means that if you make this bet you can win it approximately one-third of the time. So if you play 3 times you have a chance to win *one* time and lose two times. So two losses to *one* win means that you should get odds of 2 to 1. That is, for this to be a fair game, you should demand that you collect $2 each time you win and pay only $1 each time you lose.

What if you bet there will be three heads and three tails? Is that an even money bet? After all, you would get half heads and half tails. No, it is worse! Because there are 20 ways the toss will be exactly 3 heads and 3 tails, all the other tosses are not 3 and 3.

$$\text{Probability of 3 heads and 3 tails } = \frac{20}{64} = 0.3125$$

Of course, modern probability theory is applicable to more than just gambling. We encounter it in politics, sports, business, science and even quantum mechanics.

The Pascal (or Pingala) triangle works fine if we want to expand $(x + y)^n$ for small positive integers, like $n = 1, 2, 3, 4$. But for modern applications, we will need a general formula in order to find the coefficients when $n = 100$, or $n = 1,000,000$. We even need to find the coefficient in cases where n is negative, such as $(x + y)^{-1}$ and for fractional powers, such as $(x + y)^{1/2}$. The triangle does not work for negative or fractional powers.

In the next section, we discuss the formula for $(x + y)^n$ when n is any positive integer.

What if n is any positive integer?

We used Pascal's triangle to find the expansion of $(x + y)^n$ for cases in which n is a small non-negative integer. In order to find the expansion for larger integers we can use the following general expression which can be proved by mathematical induction. We omit the proof here because it is tedious but you can prove it for yourself.

Theorem 1: *If each of x and y is a number, and n is any nonnegative integer (whole number) then*

$$\begin{aligned}
(x + y)^n &= x^n + nx^{n-1}y + \frac{n(n-1)}{2}x^{n-2}y^2 + \frac{n(n-1)(n-2)}{2 \times 3}x^{n-3}y^3 \\
&+ ...\frac{n(n-1)...(n-k+1)}{1 \times 2 \times 3 \times ...k}x^{n-k}y^k + ... + y^n
\end{aligned} \qquad (6.1)$$

This is the general binomial theorem. The coefficients are a little difficult to compute, but if you need any one term, you can use the kth term, as shown, to calculate it.

Example:

Suppose $n = 100$ and you need the $74th$ term, then using $n = 100$, $k = 74$, the exponent on x will be $n - k = 100 - 74 = 26$, and the exponent on y will be $k = 74$. The coefficient of $x^{26}y^{74}$ will be

$$\frac{100 \times 99 \times 98 \times \ldots \times 27}{1 \times 2 \times 3 \times \ldots \times 74}$$

In the numerator, you fill in the ellipses,... with all the integers between 98 and 27, and in the denominator with all integers between 3 and 74.

Fortunately, you have a calculator that can help you do these multiplications.

Factorials

A factorial is a shorthand way of writing the products of integers, don't be surprised ! when we tell you that the notation for a factorial is the exclamation mark, !. Roughly speaking $n!$ is the product obtained by multiplying all of the integers from n down to 1.

Now-a-days, in some calculators there is a [!] button. It is called a factorial button and you can use it to compute $n!$, whose formal definition for any non-negative integer, is as follows

$$\begin{aligned} 0! &= 1 \\ n! &= n \times (n-1)! \end{aligned}$$

This is a "recursive" definition, in that the meaning at $n+1$ depends upon the meaning at n.

Example: To get 5!, you need to know 4! and to get this, you need to know 3!, and so forth. In other words you can write

$$\begin{aligned} 5! &= 5 \times 4! \\ 5! &= 5 \times 4 \times 3! \\ 5! &= 5 \times 4 \times 3 \times 2! \\ 5! &= 5 \times 4 \times 3 \times 2 \times 1 \\ &= 120 \end{aligned}$$

Problems (Fun with factorials)

1. Show that 6! = 720 and find 7!
2. Show that $110! = 110 \times 109 \times 108 \times 107!$
3. Show that $\frac{11!}{9!} = 110$
4. If the number A is the product $1 \times 2 \times 4 \times 6 \times 7 \times 8$, what numbers do you need to multiply A by in order to get 8 factorial?
5. Show that $2^3 \times (1 \times 3 \times 5) = 5!$
6. If B is the product of odd numbers $1 \times 3 \times 5 \times 7$, and C is the product of even numbers $2 \times 4 \times 6$, show that $B \times C = 7!$
7. Show that

$$\frac{50 \times 49 \times 48 \times 47 \times 46}{5!} = \frac{50!}{5! \times 45!}$$

Solutions:

1. By definition $7! = 7 \times 6!$, therefore $7! = 7 \times 720 = 5040$
2. Recursively, $110! = 110 \times 109! = 110 \times 109 \times 108! = 110 \times 109 \times 108 \times 107!$
3. $\frac{11!}{9!} = \frac{11 \times 10 \times 9!}{9!} = 110$
4. $A = 1 \times 2 \times 4 \times 6 \times 7 \times 8$ is not 8! because it is missing the factors 3 and 5, so $15 \times A = 8!$

5. If $A = (1 \times 3 \times 5)$, then it needs the factors 2 and 4 to be promoted to 5!, so

$$\begin{aligned} A &= 1 \times 3 \times 5, \text{ given} \\ A \times 2 \times 4 &= 1 \times 2 \times 3 \times 4 \times 5 \\ A \times 2^3 &= 5! \text{ Beautiful!} \end{aligned}$$

6. Given $B = 1 \times 3 \times 5 \times 7$, and $C = 2 \times 4 \times 6$, then

$$\begin{aligned} B \times C &= (1 \times 3 \times 5 \times 7) \times (2 \times 4 \times 6) \\ &= 1 \times 2 \times 3 \times 4 \times 5 \times 6 \times 7 \\ &= 7! \end{aligned}$$

7. To change the numerator into 50! we need the product of all the numbers from 45 down to 1. So, we multiply numerator *and* denominator by 45! getting

$$\frac{50 \times 49 \times 48 \times 47 \times 46}{5!} \times \frac{45!}{45!}$$

$$= \frac{50!}{5! \times 45!}$$

Writing binomial coefficients in terms of factorials

In the above example where we wanted to write the 74*th* term in the binomial $(x + y)^{100}$, we found that the coefficient of $x^{26} y^{74}$ was

$$\frac{100 \times 99 \times 98 \times \ldots \times 27}{1 \times 2 \times 3 \times \ldots \times 74}$$

which is

$$\frac{100 \times 99 \times 98 \times \ldots \times 27}{74!}$$

Now, multiply numerator and denominator by by 26! then the numerator becomes 100! because it is the product of all integers from 100 down to 1. Thus,

$$\frac{100 \times 99 \times 98 \times \ldots \times 27}{74!} = \frac{100!}{74! \times 26!}$$

So, in the binomial expression $(x + y)^{100}$, the coefficient of $x^{26} y^{74}$ is

$$\frac{100!}{74!26!}$$

and let it go at that. If you really needed the exact number, use a computer!

The coefficient, $\frac{n!}{(n-k)!k!}$, of powers of x and y in Equation (6.1) have a special notation, $\binom{n}{k}$, called the *binomial coefficient*. It is defined as follows. If k and n are non-negative integers and $0 \le k \le n$, then

$$\binom{n}{k} = \frac{n!}{(n-k)!k!}$$

which is equal to

$$\frac{n \times (n-1) \times (n-2) \times ...(n-k+1)}{k!} \frac{(n-k)!}{(n-k)!}$$

The binomial coefficient $\binom{n}{k}$ is used to determine the number of ways that k things can be chosen from a set of n things. We read $\binom{n}{k}$ as "n choose k".

Notes:

The *n choose k* formula automatically accounts for the two cases in which you may want to choose *none* or *all* of the n things. Because when $k = 0$, $\binom{n}{k} = 1$, and when $k = n$, $\binom{n}{k} = 1$.

Proof: Given a collection of n items, there is only one way, $\binom{n}{n}$ to chose all n of them and there is only one way, $\binom{n}{0}$ to choose none of them. This is because 0! is defined as 1, so

$$\binom{n}{n} = \frac{n!}{(n-n)!n!} = \frac{n!}{n!} = 1, \text{ and}$$

$$\binom{n}{0} = \frac{n!}{(n-0)!0!} = \frac{n!}{n!} = 1$$

Using the n chose k notation, $\binom{n}{k}$ as the binomial coefficient, we can compactly write the binomial theorem given in Equation (6.1) as follows.

$$(x+y)^n = \sum_{k=0}^{k=n} \binom{n}{k} x^{n-k} y^k \tag{6.2}$$

The summation on the right-hand-side of Equation (6.2) can be read as: "The sum of $\binom{n}{k}$ times x raised to the power $n - k$, times y raised to the power k, from $k = 0$ to n."[2]

The induction proof of Equation (6.2) depends upon multiplying both sides by $(x+y)$

$$(x+y)(x+y)^n = (x+y) \sum_{k=0}^{k=n} \binom{n}{k} x^{n-k} y^k$$

$$(x+y)^{n+1} = \sum_{k=0}^{k=n} \binom{n}{k} x^{n-k+1} y^k + \sum_{k=0}^{k=n} \binom{n}{k} x^{n-k} y^{k+1}$$

Then re-arranging the terms, we get the coefficient, $\binom{n}{k}$ in the first sum and $\binom{n}{k-1}$ of

$$x^{n+1-k} y^k$$

in the second sum. When we add their coefficients, we need to show that

$$\binom{n}{k} + \binom{n}{k-1} = \binom{n+1}{k}$$

---

[2] The usual way to read an exponential expression such as y^k is "y to the *kth* power". But when I served in the U.S. Peace Corps in 2000, my English speaking Zimbabwe students insisted on saying "y to the power k." I like their phraseology better.

This can be done by adding the fractions, getting

$$\frac{n!}{(n-k)!k!} + \frac{n!}{(n-k+1)!(k-1)!} = \frac{(n+1)!}{(n+1-k)!k!}$$

which is the coefficient of the $x^{n+1-k}y^k$ term in the expansion of $(x+y)^{n+1}$. In other words,

$$(x+y)^{n+1} = \sum_{k=0}^{n=k+1} \binom{n+1}{k} x^{n+1-k}y^k$$

This last equation completes the induction proof; thus, we can say that Equation (6.2) is true for any positive integer n.

Application to combinatorics

In addition to its use as the binomial coefficient, the expression$\binom{n}{k}$ has practical applications in a mathematical field called *combinatorics*, where we study the number of different ways that elements of a set can be selected, arranged and counted.

Example: Suppose your chess club has 7 equally strong players, how many ways can you select 5 of them as a team to play in a match against another club? It is 7 *choose* 5.

Solution:

$$\binom{7}{5} = \frac{7!}{2! \times 5!} = \frac{7 \times 6 \times (5 \times 4 \times 3 \times 2 \times 1)}{(2 \times 1) \times (5 \times 4 \times 3 \times 2 \times 1)} = \frac{7 \times 6}{2 \times 1} = 21$$

You have the option of naming twenty-one different five-player teams to send off to the match. It is the same as the number of ways that you can leave two of the players off a team, seven *choose* two.

$$\binom{7}{2} = \frac{7!}{5! \times 2!} = 21$$

This number, 21, is the coefficient of both x^5y^2 and x^2y^5 in the expansion of the binomial $(x+y)^7$.

$$(x+y)^7 = x^7 + 7x^6y + 21x^5y^2 + 35x^4y^3 + 35x^3y^4 + 21x^2y^5 + 7xy^6 + y^7$$

The binomial theorem, generalized

The formula for $(x+y)^n$, the *nth* power of the binomial, discussed in the last section requires that n be a positive integer. In the late 17th century Isaac Newton made some brave and daring assertions. In his book, on the convergence of "infinite equations", *De analysi per aequationes numero terminorum infinitas*, published in 1669, Newton introduced the idea of using fractions and negative numbers for n in the binomial theorem. The results were applied to finding relationships between various functions and to computing highly accurate approximations to irrational numbers. Specifically, he said that we could let n be a fraction, say $n = \frac{1}{2}$ or a negative number, such as $n = -1$.

The beginning of infinite series

What would happen if we did try to let $n = \frac{1}{2}$, as a power of the binomial $(a+b)$? Do you think we would get $a^{\frac{1}{2}}$ followed by a bunch of Pascal triangle coefficients times powers of a and b, then end up with $b^{\frac{1}{2}}$ at the end? Maybe or maybe not. Let us go back to the 1670's and pretend that we are Isaac Newton about to expand the binomial theorem for Equation (6.1)

$$(a+b)^n = a^n + na^{n-1}b + \frac{n(n-1)}{2}a^{n-2}b^2 + ... a^1 b^{n-1} + a^0 b^n$$

for some number n other than a positive integer.

Finding $(a+b)^{\frac{1}{2}}$

Holding our breath, we begin with $n = \frac{1}{2}$ instead of a whole number, in the binomial theorem. The result will start out looking like the binomial theorem, the first term being $a^{\frac{1}{2}}$ alright, but there will not be a last term that ends up as $b^{\frac{1}{2}}$. In fact, the series will not stop at all; it becomes infinite.

$$
\begin{aligned}
(a+b)^{\frac{1}{2}} \;=\; & a^{\frac{1}{2}} + \frac{1}{2}a^{\frac{1}{2}-1}b + \frac{\frac{1}{2}(\frac{1}{2}-1)a^{\frac{1}{2}-2}b^2}{2} + \\
& \frac{\frac{1}{2}(\frac{1}{2}-1)(\frac{1}{2}-2)a^{\frac{1}{2}-3}b^3}{6} + ... \text{ forever}
\end{aligned}
\qquad (6.3)
$$

The exponents for a are $\frac{1}{2}$, $\frac{-1}{2}$, $\frac{-3}{2}$, $\frac{-5}{2}$, ...getting more and more negative–never becoming zero, and the powers on the b are whole numbers increasing without bound. Does this series actually add up to a number if we let it go on for ever? The answer is *not always*. We will see that it depends upon whether $\left|\frac{b}{a}\right| < 1$ or $\left|\frac{b}{a}\right| \geq 1$. Look at the powers of a and b in the $(k+1)$ term.

$$a^{\frac{1}{2}-k}b^k$$

This can be written as

$$a^{\frac{1}{2}}\left(\frac{b^k}{a^k}\right) = \sqrt{a} \times \left(\frac{b}{a}\right)^k$$

If $\left|\frac{b}{a}\right| > 1$, then for positive integers, k, $\left|\frac{b}{a}\right|^k$ gets large and larger as k gets large. We say it "approaches infinity" when it increases without bound. Symbolically, this can be written as

$$\left|\frac{b}{a}\right|^k \to \infty, \text{ when } k \to \infty$$

So, there is no way Equation (6.3) can *converge* (add up to a finite number) if $|b| > |a|$. Furthermore, when $b = a$, a series like this does not usually settle down to a finite number. We will see why this is true later.

Let us consider a case in which the series does converge; it will be one in which $|b| < |a|$.

Example: Let $a = 25$, and $b = -2$, then $\left|\frac{b}{a}\right| = \left|\frac{-2}{25}\right| < 1$. We are looking for

$(25 - 2)^{\frac{1}{2}}$, or $\sqrt{23}$.

$$(25 - 2)^{\frac{1}{2}} = 25^{\frac{1}{2}} + \frac{1}{2}25^{\frac{1}{2}-1}(-2) + \frac{\frac{1}{2}(\frac{1}{2} - 1)25^{\frac{1}{2}-2}}{2}(-2)^2 +$$

$$\frac{\frac{1}{2}(\frac{1}{2} - 1)(\frac{1}{2} - 2)25^{(\frac{1}{2}-3)}}{6}(-2)^3 + ...$$

$$(25 - 2)^{\frac{1}{2}} = 5 - 25^{-\frac{1}{2}} - \frac{1}{2}\frac{1}{2}\frac{1}{2}25^{\frac{-3}{2}}4 - \frac{1}{2}\frac{1}{2}\frac{3}{2}25^{\frac{-5}{2}}\frac{2^3}{6} + ...$$

Doing the arithmetic, we can reduce this to

$$(25 - 2)^{\frac{1}{2}} = 5 - \frac{1}{5} - \frac{1}{2} \times \frac{1}{125} - \frac{1}{2} \times \frac{1}{3125} + .. \tag{6.4}$$

$$\sqrt{23} = 5 - \frac{1}{5} - \frac{1}{250} - \frac{1}{6250} + ... \tag{6.5}$$

This is an *infinite* series, but if we just add the first four terms, namely,

$$+5, \frac{-1}{5}, \frac{-1}{250}, \frac{-1}{6250}$$

We get

$$\sqrt{23} \approx 4.79584$$

which is a pretty good approximation to $\sqrt{23}$, since $(4.79584)^2 = 23.00008131$.

BUT WAIT! Why couldn't we reverse the subtraction and use the one-half power of the binomial $(-2 + 25)^{\frac{1}{2}}$ and also get this good answer? After all addition satisfies the commutative law, $a + b = b + a$, doesn't it?

Why should we get an entirely different infinite series when we compute $(b+a)^{\frac{1}{2}}$ than we do when we compute $(a + b)^{\frac{1}{2}}$?

$$(b + a)^{\frac{1}{2}} = b^{\frac{1}{2}} + \frac{1}{2}b^{\frac{1}{2}-1}a + \frac{\frac{1}{2}(\frac{1}{2} - 1)b^{\frac{1}{2}-2}a^2}{2} +$$

$$\frac{\frac{1}{2}(\frac{1}{2} - 1)(\frac{1}{2} - 2)b^{\frac{1}{2}-3}a^3}{6} + ... \text{ forever}$$

The answer is: the binomial theorem was proved under the assumption that n was a nonnegative integer, so yes, it is true that $(a + b)^n = (b + a)^n$. But assuming n could be something other than a nonnegative integer is tantamount to creating a new definition of multiplication, like we did in defining complex numbers or quaternions. As we found out above, any time we want the expansion of $(a+b)^{\frac{1}{2}}$, to converge, we must require that $\left|\frac{b}{a}\right| < 1$, that is not too much to ask, is it? You may, however, wish to know, "Why do we even care whether or not a given series converges?" An answer is that infinite series gives us an easy way to compute approximations to difficult numbers like $\sqrt{23}$.

Example Here is what happens when we ignore the requirement that $\left|\frac{b}{a}\right| < 1$. Let $a = -2$ and $b = 25$, then

$$\left|\frac{25}{-2}\right| > 1$$

Now try to expand $(-2+25)^{\frac{1}{2}}$

$$
\begin{aligned}
(-2+25)^{\frac{1}{2}} &= (-2)^{\frac{1}{2}} + \frac{1}{2}(-2)^{\frac{-1}{2}}(25) + \frac{\frac{1}{2}(-\frac{1}{2})(-2)^{\frac{-3}{2}}(25)^2}{2!} + \dots \\
&= i\sqrt{2} + \frac{1}{2}\frac{(25)}{i\sqrt{2}} - \frac{1}{4}\frac{(25)^2}{(i\sqrt{2})^3} + \dots
\end{aligned}
\qquad (6.6)
$$

In this series, the power on 25 increases to infinity and powers on $(-2)^{\frac{1}{2}}$, that is, $(i\sqrt{2})$, in the denominator, progresses as increasing odd numbers. This series cannot converge to the real number, $\sqrt{23}$. Not only is the series in Equation (6.6) divergent, it is also ugly.

How to fix the $(a+b)^p \neq (b+a)^p$ problem

One way we can help to bring order to the non-integral powers of the binomial $(a+b)$ is to agree that $a > 0$ and larger in absolute value than the absolute value of b. This will help us insure that we are using the increasing powers on the smaller number. Thus, we are assuming that $-a < b < +a$, then, dividing by a, which we can do without reversing the inequality because a is positive we get $-1 < \frac{b}{a} < +1$. By the distributive law we can re-write $(a+b)$

$$
(a+b) = a(1 + \frac{b}{a})
$$

Any binomial can be factored like this. So, let $\frac{b}{a} = x$, we get $(a+b) = a(1+x)$, where $|x| < 1$. Therefore,

$$
(a+b)^n = a^n(1+x)^n, \text{ with } |x| < 1
$$

Then

$$
\begin{aligned}
(25-2)^{\frac{1}{2}} &= (25)^{\frac{1}{2}}(1 - \frac{2}{25})^{\frac{1}{2}} \\
&= 5\left(1^{\frac{1}{2}} + \frac{\frac{1}{2}(\frac{-2}{25})^1}{1!} + \frac{\left(\frac{1}{2}\right)\left(\frac{-1}{2}\right)\left(\frac{-2}{25}\right)^2}{2!} + \frac{\frac{1}{2}(\frac{-1}{2})(\frac{-3}{2})(\frac{-2}{25})^3}{3!} \dots\right)
\end{aligned}
$$

and Equation 6.4 becomes

$$
= 5\left(1 - \frac{1}{25} - \frac{1}{2} \times \frac{1}{(25)^2} - \frac{1}{2} \times \frac{1}{(25)^3} \dots\right)
\qquad (6.7)
$$

The first four terms of which add up to

$$
5(0.959168) = 4.79584
$$

Which is the same answer we got before. If you want a closer approximation, just use a few more terms of the series. From now on, when we want to compute a power, n, of a binomial $(a+b)$ in which n is not a positive whole number, then we will use $(1+x)$, where $|x| < 1$.

A calculator's square root key ☑

Any electronic calculator can add, subtract, multiply and divide, and that is all it can do, period. So to get the square root of 23, it has to use a few terms of Equation (6.7). But how does a "scientific" calculator, that has a $\left[\sqrt{}\right]$ button work? It still only adds, subtracts, multiplies and divides. In fact, you can use a "regular" pocket calculator to find the square root of any positive number. Simply insert *yourself* as the missing button; use an infinite series and just do these four simple operations, yourself.

To be fair, there is one more task a calculator can do, and that is recognize, by looking at the binary digits, 1's and 0's of any numbers you enter to see whether or not any calculations result in a positive or negative number. This determines when one input is greater than or less than (or equal to) another input. But you, as a thinking human being, have the advantage of just looking at the numbers to see which one is larger (or you could subtract one from the other to see whether you get positive, negative or zero).

Problem: Find $\sqrt{50}$ without a square root key.

Solution: Let's start with $a^2 = 49$, the closest perfect square to 50. So, $a = 7$. Let d be the difference between 50 and 49, that is $d = 1$. Write 50 as $(49+1)$, or $49(1 + \frac{1}{49})$

$$
\begin{aligned}
\sqrt{50} &= \sqrt{49+1} \\
&= \sqrt{49(1 + \frac{1}{49})} \\
&= 7\sqrt{(1 + \frac{1}{49})} \\
&= 7(1 + \frac{1}{49})^{\frac{1}{2}}
\end{aligned}
$$

Now calculate a few terms of the following infinite series.

$$
\begin{aligned}
(1 + \frac{1}{49})^{\frac{1}{2}} &= 1 + \frac{1}{2} \times \frac{1}{49} - \frac{\frac{1}{2} \times \frac{1}{2} \times \left(\frac{1}{49}\right)^2}{2!} + \ldots \\
&= 1 + \frac{1}{98} - \frac{1}{8 \times (49)^2} + \ldots \\
&= 1.01015202\ldots
\end{aligned}
$$

Now multiply this by 7 and you get

$$7 \times 1.01015202\ldots = 7.07106414$$

as your approximation for $\sqrt{50}$. Check it out.

$$(7.07106414)^2 = 49.999948\ldots$$

What if $n = -1$? Finding $(1 + x)^{-1}$

Now let us see what happens if we try to use a negative number like -1 for n. Compute the binomial expansion of $(1 + x)^{-1}$.

$$
(1 + x)^{-1} = 1 + (-1)x + (-1)(-2)x^2\frac{1}{2} + (-1)(-2)(-3)x^3\frac{1}{6} +
$$
$$
+ (-1)(-2)(-3)(-4)x^4\frac{1}{24} + ...
$$
$$
(1 + x)^{-1} = 1 - x + x^2 - x^3 + x^4 - x^5 + ... \tag{6.8}
$$

The alternating series pattern in Equation (6.8) continues forever. It adds up to a finite number whenever x is any number between -1 and $+1$, *not* including $+1$ or -1.

Question: Since it looks like it would be OK for x to be equal to 1 on the left hand side of Equation (6.8), why can't we let $x = 1$ in the series on the right hand side?

Answer: Great question! Yes, on the left side, when $x = 1$, then $(1 + x)^{-1}$ becomes $2^{-1} = \frac{1}{2}$. But the series on the right

$$
1 - 1 + 1 - 1 + 1 - 1 + ...
$$

does not settle down to any one single finite number. The first term is just 1, but the *sum* of the first two terms is 0, then the sum of the first 3 terms is 1 again, and the sum of the first four terms is 0, and this continues forever. Neither 0 nor 1 is a good approximation for $\frac{1}{2}$. An infinite series, the sum of whose terms jump back and forth like this without settling down to a finite fixed number is called a *divergent* infinite series. Furthermore, any infinite series in which the sums increase without bound is also call divergent. We will discuss such *numerophobic* infinite series later when we discuss *partial sums*.

Example Use Equation (6.8)to find $(1 + x)^{-1}$ if $x = \frac{3}{13}$

$$
(1 + \frac{3}{13})^{-1} = 1 - \left(\frac{3}{13}\right) + \left(\frac{3}{13}\right)^2 - \left(\frac{3}{13}\right)^3 + \left(\frac{3}{13}\right)^4 - \left(\frac{3}{13}\right)^5 ...
$$

We can check it out. The sum of the first six terms of the above series is approximately 0.8124. But the number $(1 + \frac{3}{13})^{-1}$, itself, is $\frac{13}{16}$ or 0.8125

Although Equation (6.8) converges for all x between -1 and 1,something else happens when we turn the binomial around, $(x + 1)$, and look at the series for $(x + 1)^{-1}$.

$$
(x + 1)^{-1} = x^{-1} + -1x^{-2} + \frac{(-1)(-2)x^{-3}}{2!} + \frac{(-1)(-2)(-3)x^{-4}}{3!} + ...
$$
$$
= \frac{1}{x^1} - \frac{1}{x^2} + \frac{1}{x^3} - \frac{1}{x^4} + ... + (-1)^{n+1}\frac{1}{x^n} + ... \tag{6.9}
$$

The Equation (6.9) *diverges* (does not converge) for the numbers between -1 and 1. BUT it converges when $|x| > 1$ that is, for all $x > 1$ and all $x < -1$. Again we can not use $x = 1$.

Here is a wonderful, whimsical and maybe even useless, application of Equation 6.9.

Example: Suppose you are traveling from San Antonio to Austin, but when you get half way, you decide to turn around and travel back toward San Antonio. But again, when you get halfway back you turn back to the opposite direction. If you keep reversing your direction each time going half of the previous distance, what fraction of the distance from San Antonio to Austin will you finally end up in?

Hint: Use $x = 2$ in Equation (6.9).

By the way, the series in Equation(6.8) and, also the one in (6.9) can be obtained by using the algorithm for "long division" that we learned in elementary school. Here is how to get Equation (6.8) by dividing $(1 + x)$ into 1.

$$(1 + x)^{-1} = \frac{1}{(1 + x)}$$

$$
\begin{array}{r}
1 - x + x^2 - x^3 \\
\hline
1 + x{\overline{\smash{\big)}\,1}} \\
\underline{1 + x} \\
-x \\
\underline{-x - x^2} \\
x^2 \\
\underline{x^2 + x^3} \\
-x^3 \\
\cdots\cdots
\end{array}
$$

Problems:

Show that you can get the non-alternating infinite series:

1. $(1 - x)^{-1} = 1 + x + x^2 + x^3 + x^4 + ...$ and
2. $(x - 1)^{-1} = \frac{1}{x^1} + \frac{1}{x^2} + \frac{1}{x^3} + \frac{1}{x^4} + ...$

from either the binomial expansions or by the long division algorithm.

Here, Problem 1 converges for $|x| < 1$ and Problem 2 converges for $|x| > 1$.

Other powers of $(1 + x)$ or $(x + 1)$

If we find that the $(1 + x)^p$ for some number p, results in an infinite series that can be proved to be convergent, for some set of values of x, then we have shown that $(1 + x)^p$ exists as a function of x and may be used in some mathematical application.

Example: If $n = \sqrt{2}$, does $(1 + x)^{\sqrt{2}}$ converge?

$$(1 + x)^{\sqrt{2}} = 1^{\sqrt{2}} + \left(\sqrt{2}\right)1^{(\sqrt{2}-1)}x + \frac{(\sqrt{2})(\sqrt{2} - 1)1^{(\sqrt{2}-2)}x^2}{2!} + ...$$

How do we actually test some unknown series such as this one to determine whether or not it converges? In many cases, there is a test called the "ratio test," in which we calculate the ratio of two consecutive terms of a series in order to find the range of values, if any, of x, for which the series converges. We will not be able to discuss every possible test for convergence, but we will cover several very useful ones, such as the comparison test and the p-series tests.

The function e^x

In addition to expanding $(1 + x)^n$ for fractional powers and negative powers, the most natural idea is to assume you can let n be a positive integer as large as you want. Find out what happens to $(1 + x)^n$ if you let $n \to \infty$. If we combine this with using $\frac{x}{n}$, instead of just x, we get an interesting result from this titanic clash between one number n, going to infinity and another number, $\frac{x}{n}$, going to zero. In other words, we seem to be on the precipice of insanity if we want to find:

$$(1 + \frac{x}{\infty})^{\infty}$$

The only way we can do this is to stop short of infinity and use a very large n that is on its way to infinity and compute,

$$(1 + \frac{x}{n})^{n}$$

Then try to do some sort of *algebraic simplifying*; this means cancelling terms or factors that can be cancelled and then let n continue on its trip to infinity. One of the important consequences that comes out of this daring effort is a function we call *the exponential function* or e^x. You will probably find the $[e^x]$ key on your calculator; it is usually paired with its inverse called the [ln] key, the "natural logarithm of x." We are going to show you how we get an equation for e^x from the following definition.

Definition: For any number x, the number e^x is defined by the equation:

$$e^x = \lim_{n \to \infty} (1 + \frac{x}{n})^{n} \qquad (6.10)$$

To see what is happening here, we let x be any number not zero, then starting with $n = 1$, $n = 2$, $n = 3$, we compute $(1 + \frac{x}{n})^n$ for each n.

$$
\begin{aligned}
\left(1 + \frac{x}{1}\right)^1 &= 1 + x \\
\left(1 + \frac{x}{2}\right)^2 &= 1 + 2\frac{x}{2} + \left(\frac{x}{2}\right)^2 = 1 + x + \frac{x^2}{4} \\
\left(1 + \frac{x}{3}\right)^3 &= 1 + 3\frac{x}{3} + 3\left(\frac{x}{3}\right)^2 + \left(\frac{x}{3}\right)^3 \\
&= 1 + x + \frac{x^2}{3} + \frac{x^3}{27}
\end{aligned}
$$

Unfortunately, these small numbers for n do not reveal the deep secret locked inside of Equation (6.10). It is only after we start taking larger values of n that we can coax (6.10) to surrender its beautiful pattern, not detectable from $n = 1$, or $n = 2$. Here we go; still keeping x fixed at some number, let's try using a larger number for n, such as $n = 100$.

$$
\begin{aligned}
(1 + \frac{x}{100})^{100} &= 1 + 100\frac{x}{100} + \frac{100 \cdot 99}{100 \cdot 100}\frac{x^2}{2!} + \frac{100 \cdot 99 \cdot 98}{100 \cdot 100 \cdot 100}\frac{x^3}{3!} + ... + \left(\frac{x}{100}\right)^{100} \\
&= 1 + x + 0.99\frac{x^2}{2!} + 0.9702\frac{x^3}{3!} + ... + \frac{x^{100}}{100^{100}}
\end{aligned}
$$

In the expansion of the binomial $(1 + \frac{x}{n})^n$ the kth term is

$$
\begin{aligned}
\binom{n}{k}\left(\frac{x}{n}\right)^k &= \frac{n!}{(n-k)!\,k!}\frac{x^k}{n^k} \\
&= \frac{n!}{(n-k)!\,n^k}\frac{x^k}{k!} \qquad (6.11)
\end{aligned}
$$

In Equation (6.11), look at the coefficient of $\frac{x^k}{k!}$. Here is where we examine the consequence of letting n approach infinity.

$$\frac{n!}{(n-k)!n^k} = \frac{n(n-1)(n-2)...(n-k+1)}{n^k}$$

Ask what happens as $n \to \infty$? You will find that, after some work,

$$\lim_{n \to \infty} \frac{n(n-1)(n-2)...(n-k+1)}{n^k} = 1 \qquad (6.12)$$

Here is how we get that. First, we have to sort out the many "n's", each being a candidate going to infinity in both the numerator and denominator, and simplify the expression to see how these *incipient* infinities cancel each other out. Here is a non-rigorous argument that justifies the limit shown in Equation (6.12).

$$\begin{aligned}
\frac{n!}{(n-k)!n^k} &= \frac{n}{n}\frac{(n-1)}{n}\frac{(n-2)}{n}...\frac{(n-k+1)}{n} \\
&= 1 \times \left(1 - \frac{1}{n}\right) \times \left(1 - \frac{2}{n}\right) \times ... \left(1 - \frac{k-1}{n}\right)
\end{aligned} \qquad (6.13)$$

If a fraction is one in which some specific positive number, k is divided by a varying number n that approaches infinity, then that fraction will approach zero.[3] That is, for any fixed k,

$$\lim_{n \to \infty} \frac{k}{n} = 0$$

Now, as $n \to \infty$ then

$$\begin{aligned}
\left(1 - \frac{1}{n}\right) &\to 1 - 0 = 1 \\
\left(1 - \frac{2}{n}\right) &\to 1 - 0 = 1 \\
&... \\
\left(1 - \frac{k-1}{n}\right) &\to 1 - 0 = 1
\end{aligned}$$

So the entire expression in (6.13) approaches $1 \times 1 \times ... \times 1 = 1$, verifying the limit in Equation (6.12).

Thus, we can write the following limit.

$$\begin{aligned}
\lim_{n \to \infty} \left(1 + \frac{x}{n}\right)^n &= \lim_{n \to \infty} \left(1 + x + ... \frac{n!}{(n-k)!n^k}\frac{x^k}{k!} + ...\right) \\
&= 1 + x + 1\frac{x^2}{2!} + 1\frac{x^3}{3!} + ... + 1\frac{x^k}{k!} + ... \text{ forever}
\end{aligned}$$

In other words,

$$e^x = 1 + x + \frac{x^2}{2!} + \frac{x^3}{3!} + ... + \frac{x^k}{k!} + ...\text{forever}$$

---

[3]See Chapter 4, Axiom 16, the well-ordering principle which implies that for any positive numbers ρ and κ there is an integer n such that $\frac{\kappa}{n} < \rho$.

Hooray! We got an equation for e^x from the definition in (6.10).

We write this equation compactly as

$$e^x = \sum_{k=0}^{k=\infty} \frac{x^k}{k!} \tag{6.14}$$

Beautiful!

This equation gives you the number e raised to *any* power.

Examples: What is e^1?

$$
\begin{aligned}
e^1 &= 1 + 1 + \frac{1}{2} + \frac{1}{3!} + \frac{1}{4!} + \frac{1}{5!} + \frac{1}{6!} + \dots \\
&= 2.71828\dots
\end{aligned}
$$

What is $e^{0.04}$?

$$
\begin{aligned}
e^{0.04} &= 1 + 0.04 + \frac{0.04^2}{2!} + \frac{0.04^3}{3!} + \frac{0.04^4}{4!} + \dots \\
&= 1.04081\dots
\end{aligned}
$$

Base e logarithm

Earlier, when we defined logarithms to the base b, we wrote $y = \log_b(x)$ which means that $b^y = x$. When the base $b = 10$, the logarithms are called *common logarithms* and used in various applications. When the number e is used as the base, the logarithms are called *natural logarithms*, which are predominantly used in mathematics, physics and engineering.

The graphs of the equation $y = e^x$ and $y = e^{-x}$ are shown in Figures 6.1 and 6.2. They are known as the *exponential growth* function and the *exponential decay* function, respectively.

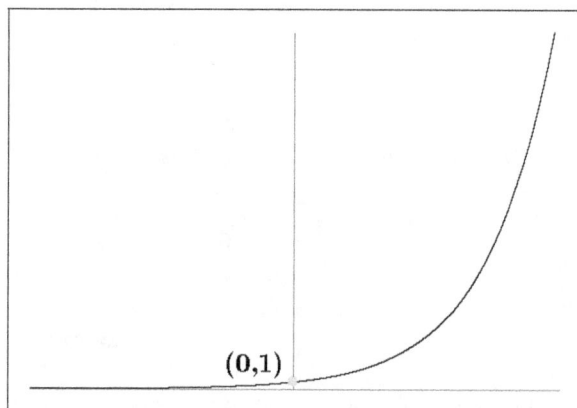

FIGURE 6.1 $y = e^x$

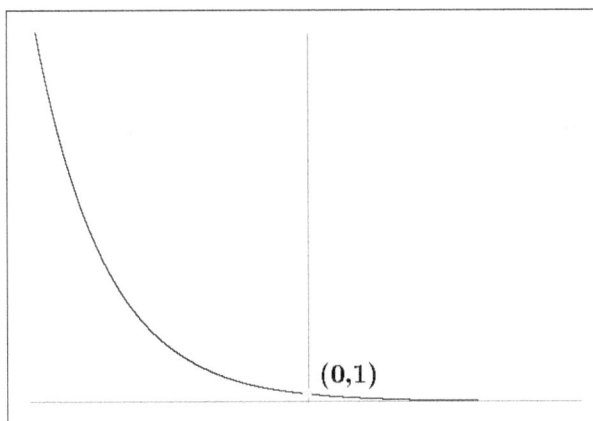

FIGURE 6.2 $y = e^{-x}$

The exponential growth equation, Figure 6.1 has applications in studies about population, epidemics and finance. The notation for logarihtms to the base e, is $y = \ln(x)$ instead of $y = \log_e(x)$. That is,

$$\ln(x) = y \text{ means } e^y = x$$

The exponential decay equation, Figure 6.2 is applied to problems involving radioactive decay, carbon dating, and thermodynamics. We will take a look at the use of the exponential function in compound interest rates.

Continuously compounded interest

Suppose you want to know how much interest you can earn on a savings account in which the interest is compounded quarterly, four times a year. How would that be different from one compounded monthly? Or weekly? If $x = $ APR, the *annual percentage interest rate*, and n is the number of conversion periods (12 (months), 52 (weeks), 365 (days), etc.) then the effective interest rate is

$$(1 + \frac{x}{n})^n - 1$$

So, for example, if the APR is 5.5% and the interest is compounded *monthly,* then $x = 0.055$ and $n = 12$, so

$$(1 + \frac{0.055}{12})^{12} = 1.05640786...$$

therefore, the effective rate is about 0.05641 or 5.641%.

What if the 5.5% rate is compounded weekly? Then $n = 52$.

$$(1 + \frac{0.055}{52})^{52} = 1.0565099$$

The effective rate is 5.651%. These were easy to compute by your calculator.

But what about *instantaneous* (also called *continuous*) compounding? What does that even mean? Compounding every instant? That would be infinitely many times a year! You might say, "Oh, that is easy just let $n = \infty$, infinity". Unfortunately, you cannot use $n = \infty$, because then

$$(1 + \frac{0.055}{\infty})^{\infty}$$

a meaningless statement, if anyone ever saw one.   But there is a way to introduce continuous compounding, we just say for n payment periods, letting n approach infinity. Thus, the continuous compounding is found from the limit of

$$(1 + \frac{0.055}{n})^{n}$$

as n approaches infinity. In other words we want

$$\lim_{n \to \infty} (1 + \frac{0.055}{n})^{n} = e^{0.055}$$

which, from your calculator is

$$e^{0.055} = 1.056540615...$$

or an interest rate of, approximately, 5.654%. Therefore, for a given principal of \$1000 the amount A after continuous compounding for a year would be

$$A = \$1000 \times e^{0.055} = \$1056.54$$

Doubling your money–the rule of 70

If a principal amount P is invested at an annual interest rate r, continuously compounded, for n years, then amount A will be

$$A = P \times e^{rn}$$

When will you double your money?

$$
\begin{aligned}
A &= 2P \\
2P &= P \times e^{rn} \\
2 &= e^{rn}
\end{aligned}
$$

That is, we want to find what number x makes $e^x = 2$?   We can solve for x by using the base e logarithm, $x = \ln(2) \approx 0.693$. You can check it out by computing a few terms of $e^{0.693}$ in Equation (6.14).

$$
\begin{aligned}
e^{0.693} &= 1 + 0.693 + \frac{0.693^2}{2!} + \frac{0.693^3}{3!} + \frac{0.693^4}{4!} + \frac{0.693^5}{5!} \\
&= 1.999
\end{aligned}
$$

Close enough, call it 2.

So, if you have an interest rate of r, then the time, rn that it takes to double you money is $rn = 0.693$, or

$$n = \frac{0.693}{r} = \frac{69.3}{100 \times r}$$

In the business department of your local university, they round off the number 69.3

to 70,

$$n = \frac{70}{100 \times r}$$

and they teach it as the "*Rule of* 70," meaning if your interest rate is written as a number between 0 and 100 (say 5.5), then you divide that number *into* 70, to find how many years it would take for you to double your money.

Example:

If the interest rate is 5.5%, then $\frac{70}{5.5} = 12.72$, or it would take about 12 years and 9 months to double your money. If the interest rate is 8%, then it takes $\frac{70}{8} = 8.75$, or 8 years and 9 months. If your interest rate is $\frac{1}{2}$ of 1%, then it takes $\frac{70}{1/2} = 140$, about 140 years.

Other types of infinite series

Not all infinite series come from the Binomial Theorem. Loosely speaking, any of the commonly used mathematical functions, such as: $\tan(x)$, $\cos(x)$, $\ln(x)$, and their inverses can be expressed as an infinite series of simple terms, powers of x, thus giving engineers and other consumers of mathematical formulas, quick, easy, and accurate numerical values to use in their calculations. This very happy circumstance for mathematicians, scientists, and economists evolved from Newton's work in calculus. And this is no surprise, because he was a key player in the invention and application of ideas in calculus.

In the next section we will take the time to develop the concept of infinite series in theoretical detail, so hold on to your hats!

Some infinite series used in calculus

The infinite series for the function e^x is one of the formulas we use in calculus to derive infinite series for other functions, such as $\cos(x)$ and $\sin(x)$.[4]

Do you recall that we used Euler's Equation,

$$e^{it} = \cos(t) + i\sin(t).$$

when we wrote the polar coordinate version of the complex number $x + iy$? Well, even if you don't, it is still true. The point is that substituting the imaginary number it in for x in the infinite series Equation (6.14), gives us

$$
\begin{aligned}
e^{it} &= 1 - \frac{t^2}{2!} + \frac{t^4}{4!} - \frac{t^6}{6!} + ... + \\
&\quad it - \frac{it^3}{3!} + \frac{it^5}{5!} - ...
\end{aligned}
$$

Therefore,

$$
\begin{aligned}
\cos(t) &= 1 - \frac{t^2}{2!} + \frac{t^4}{4!} - \frac{t^6}{6!} + ... \\
\sin(t) &= t - \frac{t^3}{3!} + \frac{t^5}{5!} - \frac{t^7}{7!} + ...
\end{aligned}
$$

---

[4]We expect readers who don't know calculus to just look at and marvel at the resulting series for $\ln(x), \cos(x)$, and $\sin(x)$. See Chapter 11 in this book, for a brief treatment of calculus.

OK, Nuff said. A proof of these formulas actually requires calculus, but we will just assume them here. They are keys to getting infinite series expansions for other calculus functions and equations such as: $y = \ln(t)$, $y = \tan(t)$ and its inverse, $t = \arctan(y)$, and others.

Trigonometry without a scientific calculator

As we mentioned, ordinary calculators do not have buttons that you can push to find square roots, trigonometric functions or logarithms. But, you can never-the-less solve any such "higher math" problems, without needing any calculator with those built-in programs. If you know the infinite series for a trig function such as $\cos(x)$, for example you can find any cosine just by adding, subtracting, multiplying and dividing numbers.

Example: Given the right triangle $\triangle ABC$, with right angle at the vertex B and with $\theta = 36°$ being the measure of the angle $\angle BAC$. Find the $\sin(\theta)$ and the $\cos(\theta)$.

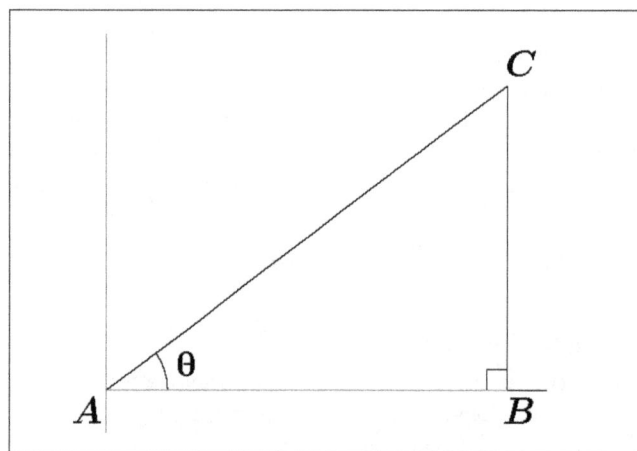

FIGURE 6.3 FIND $\cos(36°)$

First, we need to convert the degree measurement of an angle into a radian measure, because the infinite series for the cosine is in real numbers, and this means radian measure for angles. We can use Equation (5.2) from Chapter 5 of this book to get:

$$\pi \text{ radians} = 180°$$

So, $36° = \frac{\pi}{5}$ or about 0.628318... radians. The $\cos(t)$ series is

$$\cos(t) = 1 - \frac{t^2}{2!} + \frac{t^4}{4!} - \frac{t^6}{6!} + ...$$

Just use the first 3 terms, and substitute in $t = 0.628318$

$$
\begin{aligned}
\cos(0.628318) &= 1 - 0.197392 + 0.006494 - 0.000085 \\
&= 0.809017...
\end{aligned}
$$

which is what you would have gotten if you had used the cosine key on a scientific calculator. To get the $\sin(36°)$, just use 0.628318 in a few terms of the series for $\sin(x)$.

But you may ask where could you get an approximation for π with a non-scientific calculator? Hah! There are more than one infinite series that will give you π to any accuracy you want. Here is a glacially slow one, discovered in the early 1400's by the Indian mathematician Madhava Sangamagrama and later, in 1674, by the German mathematician, Gotfried Leibnitz (one of the inventors of calculus)

$$\pi = 4 - \frac{4}{3} + \frac{4}{5} - \frac{4}{7} + \frac{4}{9} - \frac{4}{11} +$$

Here, you take 4 divided by the odd numbers with alternating signs (adding and subtracting the terms). Unfortunately, it is painfully slow to converge. For example, it takes about three hundred terms to get π to two decimal place accuracy, nowhere as good as the elementary school approximation, 22/7, which is good to three decimal places. On the other hand, a blazingly fast one, invented by the Indian mathematician Srinivasa Ramanujan at the beginning of the 20th century, which we will show here as Equation (6.21), at the end of this chapter. It gives us π to a very high degree accuracy in just one or two terms.

Well-behaved and pathological functions

Mathematical functions whose graphs are continuous and have tangent lines whose slopes can be computed at each of their points are often called *well-behaved*, or *smooth*. See Figure 6.4.

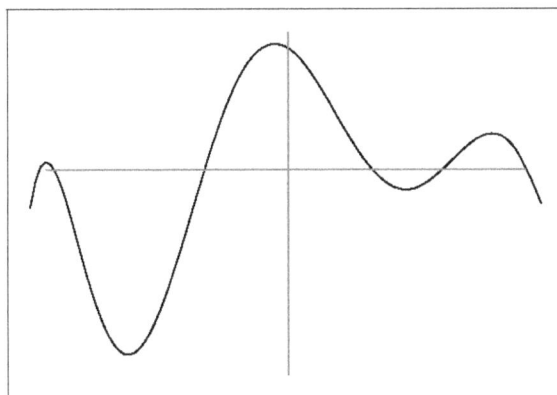

FIGURE 6.4 A SMOOTH FUNCTION

It is fair to say that most applications in physics, engineering, and economics make use of such functions more often than they do of another type called *pathological* functions.

There are uncountably many pathological functions and it is important to understand how to handle them in the most general way. These are functions which have graphs that are discontinuous, or have cusp points (those at which no line is tangent to the curve). Usually they can not be expanded into infinite series whose terms are simply powers of x.

Still modern physics, and other sciences such as meteorology, especially weather models, make use of these functions. Many of them deal with topics such as *chaos* and *fractional dimensions*, producing graphic results sometimes called *fractals*. And fractals are used to study weather patterns as well as to create graphics representing landscape and clouds in science fiction movies.

Many of these functions arose, originally, from attempts to solve problems by the use of infinite series. They did not exist before someone first wrote down the series that defined them. When we expand these new functions into infinite series, we often find that they are not just infinite series of powers of x, but rather, infinite series of trigonometric functions of x.

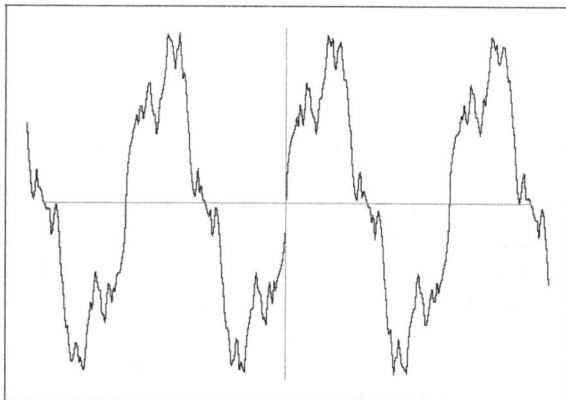

FIGURE 6.5 THE CRINKLY FUNCTON

One example is a function that is continuous at every one of its points and does not have a tangent line at any of its points. It was first defined in 1872 by Karl Weierstrass. The Weierstrass function, Equation (6.15), below, is so un-smooth that it is known as a *crinkly function*.

$$W(x) = \sum_{k=1}^{k=\infty} \frac{\sin(k^2 \pi x)}{k^2 \pi} \tag{6.15}$$

Figure 6.5, above, is a plot of just the first 40 terms of Equation (6.15).

In 1807, the French mathematician Jean Baptiste Joseph Fourier introduced infinite series of trigonometric functions that could be used to solve partial differential equations needed in the study of heat and wave equations. Later in the 1800's, gradual progress was being made in the applications of various infinite series all over the world.

Theoretical considerations

How does infinity fit in with finite numbers? The concept of infinity may seem to be a great mystery. Is it a number? a place? or just a distant horizon that we can never get to? Believe-it-or-not, there are several different types of infinity, most of them were discovered at the end of the 1800's. We will discuss them in a later chapter in this book. But right now we will be using only the "lowest" level of infinity which corresponds to the positive integers, that is, the counting numbers, $\{1, 2, 3, ..., n, ...\}$.

What keeps an infinite series from converging?

Is it true that just any old infinite set of numbers added together add up to finite number? The answer is NO, and thank goodness for that! Otherwise, we would be drowning in a sea of infinities disguised as finite numbers!

We need to define what we mean by saying something is *finite* and what we mean when we say *infinite*.

Definition: A positive number X is *finite* means that there is a positive integer n such that $n > X$.

A set S is finite means that there is a positive integer n such that S does not have n elements in it.

Examples:

The number π is a finite number because the positive integer $4 > \pi$. Any time you give me a number, if I can find a number that is bigger than yours, then your number is finite.

A googol is finite because $10^{100} + 1 >$ a googol. What is a googol? In 1945, the American mathematician Edward Kasner, asked his 9 year old nephew what the number 10^{100} was. The young boy made up an answer and said "a googol." Kasner used that word in a paper he published and the name became a standard in mathematical literature.

Later, when the internet came into existence and a certain new search engine company started up, it is alleged that the founders wanted to use the name <u>Googol</u> so as to demonstrate the vast number of sites it could access, but somebody misspelled it and named the company, <u>Google</u>. When this company first became widely known, it seemed to have been named after *Barney Google* from the newspaper comics, "Barney Google and Snuffy Smith."

When does an infinite series converge and when does it not?

The main question we want to answer is: "When we are confronted with some hitherto unknown newly minted infinite series, how do we know whether it will or will not add up to a finite number?"

Example: Consider the infinite series

$$\frac{1}{2} + \frac{1}{2} + \frac{1}{2} + \frac{1}{2} + ...$$

Does it add up to a finite number? Just suppose there *is* some finite number N such that $N = \frac{1}{2} + \frac{1}{2} + \frac{1}{2} + \frac{1}{2} + ...$then we could add $\frac{1}{2}$ to both sides and get $N + \frac{1}{2}$ on the left and just N on the right because adding one more $\frac{1}{2}$ to an infinity of $\frac{1}{2}$'s is still just an infinity of $\frac{1}{2}$'s–nothing is changed–the right hand side is still N.

$$N \;=\; \frac{1}{2} + \frac{1}{2} + \frac{1}{2} + ... + \frac{1}{2} + ...\text{Supposition} \tag{6.16}$$

$$N + \frac{1}{2} \;=\; \frac{1}{2} + \frac{1}{2} + ... + \frac{1}{2} + \frac{1}{2} + ...\text{Adding } \frac{1}{2}$$

This means that $N + \frac{1}{2} = N$, so $\frac{1}{2} = 0$ which affronts your common sense. But even more importantly, if you multiply both sides by 2, you get $1 = 0$, which violates Axiom 9 in Chapter 4. This contradiction means that the series in (6.16) cannot add up any finite number N.

Ratio Test

For an infinite series to be able to add up to some finite number, the terms cannot fail to, eventually, get smaller and smaller, actually approaching zero, and they should do it *fast enough* to keep the sum from getting too big. Here the phrase "fast enough" has a precise definition that can be determined by a method called the "ratio test." If

$$a_1 + a_2 + a_3 + ... + a_k + a_{k+1} + ...$$

is an infinite series of positive terms, and the ratio of two consecutive terms a_{k+1} divided by a_k has a limit less than 1, then the series converges and if the limit is greater than 1, the series diverges.. In other words, let a_{k+1} and a_k be consecutive terms and find the limit

$$\lim_{k \to \infty} \frac{a_{k+1}}{a_k} = b,$$

then if $b < 1$, the series converges, but if $b > 1$, the series diverges, that is does not converge. What if $b = 1$? Then the test fails and the series might either converge or diverge.

Example:

$$A = \frac{1}{5} + \frac{1}{25} + \frac{1}{125} + ... + \frac{1}{5^k} + ...$$

In this case, the terms approach zero fast enough for the series to add up to a finite number, because by the ratio test

$$\frac{a_{k+1}}{a_k} \quad = \quad \frac{1/5^{k+1}}{1/5^k} = \frac{1}{5}, \text{ so}$$

$$\lim_{k \to \infty} \frac{a_{k+1}}{a_k} \quad = \quad \frac{1}{5}, \text{ which is less than 1}$$

Actually we already knew that this series converges because we know exactly what its sum is. It comes from

$$\frac{1}{(1-x)} = 1 + x + x^2 + x^3 + ... + x^k + ...$$

with $x = \frac{1}{5}$

$$\frac{1}{(1 - \frac{1}{5})} = 1 + \frac{1}{5} + \frac{1}{(5)^2} + \frac{1}{(5)^3} + ...$$

Therefore,

$$A = \frac{1}{(1 - \frac{1}{5})} - 1 = \frac{5}{4} - 1 = \frac{1}{4}$$

Partial sums

Let us consider any infinite set of real numbers:

$$\{a_1, a_2, a_3, a_4, ..., a_n, ...\}$$

For each positive integer, n, let a_n be a number. The numbers a_n, themselves, can be positive, negative or zero. Now we recursively define the sum S_n to be

$$S_1 \quad = \quad a_1 \text{ and}$$
$$S_n \quad = \quad S_{n-1} + a_n, \text{ for } n \geq 2$$

In other words

$$
\begin{aligned}
S_1 &= a_1 \\
S_2 &= S_1 + a_2 = a_1 + a_2 \\
S_3 &= S_2 + a_3 = a_1 + a_2 + a_3 \\
S_4 &= S_3 + a_4 = a_1 + a_2 + a_3 + a_4
\end{aligned}
$$

The number S_n is called a *partial sum,* in fact, the *nth partial sum.* It is the sum of the first n terms of the infinite series. There is no doubt about the fact that S_n is a finite number because of the axioms of the number system. We are just adding up a finite number of numbers.

The problem happens when we think we can add up *all* of the numbers, infinitely many of them.

$$a_1 + a_2 + a_3 + .. + a_n + ... \text{ forever}$$

It may turn out there is no number that can be their sum, as shown, for example $\frac{1}{2} + \frac{1}{2} + \frac{1}{2} + \frac{1}{2} + ...$ in (6.16), above. If, however, in some particular series, we can show that there is some fixed number S, such that $|S - S_n|$ approaches zero as n gets larger and larger, then we can claim that S is, indeed, the sum of that infinite series. This is written as

$$\lim_{n \to \infty} S_n = S.$$

When this happens, we say that the series *converges* to S. The formal definition of this idea is as follows.

Definition: (Converging infinite series.)

If α is the infinite series:

$$x_1 + x_2 + x_3 + ... + x_n + ...$$

then the statement that α *converges,* means that there exists a number S, such that for any positive number ϵ, there exists a positive integer N, such that $|S - S_n| < \epsilon$ for every $n > N$.

Problem: Given the infinite series:

$$\frac{1}{5} + \frac{1}{25} + \frac{1}{125} + \frac{1}{625} + ... + \frac{1}{5^n} + ...$$

show that the partial sums S_n converge to $S = \frac{1}{4}$.

Solution:

The partial sums are:

$$
\begin{aligned}
S_1 &= \frac{1}{5} \\
S_2 &= \frac{1}{5} + \frac{1}{25} = \frac{6}{25} \\
S_3 &= \frac{6}{25} + \frac{1}{125} = \frac{31}{125} \\
&\quad ...
\end{aligned}
$$

We will conclude that $S = \frac{1}{4}$, by proving that $\left|\frac{1}{4} - S_n\right| \to 0$, as $n \to \infty$. Here are some values of $\left|\frac{1}{4} - S_n\right|$, for $n = 1, 2, 3$

$$\left|\frac{1}{4} - S_1\right| = \frac{1}{20}, \text{thus } S_1 = \frac{1}{4} \pm \frac{1}{4}5^{-1}$$

$$\left|\frac{1}{4} - S_2\right| = \frac{1}{100}, S_2 = \frac{1}{4} \pm \frac{1}{4}5^{-2}$$

$$\left|\frac{1}{4} - S_3\right| = \frac{1}{500}, S_3 = \frac{1}{4} \pm \frac{1}{4}5^{-3}$$

$$\ldots$$

By mathematical induction, the well-ordering axiom in Chapter 4,

$$S_n = \frac{1}{4} \pm \frac{1}{4}5^{-n}$$

for any positive integer, n.

And, by the Dedekind cut axiom if ϵ is any positive number, then there is some positive integer n such that

$$\frac{1}{5^n} < \epsilon$$

Therefore if ϵ equals any prescribed tolerance, *viz* "error," then you can find a positive integer N large enough so that, not only will S_N differ from $\frac{1}{4}$ by less than ϵ, but this will also be true for any other S_n, when $n > N$. Every one of them will be closer to $\frac{1}{4}$ than ϵ. This is what is meant by saying

$$\lim_{n \to \infty} S_n = \frac{1}{4}$$

Other tests for convergence or divergence

Since you cannot actually do infinitely many additions, you cannot tell if a given series is convergent or divergent, just by adding a few terms. But if you compare the general term of a given unknown series with the general term of a another, known to be divergent, or known to be a convergent series, you may be able to conclude what is happening with your given unknown series.

Comparison test for divergence Working with only positive numbers, an unknown series: $y_1 + y_2 + y_3 + \ldots$ *diverges* if it becomes term-by-term *greater than or equal to* a known divergent series: $x_1 + x_2 + x_3 + \ldots$. "Term-by-term greater than or equal to" means that for all integers k after some finite number, $y_k \geq x_k$.

Roughly this says, *if your series is bigger than a divergent series, then your series diverges too.*

Naturally, there is also a comparison test for *convergent* series.

Comparison test for convergence An unknown series $b_1 + b_2 + b_3 + \ldots$ converges if, after a finite number of terms, it becomes, in absolute value, term-by-term *less than or equal to* a known converging series: $a_1 + a_2 + a_3 + \ldots$.

Roughly, this says that *if your series is smaller than a convergent series then your series must converge also.*

The proofs of these tests depend upon the Dedekind cut, Axiom 17, Chapter 4.

Problem: Suppose someone wagers you $100 that you can't tell him whether or not the following series converges. How could you win this bet?

$$1 + \frac{1}{2!} + \frac{1}{4!} + \frac{1}{6!} + \frac{1}{8!} + ... + \frac{1}{(2n)!} + ... \tag{6.17}$$

Solution:

You know from the definition of e^x, when $x = 1$, that

$$e^1 = 1 + 1 + \frac{1}{2!} + \frac{1}{3!} + \frac{1}{4!} + ... \frac{1}{n!} + ... \tag{6.18}$$

This number e^1 is the *finite* number e, so we know the series (6.18) converges.

Now back to the $100 bet series, comparing (6.17) to (6.18) we see that every term in (6.17) is less than or equal to the corresponding term in (6.18). That is,

$$1 \le 1, \ \frac{1}{2!} \le 1, \ \frac{1}{4!} \le \frac{1}{2!}, \ \frac{1}{6!} \le \frac{1}{3!}, \ ..., \ \frac{1}{(2n)!} \le \frac{1}{n!}, ..$$

So, in general, the unknown series (6.17) is term-by-term smaller than the known convergent infinite series (6.18). Therefore, you have proved that the series that was being gambled upon converges, and you can collect your *c-note*.

Is the harmonic series convergent?

The fractions $\frac{1}{2}$, $\frac{1}{3}$, $\frac{1}{4}$, etc., are called *harmonics, or harmonic frequencies.* They are the musical tones, in stringed instruments, that come from stopping a string that is $\frac{1}{2}$, or $\frac{1}{3}$, or $\frac{1}{4}$, etc. of the way along the length of the string. An infinite series we call the *harmonic* series is the sum of all the reciprocals of the integers, that is, *all* the harmonics.

$$S = 1 + \frac{1}{2} + \frac{1}{3} + \frac{1}{4} + \frac{1}{5} + ... \tag{6.19}$$

The study of the harmonic series is associated with construction of bridges because it can be used to assess *resonance* or vibrations caused by traffic on the bridge or wind or other natural causes. What is deceptive about this series is that it looks like it ought to converge because the terms $\frac{1}{2}$, $\frac{1}{3}$, $\frac{1}{4}$, ...are, indeed, getting smaller, and actually approaching zero. But does it converge? Do the terms get small enough fast enough?

If we apply the ratio test, we will not be able to tell whether or not it converges because the limit of the ratio of two consecutive terms is equal to 1.

$$\lim_{k \to \infty} \frac{1/(k+1)}{1/k} = \lim_{k \to \infty} \frac{k}{(k+1)}$$
$$= \lim_{k \to \infty} \frac{1}{\left(1 + \frac{1}{k}\right)} = \frac{1}{1+0} = 1$$

Therefore, the series might diverge or converge. Since the ratio test doesn't work here, how else can we prove whether or not it converges? From a practical point of view, we cannot just compute the sum for a whole lot of terms to get much insight. For example, even after adding up 30 terms, the series does not seem to be threatening to go to infinity, the partial sum, S_{30}, is less than 4. It does not help to remember the song in the 1960's about the "harmonic convergence of the moon and stars."

It turns out that the harmonic series does, indeed, diverge, and its divergence was first proved around 1350 by the French mathematician Nicole Oresme. Then more than 300 years later, around 1687, this was also proved by the rivalrous Swiss mathematicians, the Bernoulli brothers, Johann and Jacob, each independently.

Proof:

Oresme's proof, using only algebra, is easier to understand. He starts by separating the terms into groups of fractions such that each group adds up to $\frac{1}{2}$ or more.

$$1 \geq \frac{1}{2}$$
$$\frac{1}{2} \geq \frac{1}{2}$$
$$\left(\frac{1}{3} + \frac{1}{4}\right) \geq \left(\frac{1}{4} + \frac{1}{4}\right) = \frac{1}{2}$$
$$\left(\frac{1}{5} + \frac{1}{6} + \frac{1}{7} + \frac{1}{8}\right) \geq \left(\frac{1}{8} + \frac{1}{8} + \frac{1}{8} + \frac{1}{8}\right) = \frac{1}{2}$$
$$...$$

Notice that in the group $\left(\frac{1}{5} + \frac{1}{6} + \frac{1}{7} + \frac{1}{8}\right)$, the three fractions $\frac{1}{5}$, $\frac{1}{6}$, $\frac{1}{7}$, are all greater than $\frac{1}{8}$. That is, they are all greater than the last fraction in that grouping. In the nth grouping, for any positive integer, n, the last fraction is $\frac{1}{2^n}$; the rest of the fractions in that grouping will be greater than $\frac{1}{2^n}$ and since there are 2^{n-1} fractions in *that* grouping, the total group adds up to something more than $\frac{1}{2}$.

When we add up the terms of the harmonic series we are adding infinitely many groups, each of which is $\frac{1}{2}$ or greater. In other words, the harmonic series is greater than

$$\frac{1}{2} + \frac{1}{2} + \frac{1}{2} + \frac{1}{2} + \frac{1}{2} + ...$$

and this is the divergent series we gave in (6.16). Hence, by the comparison test, we have proved that the harmonic series, (6.19), diverges.

We have only scratched the surface of convergence tests for infinite series, but we may have given you enough experience with this to actually create your own proof or, at least, understand a proof written by someone else.

There are unsolved problems in the world of infinite series.   Bernhard Riemann considered a series which defines a function, ζ, zeta of the complex number $s = x + iy$.

$$\zeta(s) = \sum_{n=1}^{\infty} \frac{1}{n^s} \tag{6.20}$$

known as the *Riemann-Euler zeta function*. Euler's name being added because he had contributed to this same problem more than one hundred years earlier, in 1737, before complex numbers had been widely known. The Riemann zeta function is at the foundation of analytic number theory and complex transformations. These are important in a field called *conformal mappings*, as well as for applications in physics, statistics and probability theory.

Riemann posed a conjecture about the series (6.20) which is still not answered as of this year, 2020. It is called the *Riemann zeta conjecture* and it goes like this: $\zeta(s) = 0$ only if s is a negative even number or if $\frac{1}{2}$ is the real part of the complex number s. This conjecture, although unproved, has given rise to several, maybe hundreds of, results just waiting to be promoted to theorems, once this hypothesis is proved.

In the 18th and 19th centuries, Euler, Jean-Baptiste Fourier, and Brook Taylor, used trigonometric functions as well as almost every type of mathematical function as the terms of infinite series. The crinkly curve, Figure 6.5, defined by Weierstrass is an example of a Fourier series. Augustin Louis Cauchy contributed theorems providing the criteria needed to insure that any series converges or diverges. Other tests, the ratio and the root tests were developed by Carl Friedrich Gauss, and Riemann. Sonya Kovalevsky and others solved partial differential equations and integral equations by using various infinite series that they discovered, or invented, in the late 1800's.

In early 1900s, an astonishing young mathematician, previously unknown in the western world appeared on the scene to make giant leaps to new levels in the study of infinite series.

Srinivasa Ramanujan

In 1913, the 25 year-old Srinivasa Ramanujan, sent a letter to the British mathematician G. H. Hardy. In this letter, Ramanujan explained that he was a poor clerk working in an accounting office in India and he had no formal training in mathematics but he had a passion for the subject. He went on to say that he had worked out a few results that apparently astounded some of the local Hindu mathematicians, and he wanted to know whether or not Hardy would be interested in looking at them. He concluded the letter as follows:

> " I would request you to go through the enclosed papers. Being poor, if you are convinced that there is anything of value, I would like to have my theorems published. I have not given the actual investigations nor the expressions that I get but I have indicated the lines on which I proceed. Being inexperienced, I would very highly value any advice you give me. Requesting to be excused for the trouble I give you, I remain, Dear Sir, Yours Truly S. Ramanujan."

There were 120 theorems (without proofs) attached to this letter. At first, Hardy didn't know how to react to receiving this letter from an unknown Hindu clerk. He looked over the list and tried to see if he could recognize anything. He did notice one or two theorems that were somewhat like ones that he, himself, had proved. He was familiar with another two formulas that were classic theorems proved by the German mathematician Carl Jacobi around 1840. There were two others that were in Hardy's field of expertise and he thought he could prove them. He had some trouble doing so, but he did manage to construct proofs for these two new theorems. There were four other formulas that Hardy found to be much more intriguing and he decided that they must have come from some very general theorem that Ramanujan was keeping a secret. Here is what Hardy had to say about some of the items in this remarkable list of theorems.

> "The formulae (1.10)-(1.13) are on a different level and obviously both difficult and deep. An expert on elliptic functions can see at once that (1.13) is derived somehow from the theory of complex multiplication, but (1.10)-(1.12) defeated me completely. I had never seen anything in the least like

them before. A single look at them is enough to show that they could only be written down by a mathematician of the highest class.

After some time, Ramanujan convinced his mother that he should accept Hardy's offer to bring him to England. So, in 1914, Ramanujan arrived in England and became a student, or more accurately a collaborator with Hardy. While it was a struggle for Ramanujan to survive in Britain because of problems he had in maintaining his cultural, societal and dietary needs, these two mathematicians produced a large number of papers. Ramanujan's published works were very deep and ranged widely from continued fractions, elliptic integrals and his own theory of divergent infinite series.

He developed several algorithms and formulas for calculating certain transcendental numbers[5] to high degrees of accuracy. It was one of these Ramanujan-type of formulas which was used by William Gosper in 1985 to compute the first 17 million digits of π and, later, by an American team, to compute, for the first time, π to one billion decimal places. More powerful computers and modification of algorithms of the Ramanujan type have currently (in 2016) resulted in π being computed to over 12 trillion places. Here is one of his infinite series (6.21) that needs only *one* term (using $n = 0$) to get 7 decimal places of π and *two* terms(using $n = 1$) to get π accurate to 15 decimal places

$$\frac{1}{\pi} = \frac{2\sqrt{2}}{9801} \sum_{n=0}^{\infty} \frac{(4n)!(1103 + 26390n)}{(n!)^4(366)^{4n}} \tag{6.21}$$

Ramanujan died of tuberculosis in 1920. No one knows how he arrived at some of his results many of which are still unproven today.

---

[5]Transcendental numbers have nothing to do with psychology. These are irrational numbers that are not the solutions to any algebraic equations.

CHAPTER 7
LOGIC

We are mathematicians! First, we are humans, therefore we are logical creatures. Several recent psychological studies have confirmed that human children, at a very young age, exhibit the ability to reason logically. The standard protocol in an experiment with a pre-verbal infant is for the scientist to note the infant's duration of and intensity of looking at a new situation and physically reacting in a meaningful way. For example, the March 16, 2018 issue of *Science,* the journal of the American Association for the Advancement of Science (AAAS), reports an interesting experiment with infants under the age of 19 months. Nicolo Cesana-Arlotti, *et al.* performed experiments showing that babies can use the process of elimination of false cases in order to infer a logical outcome.

His AAAS article reports that, infants were shown two objects that were then hidden behind a screen. The screen was partially removed showing one of the original objects. But when the other object was revealed, the infant would register acceptance, if it was the original one, or rejection if it was something else that the experimenters had secretly substituted while it was hidden from view. Here are partial conclusions derived from this experiment.

> *Infants are able to entertain hypotheses about complex events and to modify them rationally when faced with inconsistent evidence. These capacities suggest that infants can use elementary logical representations to frame and prune hypotheses.*

Babies can do logic. Hooray! So, how is it that we, as ex-babies, can ever face a road-blocking problem in logic? OK, let us look at some paradoxes that seem to show that mathematics is plagued by illogical troubles.

Paradoxes
Self-referent paradoxes

Have you ever heard of a statement that leads to the conclusion that it is false when you assume it is true and leads to the conclusion that it is true when you assume that it is false? Such a statement is called a "paradox." One of the earliest paradoxes known in mathematics was given by the ancient mathematician, Epimenides, the Cretan, who said: "All Cretans are liars." Think about this! If he is speaking the truth then he himself is a liar. You might resolve this by saying, "Well, he didn't mean that they lied all the time, so if they lie sometimes and tell the truth sometimes, then he could make this statement." OK, but now suppose that Mr. E. says to you "I am telling a lie right now." Your problem is to determine whether or not Mr. E. is telling a lie with that statement.

    1. Assume that this statement is true, then he is telling the truth and not lying as he says; so the statement is false.

    2. Assume the statement is false then he is not lying right now; so the statement is a lie; therefore the statement is true.

There are several types of paradoxes; sometimes a statement can be a paradox in one situation but not in another. For example, there is a popular bumper sticker that reads, "The Worst Day Fishing Is Better Than The Best Day Working." We can assume that the person who displays this hates his (or her) job and feels that it will never be as desirable as fishing, no matter how poor the fishing is. In such cases the bumper sticker is not a paradox but merely a wistful expression. If, on the other hand, the person is an artist or a musician or someone who is very likely to enjoy the work, then the statement is just a lie or, at best, a frivolous remark.

There is one situation, however, in which this particular bumper sticker is a genuine logical paradox and that is, when the people displaying it are professional fisherpersons. In this case they are fishing when they are working. For them, everyday working is a day fishing. So how can their worst day working be better than their best day working?

The Working *vs.* Fishing example given above is a "self-referent" paradox, a type of paradox that has been the source of trouble for the foundations of mathematics. The difficulty is that the paradox comes from a statement that contains a reference to its own truth, and under certain conditions (for example when working is the same as fishing) the statement is self-contradictory.

Two of the most troublesome self-referent paradoxes are The *Russell Paradox* in set theory and *Gödel's Proof* of the incompleteness of mathematics. These two paradoxes have had serious effects on mathematics, creating doubt and uncertainty, as we shall see when we study them in Chapter 12.

Examples:(Self-referent paradoxes)

1. Never take a stranger's advice said to you by a stranger.

2. The worst day skiing is better than the best day working said to you by a professional skier.

3. In a town, there is a barber (himself a man) who shaves every man who does not shave himself and only those men.

Who shaves the barber?

4. Paradox of the hanging bridge. In *Don Quixote*, the author, Cervantes, tells the story of a bridge which has a gallows at one end and four judges sitting at the other end. It was the judges job to execute the following law:

"Whoever intends to pass from one End of this Bridge to the other, must first upon his Oath declare wither he goes, and what his Business is. If he swear Truth, he may go on; but if he swear false, he shall be hang'd, and die without Remission upon the Gibbet at the End of the Bridge."–*Part II, Book IV, Chapter LI.*

Several people had come to this bridge, those who lied were hung and those who told the truth were allowed to pass. But one day a person came to the bridge and, "...declared by the Oath he had taken, he was come to die upon the Gallows, and that was all of his Business." The judges are still pondering what to do with that person.

5. If Zeus is all powerful can he make a rock he cannot lift? A question asked by an ancient Greek philosopher.

6. The following pair of statements:

A: "THE FOLLOWING STATEMENT IS TRUE."

B: "THE STATEMENT ABOVE IS FALSE."

These self-referent paradoxes would be a good way to flummox a so-called lie detector. See Figure 7.1

FIGURE 7.1 DISCOMBOBULATING A LIE DETECTOR

Reversal Paradoxes

Unlike the self-contradictory paradoxes of the previous section, the problems in this section involve some interesting paradoxes on "combined" data. Here we will encounter some statements that are not true paradoxes, but rather are "surprises." The surprise is that separate data, from various sources, seems to indicate some definite conclusion, but combining the data yields just the opposite conclusion!

These *reversal paradoxes* (also called "pooled data" paradoxes) are known as "Simpson's Paradox" named after the American statistician E. H. Simpson, who discussed it in a paper in 1951. Credit must also be given to Karl Peterson who mentioned it in 1895 and UdneYule who discovered it in 1903. The paradox can be resolved by examining the basis of each calculation and noticing that comparison of best and worst results can shift from one base to another.

This type of paradox can occur, inadvertently, in statistical analysis (see the graduate school example below); or, perhaps, deliberately in a statement of a political viewpoint (see the Income tax example, below).

Example: (Graduate School Enrollment).

This example is based upon the pooled data paradox described by Simpson, in 1951.

A certain graduate school has two departments, A and B. The enrollment policy came into question as being biased against women when it was discovered that of the 300 men and 300 women who applied to the school, 120 men were accepted, but only 93 women were accepted. Both departments A and B claimed that they had not discriminated against women, in fact they both claimed to have accepted a higher proportion of women than men. A study of their applications and admissions records revealed that Department A had 200 applications from men and they accepted 100 of them and they had 100 applications from women and they accepted 51 of them. In Department B, the figures were: 100 men applicants, 20 were accepted, and of the 200 women applicants 42 were accepted. See the table below.

Dept. A	Applicants	Accepted	Pct.
Men	200	100	50%
Women	100	51	51%
Dept. B	Applicants	Accepted	Pct.
Men	100	20	20%
Women	200	42	21%
School	Applicant	Accepted	Pct
Men	300	120	40%
Women	300	93	31%

School Enrollment Paradox

We can see Department A accepted a higher percentage of women applicants, as did Department B, but in the whole school (Combining both departments), the percent of women accepted was less than the percent of men. This is a simplified example of the study done by Simpson for two departments in the Stanford Graduate school.

Notice here the admission records for the two departments is not <u>averaged</u>, but rather <u>pooled</u>.

$$\frac{100}{200} \; < \; \frac{51}{100}, \text{ Department A, Pct. of Men} < \text{Pct. of Women}$$

$$\frac{20}{100} \; < \; \frac{42}{200}, \text{ Department B, Pct. of Men} < \text{Pct. of Women}$$

$$\frac{120}{300} \; > \; \frac{93}{300}, \text{ Whole School, Men's Pct.} > \text{Women's Pct.}$$

We know that pooling the fractions $\frac{100}{200}$ and $\frac{20}{100}$ into the fraction $\frac{100+20}{200+100}$ is not the normal way to add fractions. The basis for calculating the percentages was switched. Fewer women applied for the department that was easiest to get into (higher acceptance rate) and more women applied to the department that was harder to get into.

Example: (Average rate). What is the average of 10 plus 20? Are you sure? A girl rides up a one-mile hill on her bicycle at a speed of 10 miles per hour and she rides back down the hill at a speed of 20 m.p.h. The average speed for the whole round trip is *not* 15 m.p.h. Why not? Here, you need to know the time it took for the uphill trip and the time it took for the downhill trip. Uphill she used 6 minutes.

$$\frac{1 \text{ mile}}{10 \text{ m.p.h.}} = \frac{1}{10} \text{ hour} = 6 \text{ minutes}$$

Downhill she used

$$\frac{1 \text{ mile}}{20 \text{ m.p.h.}} = \frac{1}{20} \text{ hour} = 3 \text{ minutes}$$

The whole 2 mile trip took 9 minutes, or $\frac{9}{60}$ of an hour, so rate $= \frac{\text{distance}}{\text{time}}$, or

$$\frac{2 \text{ miles}}{(9/60) \text{ hour}} = 13.33 \text{ m.p.h. not } 15 \text{ m.p.h.}$$

Example: (Income Tax) In a town with a population of 100, there are 99 people who make $100 per year and 1 person who makes $1,000,000 per year. The government of this town collects approximately $1,000 in taxes and the taxes are paid as follows: The person making a million dollars pays $150 and each of the other citizens pay $8.59. The millionaire complains that he is paying too much in taxes. He asserts that "I am the person with the top 1% in income, but I pay 15% of the all the taxes collected. Here are the facts:

a. The total tax paid is $1,000.41. That is, $99 \times \$8.59 + 1 \times \$150 = \$1,000.41$.

b. The millionaire's assertion is true; his $150 is approximately 15% of all of the taxes collected.

c. The 99 people, each paid 8.59% of their income.

d. The millionaire paid $\frac{1}{100}$ of 1.5% of his income.

This is an example of switching the base from the tax rate on individual incomes to a percentage of all taxes paid. A correct comparison should be to compare tax rate on individual incomes for both groups. In that case the fair tax on the millionaire would have been $0.0859 \times \$1,000,000 = \$85,900$, instead of $150.

Example: (Baseball)

There are other real-life examples of *pooling* data as opposed to *averaging* data. In baseball, the batting averages of two players might show that one player is better than the other in each of two seasons, but is worse when you consider the over-all two season history of the two players. A representational example here that lets us explain how numbers work in Simpson's paradox .

Year	Batter A	Batter B
1998	$127/410 = .310$	$204/684 = .298$
1999	$168/663 = .254$	$77/318 = .242$
Combined	$295/1073 = .274$	$281/1002 = .280$

Two Major League Batters

Here, A has a better average than B in both 1998 and 1999, but B has a better average than A for the two-year total. How a real situation like this could happen is baffling, but it is quite easy to see how the numbers do their part.

Without changing a single calculation, we can ransack this table in a very natural way, that will dispel the surprise. It is this: Compare A's best average, .310 with B's worst average, .242. Then, compare B's best average, .298 with A's worst average, .254. When you do this, neither one wins both comparisons; one wins once and the other wins once. Therefore there is no shock in recognizing that one or the other could have a better combined score. This assuages the anxiety you felt when you learned that one person is the winner each of the two years, but at the same time that same person was the loser over the two year period..

Example: (Medical)

In 1986 C. R. Chaig did a study of two different treatments for disposing of kidney stones that yielded the following data.

Stone size	Treatment A	Treatment B
Small	$81/87 = 93\%$	$234/270 = 87\%$
Large	$192/263 = 73\%$	$55/80 = 69\%$
All cases	$273/350 = 78\%$	$289/350 = 83\%$

Kidney Treatment Paradox

What is there to be done in these paradoxical situations? The simplest answer to say: "Be aware of them and not take sides without finding out more details."

But in the medical case, how will awareness of the paradox be useful in providing better medical service? These data were collected from a controlled experiment. It is clear that both treatments are less successful on the large (harder to treat) stones. In both cases, A is better than B, alright, but look at Treatment B's best result; it is better than A's worst. This explains the pooled data paradox here, as well as in the baseball and the graduate school reversals.

Any time each entity's *best* score is better than the other entity's *worst* score, the reversal will happen. Otherwise, the paradox cannot happen. You can prove this by writing out the appropriate inequalities in a general case.

Zeno's paradox

Around 450 BCE, Zeno of Elea, an ancient Greek philosopher and mathematician proposed a problem, generally known as "Zeno's Paradox," which turned out to be more of a misunderstanding about the nature of continuous time and distance, rather than a genuine paradox. It goes like this:

"Suppose that Achilles is in a race with a tortoise, with the tortoise starting some distance ahead of Achilles. Assume they both start simultaneously at points A and B respectively. Achilles is running after the tortoise and the tortoise is running away from Achilles. When Achilles reaches point B, the tortoise has already advanced to a point B_1 further down the road. When Achilles reaches point B_1, the tortoise has advanced further to the point B_2, thus Achilles can never catch the tortoise."

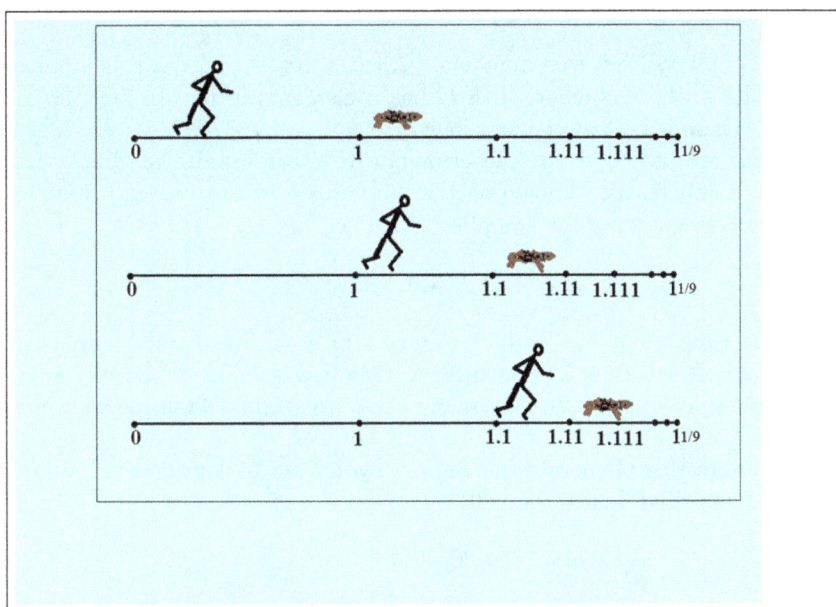

FIGURE 7.2 IT IS A TIE AT THE $1\frac{1}{9}$ MILESTONE.

Of course, to believe this you are required accept a complete misunderstanding of the continuity of time and the concept of a limiting process. It can only be true if the finish line is very close to the tortoise. But otherwise it is false. Why?

Example:

See Figure 7.2. Suppose Achilles runs ten times as fast as the tortoise and starts one mile behind him. Then when Achilles has run one mile to get to the Tortoise's starting place, then the tortoise has run $\frac{1}{10}th$ of a mile, and when Achilles gets to the 1.1 mile mark, the tortoise is at the 1.11 mile mark, eventually when Achilles reaches the 1.11 mile mark the tortoise is at the 1.111 mile mark, and so forth. But, Achilles *will* catch Tortoise at the following mile marker: 1.111111..., with the $1's$ going on forever. What is this number 1.1111...? Will it take forever for Achilles to get there? That is what Zeno thought. But, it is just the number $1\frac{1}{9}$, one and one-ninth mile. That is exactly where Achilles will catch and pass Tortoise.

How do we know that $\frac{1}{9} = 0.11111...$? Start with

$$\frac{1}{3} = 0.3333333 \text{ ...forever, Now multiply both sides by } \frac{1}{3}$$

$$\frac{1}{3} \times \frac{1}{3} = \frac{1}{3} \times (0.3333....)$$

$$\frac{1}{9} = 0.11111...$$

The seeming paradox of infinity

Reviewing how we defined finite and infinite in the previous chapter, we say a set M is *finite* if there is some whole number n, such that M does not have n elements in it. This seems like a strange way of defining a finite set, but it has an advantage when we come to defining infinite sets. A set M is *infinite* means that if n is any positive whole number, then there are n elements in M.

Example:

The set, E, of all positive even numbers, $\{2, 4, 6, 8, 10, 12, ..., 2k, ...\}$, is infinite because if n is any positive whole number, then E has n elements in it. In fact, $2n$ is the nth element of E. There is no last element in E.

The numbers we use to count the elements in either infinite or finite sets are the positive whole numbers, also known as the *counting numbers*, usually denoted by the letter $\mathbb{Z}$ for the German word for "number", "zahlen"

$$\mathbb{Z} = \{1, 2, 3, 4, 5, 6, 7, ...\}$$

Most of the time when we "count" elements in a set we merely match them with elements from $\mathbb{Z}$. It is sort of like counting people in a stadium by matching them with tickets sold, or by looking to see how many seats are empty, knowing how many seats there are.

Here is an interesting phenomenon. Suppose you want to count some infinite set such as, say, all of the positive multiples of 10.

$$M = \{10, 20, 30, 40, ... 10n, ...\}$$

You match the elements of this set with the elements in $\mathbb{Z}$.

$$
\begin{array}{ccccccc}
\mathbb{Z} & 1 & 2 & 3 & 4 & ... & n & ... \\
& \updownarrow & \updownarrow & \updownarrow & \updownarrow & \updownarrow & \updownarrow & ... \\
M & 10 & 20 & 30 & 40 & ... & 10n & ...
\end{array}
$$

It is clear that we will use up all the elements of $\mathbb{Z}$ in counting M. Every multiple of 10 has a counting number matched to it and every integer is matched with a multiple of 10. How can this be? Doesn't $\mathbb{Z}$ contain all of the elements of M (and then some) already? M is called a *proper subset* of $\mathbb{Z}$ because every element of M is an element of $\mathbb{Z}$ and there are some left-over elements of $\mathbb{Z}$, like 7 for example, that are not in M.

Another way to define an infinite set is to say that *it is one which is equally numerous with a proper subset of it.*

Examples: There are as many even numbers as there are counting numbers. They are both infinite sets. There are as many odd numbers as there are counting numbers. They are both infinite sets. If you combine the infinitely many even numbers and the infinitely many odd numbers, you don't get two infinities, but just the one infinity of the counting numbers. Infinite sets can give rise to some rather strange circumstances.

Hilbert hotel problem

The following problem was made up by David Hilbert.

A hotel with infinitely many rooms (one for each counting number $1, 2, 3, ...$) is totally occupied. Every room has a guest in it; there are NO VACANCIES. A person comes into the lobby and wants to know if he can get a room for the night. The desk clerk says, "Yes we can accommodate you." How does he do it? He cannot go to the "end" of the set of rooms because there is no end. But there is a beginning. He can have the guest in Room 1 step outside and move to Room 2, while the guest in Room 2 moves to Room 3, etc.

In general, the guest in Room n moves to Room $(n + 1)$, this goes on forever. But now Room 1 is vacant and the new guest can move right in. Everyone has his or her own room again. "There's always room for one more" is literally true in this case. We have two infinite sets, A and B, in which the elements are in a one-to-one match.

$$
\begin{array}{ccccccc}
A & 1 & 2 & 3 & ... & n & ... \\
  & \updownarrow & \updownarrow & \updownarrow & \updownarrow & \updownarrow & \\
B & 2 & 3 & 4 & ... & n+1 & ...
\end{array}
$$

This says that the sets A and B have the same number of elements. You can see how this hotel can accommodate any finite number of new guests. There's always room for 10 more or a million more, and so forth. Is there always room for countably infinitely many more? YES. Move the current guests into the odd numbered rooms, and the new guests into the even numbered rooms. You can even arrange for the guests from three or four or any other number of full hotels to be accommodated in this one single hotel. This is not a paradox, but simply an efficient arrangement of an infinity of numbers.

Aristotelian logic

In addition to his Achilles *vs* Tortoise Paradox, Zeno had another paradox about a moving arrow that never actually moves. He claimed that if you look at the arrow at any one time it is in a given fixed position, and since this occurs at every moment in time, the arrow is always in a fixed position, hence it never moves. This idea seems silly to us now, and it also made no sense to the great philosopher, physicist, and logician, Aristotle (384-322 BCE).

Around 330 BCE, Aristotle wrote a book on physics in which he defined the concept of a space-time continuum. In one chapter, titled: *On Continuous Motion and Zeno's Paradoxes*, Aristotle argued that space and time could be infinitely subdivided and that a moving object continued to move throughout a given time interval, never stopping. Furthermore, with respect to Achilles and the Tortoise, he says that Zeno was in error in assuming that the time intervals always gave the tortoise another chance to get farther ahead, and this is because time is continuous so the slower runner will be overtaken in sufficient time. He made it very clear that Zeno's paradoxes were based on fallacies which assumed that either space or time was discontinuous. By-the-way, about 100 years after Aristotle, Archimedes (287-212 BCE), developed the ideas of continuity and limits to such an extent that he is often thought of as being the inventor of calculus.

But getting back to Aristotle, among his greatest legacies handed down to us was his treatment of *mathematical logic*; he had developed a complete theory of mathematical statements, which he called *premisses*. Today we use the word *premises*. In his book, *Analytica Priora*, Aristotle gave us rules on the use of concepts such as syllogisms, hypotheses, and *negations*, which he called *objections*. He showed how to use these concepts to construct mathematical proofs either by *direct* argument or *indirect* argument, *reductio ad absurdum*.

Universal and particular premises

One of the most important ideas he introduced, and devoted several pages to was the difference between universal and a particular premise. You can think of this as the difference between the terms, *all vs some*. An example he gave was the statement,

All men are animals.

This he called a universal premise, whereas, a particular premise concerning this same subject is

Some men are animals.

His idea was this: to state a denial of the *universal* premise you must make your objection be *particular* and to state a denial of the *particular* premise, you must make your objection be *universal*. This is done, in order to come out with two statements that will cover all possibilities. You must use the word *not* and switch the words *all* and *some*.

Thus, he asserted: If it is not true that "All men are animals," then it must be true that, "Some men are not animals." It would be *incorrect* to assert that "All men are not animals." Since this latter premise means that "No man is an animal." We cannot use the two universal premises, "All men are animals" and "No men are animals" as covering all possibilities since neither of these might be true because there is a middle statement, namely some men are animals and some are not.

Law of the excluded middle

His basic rule was the law of the *excluded middle*, which says that any one meaningful premise p such as "All men are animals," or its objection, "Some man is not an animal," and must be either true or false; there is no middle ground in which they are both true or both false. The law of the excluded middle is:

One of p or not p must be true and both may not be either true or false.

Contrapositive

Closely related to the law of the excluded middle is the law of *contraposition*. We will give examples of the rule before stating the rule.

If I jumped in the water, then I got wet
means the same thing as
If I did not get wet, then I did not jump in the water.

Keeping this example in mind, study the following statement:

If p then q means the same thing as If not q then not p.

This rule holds no matter what p says. It is called a context-free law; that is, you don't need to know what p actually says, in order say the law is valid. And this, my friend, is the difficulty in teaching abstract logic. When you tell a beginner to learn this generalized law, the abstractions get in the way of understanding.

You can round out the water jumper statement this way,

If I did not get wet, then I did not jump in the water,
because, otherwise, if I had, then I would have gotten wet.

This is somewhat the way Aristotle stated it. *If p then q implies that if not q then not p, otherwise q.* Chew on that a while.

Examples:

An objection to the assertion: *All roses are red* is the statement that *there is, at least one, non-red rose.*

Problem: Consider the following three statements:

$$S_0 = \text{All roses are red.}$$
$$S_1 = \text{Not all roses are red.}$$
$$S_2 = \text{All roses are not red.}$$

Question: Are the two statements S_1 and S_2 both legitimate objections to S_0?

Answer: No. S_1 is legitimate; it says, *there is a rose that is not red*. But S_2 says *you can't find a red rose.*

Philosophical controversy

Some philosophers will say that "S_1 and S_2 do mean the same thing; it is *only grammatical*. After all, these two statements S_1 and S_2 have exactly the same words, so is it just a question of where the word *not* appears." But, in this case, grammar is everything.

This can easily be proved by using a few illustrations, called *Venn Diagrams*, depicting *sets* of things. John Venn, an English logician introduced his diagrams in order to depict logical statements as a part of set theory in 1880. He wanted to settle such disputes about logic. Other mathematicians had also been using such diagrams prior to Venn, but he is given credit for organizing the work into a coherent theory.

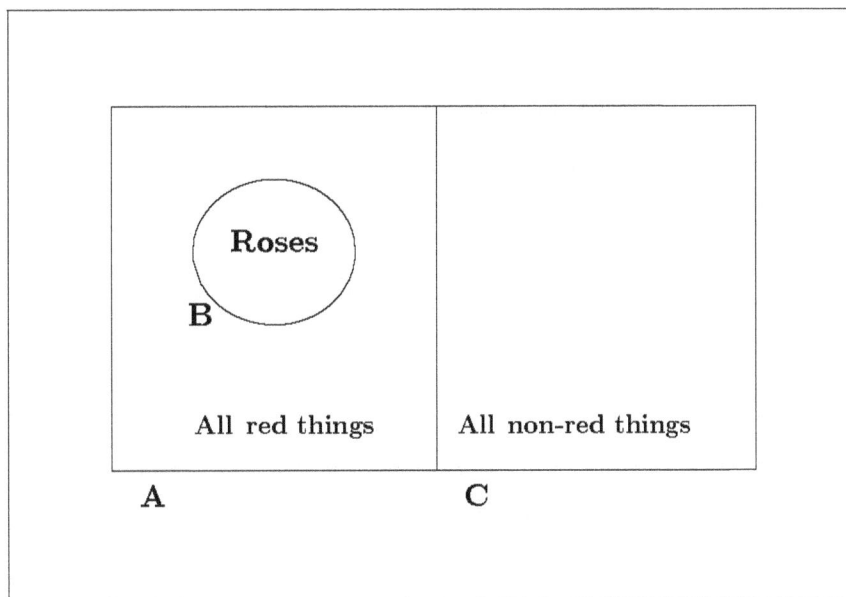

FIGURE 7.3, S_0 ALL ROSES ARE RED

Here is how we read Figure 7.3. All of the points in set A are *red things*, all of the points in set C (outside of set A) are *non-red things*, and all of the points inside of B are roses. We can see all roses are in the set of red things. This diagram depicts statement: S_0 "All roses are red."

Below in Figure 7.4, we see a Venn diagram of statement S_1 "Not all roses are red."
Which means the set of roses is not entirely within the set of red things.

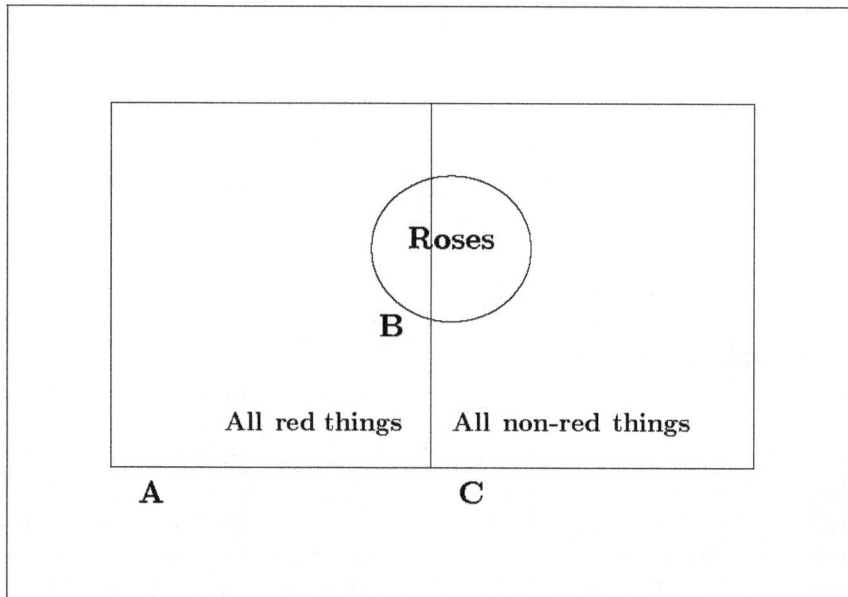

FIGURE 7.4, S_1 NOT ALL ROSES ARE RED

Figure 7.5, below, depicts the case where all roses are not red; that is, all roses are in
the set of "not red" things.

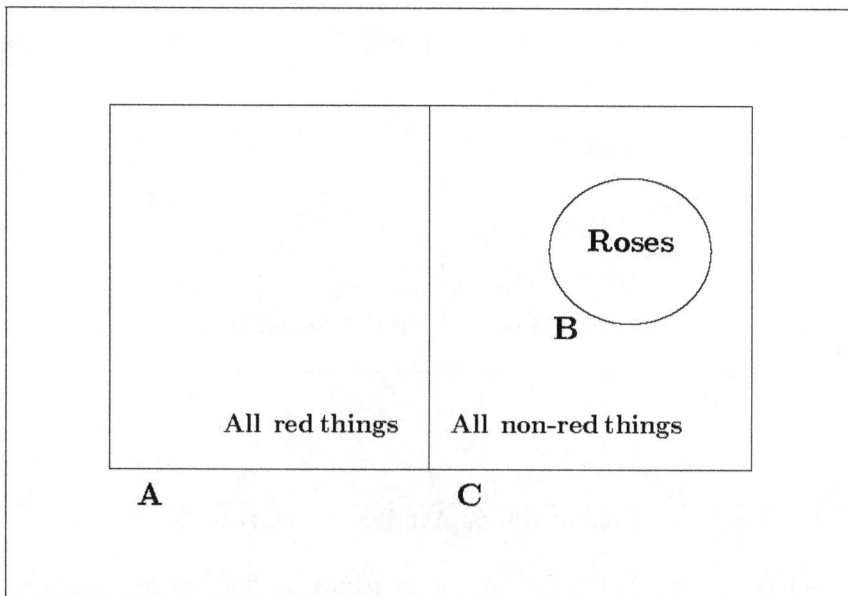

FIGURE 7.5, $S2$ ALL ROSES ARE NOT RED

Summary of the of the laws of logic

Today, here is the way we summarize Aristotle's' rules of logic.

1. If a premise is universal, its objection must be particular.
2. If a premise is particular, its objection must be universal.
3. A premise and its negation must be conjointly exhaustive and mutually exclusive.[1]

Example: (From Alice in Wonderland)

The mathematician and logician Charles Dodgson, wrote Alice in Wonderland under the pen name Lewis Carroll in 1864. He related the following conversation between Alice and a bird. Alice said that the bird was correct to say, "If you are a serpent, then you eat eggs.", but incorrect when the bird said, "if you eat eggs, then you are a serpent." Alice was quite right to correct the bird by saying:

> *If you are a serpent then you eat eggs* does not mean the same thing as *If you eat eggs, then you are a serpent,* and she continued, little girls eat eggs as well. To which the bird, unconvinced, illogically replied, "then you must be some kind of serpent."

This is an example of the logic law that the statement, "If p, then q," does not necessarily have the same truth value as its *converse*, "If q, then p."

We Want Proofs!

Mathematicians are somewhat like frustrated protestors:

"What do we want?

Proofs!

When do we want them?

Now!"

A mathematician and her husband were driving along a country road and he says to her, "Look, dear, those sheep have been shorn." She replies, "At least on this side." Why wasn't she, as a mathematician, willing to concede that the other side of each sheep was also clean shaven? It's simple; she had no *proof.* She would have needed to see the other side of each animal, or to have been apprised of an axiom which said that no sheep would be released into that pasture unless both sides had been shorn.

We will try to get a proof directly by deriving a true conclusion from the hypothesis or we get it indirectly by denying the conclusion and showing that the denial leads to a false statement.

Even in the field of statistics we employ a sort of indirect argument by the use of a device called the *null hypothesis.*

The null hypothesis in statistics

Because the field of probability and statistics deals with randomness and uncertainty, you might expect it not to require all of the trappings used in obtaining proofs; but there is one interesting case in which it is a perfect example of proofs by indirect arguments.

Suppose you want to use a sample of some population to determine whether or not a given *trait* is likely to be true about that population. You claim that the aforementioned trait *does not exist* in that population. This is your *null hypothesis,* a denial of the trait

---

[1] Conjointly exhaustive means cover all possible cases and mutually exclusive that they cannot both be true, or both be false.

in question. Then you examine the data you collected in your sample and ask: "How likely is it that I could have gotten a sample like this from a population *without* this trait?" Then after analyzing the data you might say, "Wow! Without this trait it would be very rare, in fact, less than $\frac{1}{2}$ of 1% of the time, (or some other previously chosen small percentage) that I could get a sample like this, so I reject the null hypothesis, and acknowledge that this trait is in the population."

You could be wrong in rejecting the hypothesis and that is called a Type I or a *false positive* error. Which means wrongly rejecting the negative. If, on the other hand, samples such as your's are common among populations without that trait you might be inclined to accept the null hypothesis. This could be an error too, called the Type II or *false negative* error, wrongly accepting the negative. Each such type of error has problems associated with it, such as the conclusion that a certain medication is not working when it really is or that it is working when it really isn't. The important thing is that you have new guidelines to help with your research or your analysis.

Fun with a logical Puzzle

Ms. Green, Ms. Black and Ms. Blue are out for a stroll together. One is wearing a green dress, one a black dress and one a blue dress. "Isn't it odd," says Ms. Blue, "that our dresses match our last names, but not one of us is wearing a dress that matches her own name?" "So what?" responded the lady in black.

Problem: What is the color of each lady's dress? Assume that Ms. Blue was not answering herself.

If you solved this problem you probably used the process of elimination, such as "Ms. Blue spoke, so she was not wearing a blue dress." Then you thought of some other possible contradiction, and so forth. This type of reasoning is a form of an indirect argument where you want to find the correct solution by ruling out wrong or contradictory ones. But suppose you had a logical system in which "The Law of the Excluded Middle" were not allowed? Then you would be stuck.

And this is exactly what happened in the 1960s; some mathematicians formulated a new branch of logic that required you to prove mathematical statements only by direct proofs.

The Constructivist Movement

In the 1960's a group of mathematicians started a movement that required that mathematical papers be restricted to presenting only those proofs that would provide the reader with a step by step algorithm for obtaining the conclusion. You could not establish the existence of some number simply by getting a contradiction to the assumption that such a number didn't exist. They assumed the other axioms of Archimedian logic, but not the excluded middle, which lets us say that $-(-S) = S$, the negation of (the negation of S) is S. The main proponent of this movement was the American mathematician, Erret Bishop, and his most influential work was his book, *Foundations of Constructive Analysis*, published in 1967.

While proofs obtained by constructive methods were rigorous and very sound, they did have some drawbacks. It is not that these mathematicians were wrong, they just insisted upon much more restrictive constraints. They would say that if a statement had not been proved true or false constructively, then it was not true or false.

For example, they would reject the *indirect* proof that $\sqrt{3}$ is irrational. They would expect you to show how to create a formula for a sequence of rational numbers $\frac{a_1}{b_1}, \frac{a_2}{b_{21}}, \frac{a_2}{b_3}, \frac{a_4}{b_3}, \ldots$ for each integer, n with $\left(\frac{a_n}{b_n}\right)^2 \neq 3$, and that the steps of such a proof must yield a formula in terms of n for all fractions, $\frac{a_n}{b_n}$. These types of proof are very difficult to find and were tediously long and involved. It would be like expecting a person to build a road with a pick and shovel instead of major heavy equipment.

Fortunately, we still continue to use the law of excluded middle and we have even computerized the process, teaching computers how to prove theorems.

Logical terms: and, or, not

We humans are intrinsically logical and naturally know how to decipher complicated statements. For example, we know that if a person tells you "I am going to the bank and the grocery store today," for that to be true we expect that they will go to *both* places at some time during the day. But if they say "I am going to the bank or the grocery store today," we know that they will be at one of these places, but we would not call them a liar if they went to both places. In other words, when you use the word *and*, you want both parts to be true, but when you use the word *or*, then the combined statement is true if *either* part or *both* parts are true.

Suppose you said you were going to both places, but you did not go to the bank, then your *and* statement is false.

We can formalize these concepts as follows: Let T stand for a true statement and F stand for a false statement, then

$$(F \ and \ T) \ \text{is} \ F$$

If you went to the bank, but did not go to the grocery store your combined statement $(T \ and \ F)$ is F. And if you did not go to either the bank or the grocery store, your combined statement is also false; that is, $(F \ and \ F)$ *is* F. The only the combined *and* statement that could be true is that you went to both the bank and the grocery store. $(T \ and \ T)$ *is* T. The *or* statement, on the other hand, is true if either part (or both parts) are true and can only be false if both parts are false.

Teaching computers to be logical

These concepts are natural to humans, and in a new field of computer science, Artificial Intelligence, (AI). Computers are taught to solve logical problems. If we want to use a computer this way, we must teach it how to discern when statements combining *and*, *or*, and *not* are true or false. This is done with "logic circuits" where we can turn on certain switches (allowing current to flow) and turn off others. We can actually build logical circuits that allows a computer to re-interpret **every** logical instruction into *and*, *or*, and *not* statements. In modern computers and calculators the circuitry is on a microchip. But we will illustrate this concept here by showing you some wiring diagrams depicting the flowing or non-flowing of electricity at various switches (or "gates").

The computer chips are actually tiny wiring diagrams, and the first ones were made by photographically reducing hand-drawn pictures of circuits. Credit for the invention, in 1959, of this type of technology goes to Jack Kilby of Texas Instruments and Robert Noyce of Fairchild Semiconductor Corporation.

Notation for logic statements

First, let us introduce the mathematical notation for *and*, *or* and *not*.

> The symbol $\wedge$ is *and*. $p \wedge q$ reads *p and q*.
> The symbol $\vee$ is *or*. $p \vee q$ reads *p or q*.
> The symbol $\neg$ is *not*. $\neg p$ reads *not p*.

Any *and* statement $p \wedge q$ is a connection between two statements, p, q each of which can either be TRUE or FALSE giving us four possible combinations. Likewise for any "or statement," $p \vee q$.

Logic circuits

The wiring diagrams below are called *logic circuits*. We will diagram the *and*, *or*, and *not* electrical circuits to show how a computer uses combinations of *True* and *False*. These involve *gates* or switches than can be placed in an *on* (the gate is closed) position or an *off* (the gate is open) position. When the switch is ON, the gate represents T and when the switch is OFF, the gate represents F.

AND Circuit

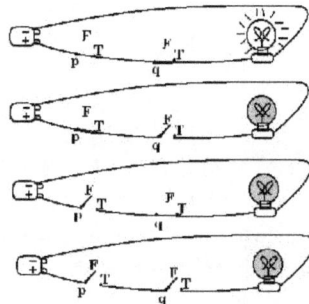

FIGURE 7.6 THE AND LOGICAL CIRCUIT

The Figure 7.6, above, the AND logical circuit, shows that the only way $p \wedge q$ can be true is that both the p gate and the q gate are closed, allowing electricity to flow.

AND Table   Pedagogically, this diagram is expressed as a table. Thus, the AND table for $p \wedge q$ is:

p	q	$p \wedge q$
T	T	T
T	F	F
F	T	F
F	F	F

AND table

What this table is saying, row by row, is that:

If p is true and q is true then the combined statement $p \wedge q$ is true.
If p is true and q is false then the combined statement $p \wedge q$ is false.
If p is false and q is true then the combined statement $p \wedge q$ is false
If p is false and q is false then the combined statement $p \wedge q$ is false.

OR Circuit

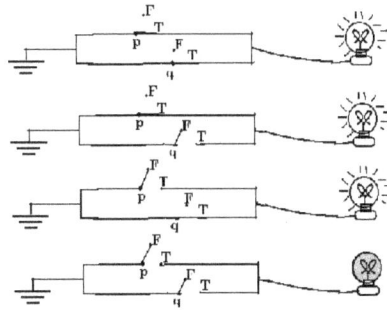

Figure 7.7 THE OR LOGICAL CIRCUIT

In Figure 7.7, above, the only way $p \vee q$ can be false is if both p is false and q is false, because if either one or the other, (or both) are closed the electricity is allowed to flow.

OR Table   In the four cases for the *or* table $(p \vee q)$, is:

p	q	$p \vee q$
T	T	T
T	F	T
F	T	T
F	F	F

OR Table

We can see that a combined statement p or q would be true if either p is true or q is true or both are true and it can only be false if both p and q were false.

NOT Circuit

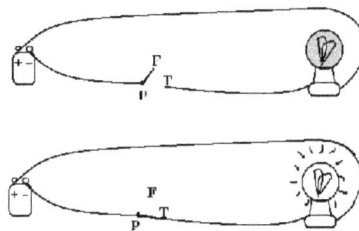

FIGURE 7.8 THE NOT CIRCUIT

In Figure 7.8, above, when p is True, and the switch is open, then the circuit is in the $\neg p$ state, no current flows. When p is False, and the switch is closed then, in the $\neg p$ state,the current flows.

NOT Table   Here is the NOT table for a single statement p that could be either true or false.

p	$\neg p$
T	F
F	T

NOT Table

In this table $\neg p$ is false only if p is true and $\neg p$ is true only if p is false.

COMBINED Tables

We will not attempt to draw the diagrams for combinations of the *and*, *not* and *or* circuits, but we can easily show you the tables for such combined circuits. The first and probably most used one is the table for *implication*, if p then q It is called the IMPLIES TABLE.

IMPLIES Table   Let us introduce one more symbol, $\Longrightarrow$, which reads, "implies". The statement $p \Longrightarrow q$, p *implies* q stands for *If p then q.*
   We can use a truth table to show that are three different equivalent versions of this *if p then q* statement. Here is a combined truth table for the three statements:

Implication p *implies* q $(p \Longrightarrow q)$

Contrapositive *not q implies not p* $(\neg q \Longrightarrow \neg p)$

Disjunctive Negation *either q or not p* $(q \vee \neg p)$

p	q	$p \Longrightarrow q$	$\neg q \Longrightarrow \neg p$	$q \vee \neg p$
T	T	T	T	T
T	F	F	F	F
F	T	T	T	T
F	F	T	T	T

The IMPLICATION truth table

   What this Implication truth table is telling you is something really remarkable.   It is saying that, for any two statements p and q, regardless of their content and truth, the following three statements are all "equivalent" $\rightleftarrows$ to each other. That is, one is true, if and only if the other is true.

$$(p \Longrightarrow q) \rightleftarrows (p \Longrightarrow q)$$
$$(\neg q \Longrightarrow \neg p) \rightleftarrows (p \Longrightarrow q)$$
$$(q \vee \neg p) \rightleftarrows (p \Longrightarrow q)$$

Examples: Jumping in the water implies that I get wet.

1. If I jump in the water, then I get wet.

2. If I didn't get wet, then I didn't jump in the water.

3. Either I got wet or I didn't jump in the water.

If you look at the implication table, you will see that all three columns $p \implies q$, $\neg q \implies p$, and $q \vee \neg p$ are the same. That is, any time one of these is false, so are the other two and any time one is true, so are the other two.

Note: There is one thing that makes the disjunctive negation important, and that is, we can use it to trick a computer into evaluating "If p then q" statements. This is because we can construct *not* and *or* circuits that allow computers to evaluate statements such as "q or not p". Voila! It can evaluate "If p then q" because it is the same thing as "q or not p."

Vacuously true statements

You will notice, in the p *implies* q table, that some strange looking results occur. For example, in the $p \implies q$ column we claim that the statement is true when p is false and q is either true or false. That is "If F then T" is a true statement and "If F then F" is a true statement. What's up with that?

Examples

If the moon is made of green cheese, then apples are oranges. This is known as a "vacuously true" statement. It is true because the moon is not made of green cheese, so assuming that it is lets you draw any conclusion.

In some small room in my house which doesn't have any horses in it, I can make this statement: "If h is a horse in this room, then h has pink ears." This is a true statement because there are no horses in the room *which* do *not* have pink ears, since there is no horse in this room. The word *which* refers to the empty set, $\emptyset$ (sometimes called "phi") of such horses, because there are no such horses. The *which* has no referent. It is called the *irreferent which*.

Another example of the *irreferent which* is: *There is no integer* which *is greater than every other integer.*

Here is a poem that my wife wrote about this mathematical situation. When I told her about this concept she must have thought I said "an irreverent witch." This is what she wrote.

The Irreferent Which (The Poem)

From: *Slices of Life*, a collection of poems by Jean Falbo.

THE IRREVERENT WITCH

An irascible old Witch
Entered my bedroom with a swish
Proclaiming
Every horse in this room has pink ears!

Whaaa? I sleepily bleated
There aren't any horses in here

Ha! She snerkled and snorted! Ha!
Vacuously true!
Vacuously true!
And she gave my love handles a poke
With a bony finger and an unmanicured nail, as she spoke

I rested my head on my pillow and I thought and I pensed
And I pensed and I thought
Only by logic can I banish this freak–and I was set to do it
But, inside I had an empty feeling I'd rue it
But, still I'd try to do it.
There, on my dresser, in my jewelry box is a horse of gold

With wings and mane and unicorn horn
That will prove you false, as sure as I'm born,
I shouted, as I leaped from my bed

That's it!
That's it!
You irreferent twit, I said

I'm outta here semper Phi
Null and void, she angrily retorted
And away she sped.

This poem and the concept of *vacuously true* statements are based upon the Aristotelean Logic, requiring every premise to have either a value of True or False.   But, in modern mathematics, there are some *non-Aristotelian* logics, ones in which there are more than two possible truth values.

Non-Aristotelian, multi-valued logic

During the years from 1950 to 2000, mathematicians invented logical systems that had more than just the two truth values "True" and "False."

Some of these systems helped to develop modern applications in computer computations; for example, a *three-valued logic* can be implemented by having a switch that can be put into one of *three* different positions, closing and opening electrical circuits over three different routes.   *Four-valued logic* has been used to increase the storage capacities of memory chips.   Various applications have been found for other finitely many valued logics.

Others require that statements take on infinitely many values, say 1 for True, 0 for False and some fraction x between 0 and 1 for any one of infinitely many different possibilities.   These systems are useful in developing probability calculations.   They help in making predictions in weather, medical research, business applications and political polls.

Kleene logic

Here is an introduction to one of the three valued logics. It is interesting to compare the truth tables for this logic to those of Aristotelian logic. This was originally published by the American logician, Stephen C. Kleene in 1952. Let any premise have one of three different values, T, U, F, standing for *True, Unknown* and *False*. The *Not Table* is as follows

p	$\neg p$
T	F
U	U
F	T

Kleene's NOT

As can be seen, above, if a statement is not true it is false, if you negate an unknown, it is still unknown, the negative of a false statement is true. Now, here are the AND, $\wedge$, and OR, $\vee$, Tables in Kleene's logic.

p	q	$p \wedge q$	p	q	$p \vee q$
T	T	T	T	T	T
T	U	U	T	U	T
T	F	F	T	F	T
U	T	U	U	T	T
U	U	U	U	U	U
U	F	F	U	F	U
F	T	F	F	T	T
F	U	F	F	U	U
F	F	F	F	F	F

Kleene's AND and OR Tables

In the AND, $\wedge$, table, F and anything is F. While all three of: $U \wedge U$, $U \wedge T$, and $T \wedge U$ are U. The only AND combination that produces T is $T \wedge T$, just as with Aristotelian logic.

In the OR table, the only way an OR statement can be false is that both p and q must be false; in all other cases we get either T or U.

In his logical system Kleene defines the implication (if then) statement to be the disjunctive negation.

$$p \implies q, \text{ means } q \vee \neg p$$

If you set up the truth table for $p \implies q$ you will see that it is compatible with Aristotelian logic. This shows that Kleene's logic is actually a generalization of Aristotelian logic, because if you omit the U possibility in all tables, you get Aristotelian. An interesting exercise is to state the denial of $p \wedge q$ in Aristotelian and compare it to the denial of $p \wedge q$ in Kleenean logic.

Hint: In regular (Aristotelian) logic, you can write the following theorem:

$$\neg(p \wedge q) = \neg p \vee \neg q$$

Have fun comparing this to $\neg(p \wedge q)$ in Kleene's logic, by using the Kleene tables for *and*, *or*, and *not*.

From logic to axioms

Aristotle's logic of 330 BCE blossomed into Euclid's geometry of 280 BCE, when Euclid published his 13 volume work, *Elements of Geometry*. What Euclid did was to organize all known geometry into a coherent collection of definitions, axioms and theorems and showed that logic could be used to prove the theorems. This began the idea that a mathematical system could be arranged in such a way that you could derive true statements in an axiomatic system.

CHAPTER 8

GEOMETRY–THE WELLSPRING OF MATHEMATICS

Geometry in prehistoric times

So far in this book we have discussed many modern topics such as group theory, real numbers, vectors, imaginary numbers, infinite series, number theory, and logic. We have been impressed with their "miraculous" just-in-time (or even a little ahead of time) appearance to solve new problems. But where did all this stuff come from? There can be no doubt that all of today's mathematics emerged from *language*, *numbers*, and *geometry*.

Speaking, counting, and measuring are common ancestors of the abstract mathematics we know and use today. It is clear that we were communicating, either by sign language or verbally before we were counting or measuring things, and it is likely that both numbers and measuring appeared simultaneously as we attempted to solve problems that allowed us to survive in dealing with nature and each other. There is a nexus between numerical and geometric concepts. Numbers can arise from geometry, measuring the sides of triangles, for example.

We can imagine that the first geometry consisted of starting from where you were and going to some place you wanted to be, walking prudently in as straight a line as possible, and as safely as experience has taught you to do. I can picture someone in a hunting party observing, "The path over there is the best way to avoid the alligators." That was probably the first definition of a straight line.

Suppose one of the early humans collected little sayings about walking around in the jungles. "If you always turn to the right at a 90° angle after walking so many steps, and you do this three times the same number of steps, you will get back to home." OK, so how did you know to actually turn 90 degrees? Maybe you had a little piece of a rock with angle markings on it, such as one of those found in 2002 dating back to 75,000 BCE in the Blombos Cave in South Africa. We are ignorant of the meanings of these marks, but from other cave drawings, we know that the early humans depicted circles and square as well as animal figures. I speculate that a square could serve as a map for stone-agers to figure out what direction was perpendicular to the current path.

Later, in 13,000 BCE, primitive geometric designs were painted in caves in Lascaux France, giving us more evidence that geometry has been around for a long long time. The

Babylonians and Egyptians had fully developed many theorems in geometry by about 3000 BCE.

Formal geometry as studied in Egypt paralleled the construction of roads, buildings and cities. To get precise right angles in building the pyramids and other structures, the carpenters, called the "rope stretchers," used a tool that consisted of a rope with twelve knots equally spaced. See Figure 8.1. Three workers could stretch the rope in such a way that it created a triangle that had one side of length 3 another of length 4 and the hypotenuse of length 5, thus creating a right triangle. This was a very early use of the Pythagorean theorem, 2500 years before Pythagoras.

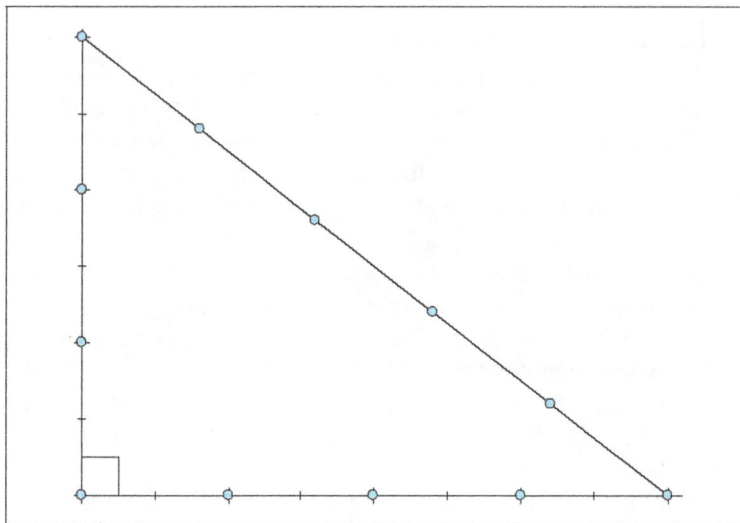

FIGURE 8.1 THE STRETCHED ROPE

Mathematics in ancient Greece

Greek Mathematics flourished from about 600 BCE to the beginning of the Common Era: Pythagoras in 540, Plato in 400, Aristotle in 340, Euclid in 300, and Archimedes and Apollonius in 200 BCE.

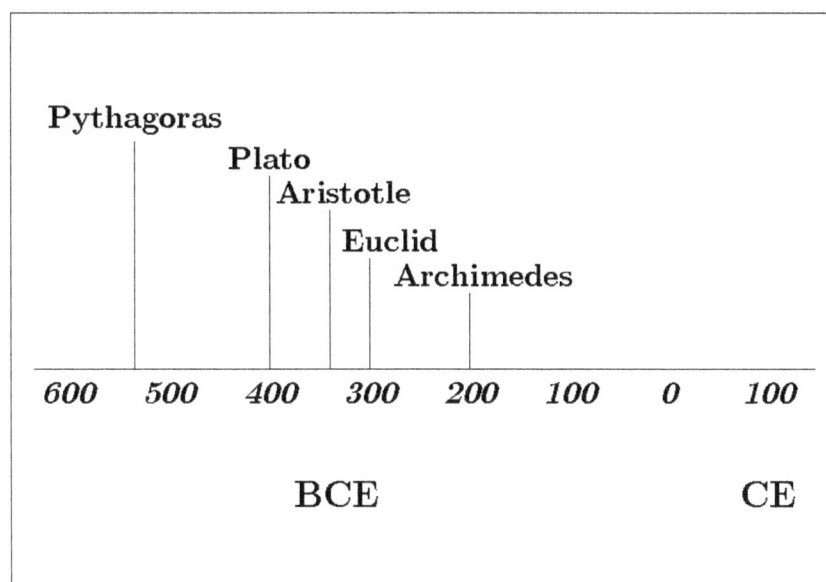

Pythagoras
Plato
Aristotle
Euclid
Archimedes

| *600* | *500* | *400* | *300* | *200* | *100* | *0* | *100* |

BCE   **CE**

GREEK MATHEMATICS BEFORE THE COMMON ERA

Plato's Academy was an important part of the development of Greek Mathematics. Plato, a student of Socrates, founded his Academy in 387 BCE and it ran for 40 years until his death. According to the mathematical historian Howard Eves:

> *Almost all of the important work of the fourth century BC was done by friends or pupils of Plato, making the Academy the link between the mathematics of the earlier Pythagoreans and that of the later, long-lived school of mathematics at Alexandria. Plato's influence on mathematics was not due to any mathematical discoveries he made, but rather to his enthusiastic conviction that the study of mathematics furnished the finest training for the mind, and hence was essential for the cultivation of philosophers and those who should govern his ideal state. This explains the renowned motto over the door of his Academy:* <u>Let no one unversed in geometry enter here.</u>

(Howard Eves: An Introduction to the History of Mathematics, Fifth Edition, Saunders Publishing, 1983)

In any case, there were hundreds of geometric "facts" known to humans for thousands of years before Euclid, but they remained a loose collection rather than a comprehensive body of knowledge. What did Euclid do that was so great? He axiomatized the subject around 280 *BCE*. What does that mean? He organized all the theorems into a sequential order: Theorem number 1, Theorem number 2, and so forth. No theorem was allowed to be proved by use of a higher numbered theorem.

To do this, he had to keep moving back to the most fundamental theorems and ended up with theorems for which he could not find any other theorems upon which to base his proof. At this point he had to assume some statements as true and which *could not* be proved. These unprovable statements he called *postulates* (these days, we call them *axioms*.) He also he had to *invent* terms and definitions in order to efficiently state the theorems.

One modern day criticism of Euclid's work is that he did not acknowledge the fact that you can't define every term. For example, when you say that ,"A point is that which has no part," you really haven't defined a point unless you say what *part* means. Everything can't be defined in terms of something else because your definitions will never end or they will circle back to previously defined terms. Today we avoid this endless or circular reasoning, by designating certain words, like *point*, as *undefined* and accept any meaning that you want to assign to them, as long as your meaning still makes sense, when you make a mathematical statement about them.

Euclid's work consisted of 13 books. At the start of the first book he had a mixture of about 25 undefined and defined terms, possibly with some unconscious assumptions thrown in, and only five clearly designated axioms (*Postulates*). From these, all the theorems (which he called *Propositions*) of plane geometry were proved.

We want to give the reader a flavor of Euclid by presenting the first few of his actual definitions, common notion, postulates, and propositions.

(Adapted from: *The Thirteen Books of Euclid's Elements* by Sir Thomas L. Heath, Dover Publications, 1956.)

DEFINITIONS

1. A point is that which has no part.

2. A line is breadthless length. This is, in modern language, a line segment.

3. The extremities of a line are points.

4. A straight line is a line which lies evenly with the points on itself.

5. A surface is that which has length and breadth only.

6. The extremities of a surface are lines.

7. A plane surface is a surface which lies evenly with the straight lines on itself.

8. A plane angle is an inclination to one another of two lines in a plane which meet one another and do not lie in a straight line.

9. When the lines containing the angle are straight, the angle is called rectilinear.

10. When a straight line set upon a straight line makes adjacent angles equal to one another, each of the equal angles is right and the straight line standing on the other is called perpendicular to that on which it stands.

11. An obtuse angle is an angle greater than a right angle.

12. An acute angle is an angle less than a right angle.

POSTULATES

Let the following be postulated:

1. To draw a straight line from any point to any point.

2. To produce a finite straight line continuously in a straight line.

3. To describe a circle with any center and any distance.

4. That all right angles are equal to one another.

5. That, if a straight line falling on two straight lines make the interior angles on the same side less than two right angles, the two straight lines if produced indefinitely, meet on that side on which are the angles less than the two right angles.

COMMON NOTIONS

1. Things which are equal to the same thing are equal to each other.

2. If equal be added to equals, the wholes are equal.

3. If equals be subtracted from equals, the remainders are equal.

4. Things which coincide with one another are equal to one another.

5. The whole is greater than the part.

PROPOSITIONS

1. *On a given finite straight line to construct an equilateral triangle.*

2. *To place at a given point (as an extremity) a straight line equal to a given straight line.*

3. *Given two unequal straight lines, to cut off from the greater a straight line equal to the less.*

Notes:

Proposition 1 says that if you are given a line segment, XY, you can make a triangle with all sides equal to length of XY.

Proposition 2 says that given a point A and a line segment XY, you can "move" XY and make A one of its end points.

Proposition 3 says that if AB is longer than XY, you can find a point C on AB such that AC is equal to XY.

The fifth postulate

From this simple beginning, there follows all the propositions of plane geometry. The first twenty-eight propositions can be proved, and were proved, using only the first four postulates. These twenty-eight theorems have been called *neutral geometry,* not requiring the fifth postulate for their proofs. After Proposition 28, the theorems start to require the fifth postulate, or other theorems requiring the fifth postulate, as part of their proofs.

Book 1 Propositions 29, 30, 31, and 32.

29. *A straight line falling on two parallel straight lines makes the alternate angles equal to one another, the exterior angle equal to the interior and opposite angle, and the interior angles on the same side equal to two right angles.*

30. *Straight lines parallel to the same straight lines are also parallel to one another.*

31. *Through a given point a straight line parallel to a given straight line can be constructed.*

32. *In any triangle, if one of the sides be produced, the exterior angle is equal to the two interior and opposite angles, and the three interior angles of the triangle are equal to two right angles.*

Euclid's proof of Proposition 29 uses postulate 5 and his proof of 30 uses proposition 29 (and, hence the fifth postulate), but his proof of 31 does not explicitly cite propositions 29, nor 30 nor the fifth postulate. The thing about Proposition 31 is that, it says there is *a* parallel line through a given external point, but it does not say that there is *only one* parallel. However, in proposition 32 and subsequently Euclid makes use, without justification, of 31 *as if the construction had yielded a unique parallel.* This is tantamount to his having assumed the following statement which is equivalent to the fifth postulate.

The Playfair axiom: Named for a Scottish mathematician John Playfair, 1795, and equivalent to the Fifth postulate.

If CD is a line and P is a point not on CD, then there is one and only one line AB containing the point P that is parallel to CD.

The Proclus axiom: Another, less wordy axiom, equivalent to the fifth postulate was stated in 400 CE by Proclus Diadochus

If a line intersects one of two parallel lines, then it intersects the other also.

Historians have noted that the neutral geometry theorems were deliberately listed first; as if Euclid was trying to avoid using the fifth postulate, thinking it could, somehow, be derived as a proposition itself. Further evidence that Euclid may have had some trepidations about the fifth postulate was the fact that it was extremely wordy and awkwardly phrased. The above two theorems, Playfair and Proclus, equivalent to and much simpler than the fifth, could have been stated instead of the fifth postulate.

Perhaps Euclid was trying to tell the world, "Look at the fifth postulate, see what you can do about it." Picking up on Euclid's clues, many mathematicians tried to prove the fifth postulate as a theorem. They either gave up after failing to do so, or they acknowledged that they needed to assume another statement to complete the proof. In some cases, they subconsciously assumed other statements which turned out to be equivalent to Euclid's fifth postulate.

Saccheri quadrilateral

Finally, in 1733 AD, one mathematician, *Girolamo Saccheri*, assumed the first four postulates and a <u>denial</u> of the fifth postulate, hoping to get a contradiction of *something*.

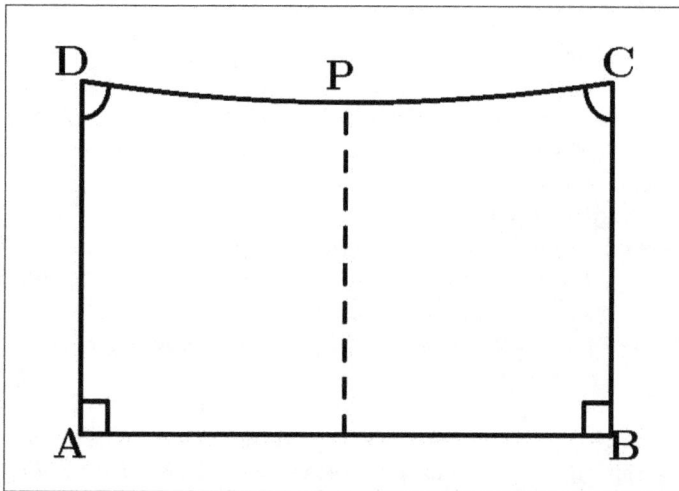

FIGURE 8.2 SACCHERI QUADRILATERAL

He derived all of the first 28 theorems, plus several other statements which were true assuming the denial of the fifth postulate but contradicted results using the Euclidean fifth postulate.

An example is the Saccheri quadrilateral, see Figure 8.2, the two sides AD and BC are congruent to one another, the two angles at A and B are right angles, and the two angles at C and D are equal to one another. This can be constructed with just the first four postulates of Euclid. If the fifth postulate of Euclid is assumed, then you can prove that the angles at C and D are right. If, however, you assume the denial of the fifth postulate then you can prove that the angles at C and D *must be acute.*

Nevertheless, he could not get a contradiction of any of the neutral geometry theorems. What he had actually done, but never realized, was to discover several theorems of a new geometric system–a Non-Euclidean Geometry. In his failure to recognize that he had discovered this new geometry, he made a feeble and transparently false claim that he had proven the fifth postulate. His work was not acknowledged as being a new contribution to mathematics at that time.

About 100 years after Saccheri, two mathematicians, the Hungarian Janos Bolyai, and the Russian Nicolai Lobachevski, working independently, and ignorant of Saccheri's work, each announced that they had discovered a geometry in which the Euclidean fifth postulate was false. When they published their works, the great German mathematician Carl Frederick Gauss, responded that, prior to 1830, he had discovered this same result, but did not announce it. Remember his motto was "few but ripe," so he held back on some of his discoveries and did not publish them until he felt they had been fully developed.

Non-Euclidean geometry
Lobachevskian geometry

So, in 1733, poor old Saccheri gave up too soon by deciding that he had a contradiction arising from the denial of the fifth postulate. But, in 1830, Lobachevky, and in 1832, Bolyai, both tried the same idea. Only this time, each one announced that a denial of the Euclidean fifth postulate was not inconsistent with the first four axioms, and that you could have a valid geometry which satisfied the first four axioms of Euclid *plus* the denial of his fifth postulate. Thus, the Euclidean fifth postulate is neither provable nor disprovable. This meant, of course, that we had two different "plane" geometries, Euclidean and non-Euclidean.

One interesting theorem in neutral geometry is that in any triangle, the sum of the angles is *less than or equal to* two right angles. This theorem can be proved by using only the first four axioms. If you assume the Euclidean fifth postulate, you can prove that the "equal to" part is true but not the "less than" part. If, instead, you assume the Lobachevskian fifth postulate, you can prove the "less than" part is true, but not the "equal to" part. Actually, this theorem follows from the Saccheri quadrilateral.

"But really!" you must be asking, "Is this other geometry a physically solid thing, or is it just a figment of a fertile imagination?" We answer that it is no less real than imagined straight lines of Euclidean geometry, how real is length without breadth? Even today, when physicists discuss the "shape" of the universe, they are willing to entertain models, that fit non- Euclidean as well as others that are Euclidean. There is a sense that tells us that geometry is local, when we are on a playing field, it is as flat as a pancake,

Euclidean Geometry. When we are on certain street intersections in San Francisco, the universe is saddle shaped (Lobachevskian). The streets go downhill in both the North and South directions and uphill in both the East and West directions. When we are on a mountain top or in an aircraft we perceive ourselves as being in a spherical universe.

In fact, in the following example we see a model of Lobachevskian geometry made up of pieces of a Euclidean geometry and if this non-Euclidean space is not valid then neither is Euclidean Space.

Felix Klein's non-Euclidean model

To illustrate non-Euclidean geometry let us introduce the following small geometric system first published in the 1870s by a German mathematician, Felix Klein. This model consists of points, line segments and a circle, all of which are in the Euclidean plane. It is said to be "embedded in" the Euclidean plane.

Just imagine that the entire universe is a flat circular disk, say a dinner plate. Call it the disk-universe. If you draw a line straight across the plate then that is a line in your universe. See the line AB or the line XY or the line CD in Figure 8.3 below.

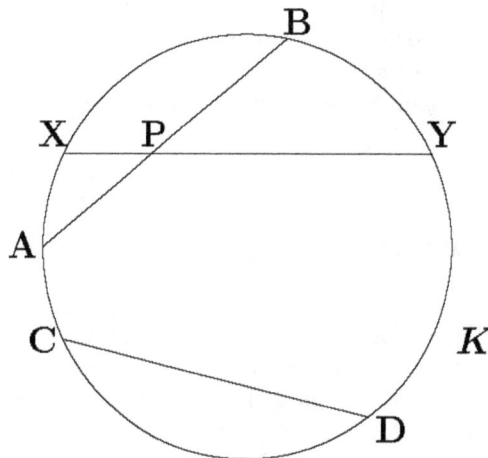

FIGURE 8.3 KLEIN MODEL OF NON-EUCLIDEAN SPACE

Now, what are parallel lines in this disk universe? They are lines that have no point in common. Do not think of them as lines that are some fixed distance apart like rails in a railroad track. They are lines that end at the boundary of the disk before they get a chance to intersect each other. In Figure 8.3 the lines AB and CD are parallel to each other (as are XY and CD) because they do not cross each other anywhere in this disk universe. On the other hand, the lines XY and AB are not parallel.

This is the way that Klein defined his Lobachevskian model. He started by saying let $\mathcal{K}$ be circular disk, meaning all of the points inside of the circle, (the German word for circle is Kreis). A line is defined as a segment drawn from one point on the circular boundary to another such point, not including the endpoints.[1] The points and lines in $\mathcal{K}$

---

[1]Measurements of length in this model are done with a "yardstick" that "shrinks" as you move

satisfy the first four postulates of Euclidean geometry, but not the fifth postulate. This model violates the Playfair axiom, for example, XY and AB are two lines that intersect in a point P, but both are parallel to the line CD. You can prove that all other neutral theorems are true in this model.

Riemannian geometry

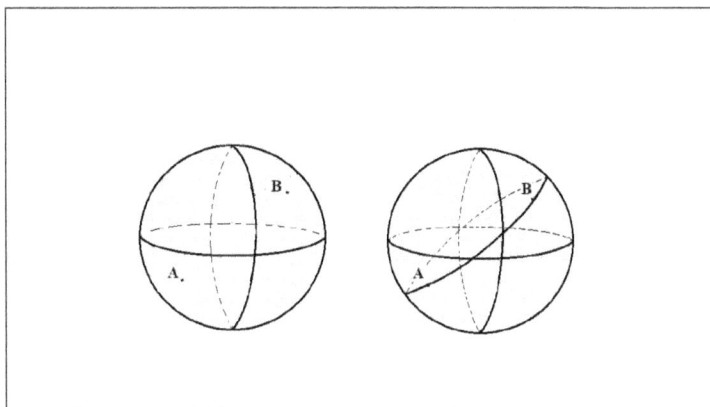

FIGURE 8.4 RIEMANNIAN GEOMETRY

To complicate matters, even further, there is another type of non-Euclidean geometry which can be obtained by replacing the fifth postulate with the assumption that there are no lines parallel to a given line through a given point, not on that line. It was invented by Bernard Riemann in 1854.

Postulates of Riemannian Geometry

To get Riemannian Geometry, we replace the first and fifth postulates of Euclidean Geometry, as follows.

Instead of saying two points determine a line, we postulate that *every pair of lines have two points in common*. And instead of saying that for any given line, l, and any point, P, not on l, there exists one and only line k through P that is parallel to l, we postulate that there are no parallel lines to l.

In this geometry, we can prove that in every triangle the sum of the angles is $> 180°$, a violation of a theorem in neutral geometry. The model for Riemannian Geometry is the surface of a sphere. A *straight line* is a "great circle," that is, any circle on the sphere whose center is at the center of the sphere. See Figure 8.4. All of the longitudes are great circles, and the *only* latitude that is a great circle is the *equator*.

If we take the Earth as a perfect sphere, then an airplane flying the shortest distance between two cities is travelling a *great circle route*. For any two points on the sphere there is exaclty one great circle through them and the shortest distance between them is along the shorter arc of this great circle. Any two great circles meet at two different points, poles apart, meaning that these points are the end points of the diameter of the circle.

You draw a great circle on the surface of the sphere from A to B to find the shortest distance from A to B. In any geometry the shortest distance between two points is called

---

towards the boundary of K. This allows you to extend a straight line indefinitely as required by the second postulate.

a *geodesic* path for that geometry. The great circles are the geodesics on a sphere. Any standard unit of measurement along a geodesic is called a *metric* for that geometry.

How non-Euclidean geometry changed mathematics

Two other specialized geometries, were created in the 20th century, Minkowsky Geometry and fractal geometry. These are actually *applications* of modern theoretical discoveries.

In 1907, the German mathematician Hermann Minkowsky, introduced a metric in Euclidean space-time geometry that proved to be an important concept later used when his famous student, Albert Einstein, developed the theory of relativity. Minkowsky helped to explain how acceleration of an object toward the speed of light affects the length of the object.

In the 1980's, Polish-born mathematician, Benoit Mandlebrot invented a geometry that can be used to represent erratic phenomena such as cloud formation, turbulence, weather patterns, and rugged geographic terrain as well as chaotic behavior.

But neither of these had the effect of the inventions of Lobachevskian and Riemannian geometries, which caused a revolution in science, especially in our understanding of how the universe is structured. And in mathematics, these discoveries went far beyond spatial considerations. As we have already seen, modern mathematical thought encompasses not only new geometries, but new logic, abstract algebra, topological spaces (so-called "rubber sheet geometry") and mathematical break through in analysis and number theory.

Why is this true? Imagine this: Once we realized that the axioms of mathematics are not intuitive self-evident truths, but are actually only arbitrarily postulated statements, then a whole new, hitherto unknown, grand universe blooms like a supernova! We can pin point the date as being around 1830 and the two seminal ideas that so radically revamped mathematical thought. These were:

1. The invention of Group Theory
2. The discovery of non-Euclidean Geometry

This would make the early 1800's the dawn of modern mathematics.

But what happened to Euclidean Geometry? Is it still good? It turns out that while the non-Euclidean geometries were tempting siren calls taking us away from Euclid, and leading us to new mountain tops, Euclidean geometry kept sailing along creating in its own wake, new fields of algebra, analytic geometry, calculus, Newtonian physics, and modern engineering.

Geometric algebra in Euclid's books

Thus, the existence of these new geometries did not obviate our dependence on good old Euclidean Geometry. The ordinary 3-dimensional space that we intuitively perceive in a small region all around us is Euclidean space. We measure distances by the Pythagorean theorem and we hit golf balls into parabolic paths. We accurately send vehicles into space and satellites into orbits.

To see the theoretical connection between Euclid's work and what is happening today, we only need to look at the end of Book I and the beginning of Book II and the first few propositions of Book VI (of his thirteen books). He introduced *geometric arithmetic*, which was a way to add, subtract, multiply and divide numbers by adding, subtracting, multiplying and dividing "lines," that is, actually, line segments.

Propositions 4, 5, 6, 7, and 8 of Book VI uses similar triangles to get algebraic calculations in terms of proportional sides. By introducing a segment of unit length, Euclid was able to do numerical problems by geometric methods. An interesting example, is the construction of a line segment that is the quotient of two given line segments.

Problem: *To divide a given line by another given line.*

Solution:

Let AC and CD be two line segments, with lengths a and b, respectively. Let AB be a line segment of unit length. That is, AB is of length 1 unit in whatever units (inches, feet, etc.) are used to measure AC and CD. See Figure 8.5. We want to find a line segment of length b/a.

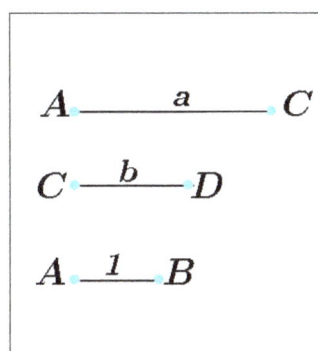

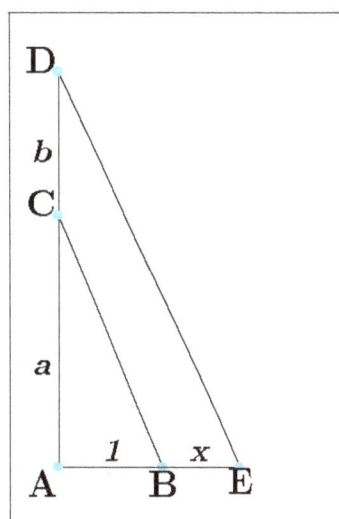

FIGURE 8.5   FIGURE 8.6

Line up AC with CD and make AB be perpendicular to AD at A. Through D draw a line parallel to CB and intersecting the line AB at E.

In Figure 8.6, the two triangles, $\triangle ABC$ and $\triangle AED$ are similar. Therefore, their corresponding sides are proportional as follows:

$$\frac{AC}{AB} = \frac{AD}{AE}$$
$$\frac{a}{1} = \frac{a+b}{1+x}$$
$$a(1+x) = 1(a+b)$$
$$a + ax = a + b$$
$$x = \frac{b}{a}$$

Thus, we have constructed the line segment BE whose length is the length of CD divided by the length of AC. If, instead, we want to multiply the two line segments we just replace x by b and b by x in this figure, and we will have constructed a segment of length ba.

Numerical manipulations of these types were picked up and further developed into *algebra* by the Persian mathematician al-Khwarizimi in 800 *CE*. But it took the next 8 centuries, for *analytic geometry,* the process of writing geometric theorems in terms of algebraic equations, to finally gobble up the world of geometric constructions.

Analytic geometry

The axiomatic approach of Euclid's to geometry was called *synthetic* geometry in the sense that you synthesized, built up, figures by citing the axioms. The new geometry was called *analytic* geometry, in the sense that you analyzed the problem by breaking down the construction of the figures and, using numerical equations, verifying the geometric truth.

In 1630, René Descartes, invented analytic geometry. Instead of points, lines and circles being the primitive terms, now we just associate every point of any geometric figure with numbers. On a line, every point is a number, x, and in the plane every point is associated with an ordered pair, (x, y), where each of x and y is a number.[2] Also, a line is an equation $ax + by = c$ instead of a geometric figure. And, a circle with center at the point (a, b) and radius r is the equation

$$(x - a)^2 + (y - b)^2 = r^2$$

which means that if (x, y) is any point on that circle, then its coordinates x and y satisfy this equation.

The reason Analytic Geometry was invented was to solve difficult construction problems. It was no longer necessary to draw a geometric figure to solve problems. Nor is it even possible to do this when solving geometric problems in four-dimensional space, or five, or n dimensional space, for any n, $(x_1, x_2, x_3, .., x_n)$. Not only do we solve geometry problems in three or more dimensions, but we apply the mathematics of n dimensional space to problems not usually associated with geometry, such as: business, gambling, economics, sociology, war and peace. In the next chapter we will discuss the applications of multidimensional vectors and matrices in these fields.

---

[2]We say *ordered* pair to emphasize that it matters as to which number is first and which is second. For example, the ordered pair $(3, 10)$ is not the same as the ordered pair $(10, 3)$.

CHAPTER 9

MATRIX APPLICATIONS

Mathematics, East and West

In the Far East, around the time of Euclid in the western world, Chinese mathematicians were solving system of linear equations. The problems they solved dealt with exchange of commodities (millet and rice), equitable taxation, profits and losses and other applications that we engage in even today. Over several centuries, the methods used to solve these problems found their way into a book, *Jiuzhang Suanshu*, or "The Nine Chapters of the Mathematical Art" dating back to as early as 300 BCE. Later, in Japan in 1683, sophisticated matrix methods were used in the work of Seki Takakazu, also known as, Seki Kowa. Surprisingly, the methods used in the Jiuzhang Suanshu book and by Seki Kowa, were basically the same ones discovered in the 1800's in the western world.

Unfortunately, these mathematical events are seldom, if ever, recorded in the western history of mathematics. The credit for these particular discoveries usually goes to Germany early in the 19th century CE when the method of row reduction[1] was invented (re-invented) by the brilliant and prolific German mathematician Carl Friedrich Gauss. Then, again, in 1888, to Wilhelm Jordan, also from Germany, who independently invented it.

Modern times

During the 19th, 20th, and beginning of the 21st centuries, engineering (bridges, dams, skyscrapers, bullet trains) experienced unbelievable growth in theory and applications. What are the mathematical tools developed to solve problems in these areas? At the risk of oversimplifying the situation, I would say: calculus, complex variables and differential equations. And what about physics, (relativity, quantum physics), biology (evolution, cell theory, genome sequencing) and chemistry (penicillins, contraceptive pills, liquid crystal displays)? The most likely "math tools" in these areas are: probability, abstract algebra and topology.

During this same time, society went through the industrial revolution, international corporations, and global trade. For the fields of economics, business, and management,

---

[1] We introduced you to this method in Chapter 2, pages 33-40.

the appropriate tools were found to be in *finite mathematics*: transition diagrams, input-output matrices, decision theory, game theory, and other applications of matrices. Finite mathematics is more concerned with whole numbers or discrete measurements, like the number of people in a study, as opposed to continuous and infinitely subdividable measurements. In the next few pages we will examine some interesting examples illustrating the applications of matrices to economics and business.

Nobel prize-winning matrices

Eugene Wigner, a Hungarian-American physicist, shared in the 1963 Nobel prize in physics. Prior to receiving this prize, Wigner gave a lecture on the Unreasonable Effectiveness of Mathematics. In it, he put forth the concept of "*the miracle of the appropriateness of the language of mathematics for the formulation of the laws of physics.*" Here we apply this same thought to *economics* in place of *physics*.

Matrix multiplication, division, addition and subtraction, as we discussed in Chapter 2, are the perfect tools for understanding how the different parts of an economy interact with each other, how they affect the current state of being and how the past got us to this place and what to expect in the future. This is an example of the power and beauty of mathematics.

In recent years, the Nobel Prize in Economics has been awarded to economists who used matrices to describe the inter-action between various segments of an economic system. Just take a quick look at the Nobel Prizes in the years 2005, 2000, 1994, 1986, and further back. You will find an impressive list of matrix topics: Matrix decision theory, Game theory, Transition matrix models, and Inter-generational matrices.

Leontief matrix

Take, for example, the work done by the 1973 Nobel Prize winner, Wassily Leontief. His research over the previous twenty years included his seminal book: *The Structure of American Economy, 1919-1939*, published by Oxford University Press, New York in 1951. In general, a Leontief matrix reveals important information about an economic system; for example, it tells us the *overall cost* to the rest of the system in producing one unit of a given good. Furthermore, if we know the *gross product* (GNP) for an economic system, then we can compute the *net product* by subtracting the overall cost from the gross product.

He used matrices that encompassed 46 segments of the economy. Since we don't relish the computation–by hand–of the inverse of a 46 by 46 matrix, we will confine our discussion here to the greatly simplified economic systems that can be described as four-by-four matrices. This is adequate to illustrate the main points.

In the Leontief model, the input-output matrix is a square matrix in which the rows and columns represent the segments of an economy.

Example: From one of Leontief's charts for 1939, we extract the following four by four table.

↗	AG	FE	NM	CON
AG	0.076	0.000	0.000	0.008
FE	0.006	0.305	0.000	0.152
NM	0.005	0.010	0.101	0.530
CON	0.026	0.004	0.002	0.000

Table 9.1 Input-Output Table

Where:

- AG is Agriculture and Fishing

- FE is Ferrous metals

- NM is Nonmetallic Minerals

- CON is Construction

The units may be in money, volume or any other measurements of value.

Example: To help you get your bearings in the table above, look at the number 0.008 in the AG row and the CON column. It is the fraction of a unit of AG used in producing a unit of CON.

What is the impact of all the segments on AG? To answer this, let us look at the whole first row to see what fraction of AG is used in producing given amounts of the four segments, including itself. The number 0.076 in the first row and first column is the fraction of a unit of the first industry's output consumed by a unit of the first industry's output. This is the percentage of AG consumed by AG, when it produces one unit. Continue along this row to compute the fraction of AG consumed by the production of one unit of FE, then by NM and finally by CON.

Setting this up as a matrix multiplication problem, let us define the following matrix M from Table 9.1

$$
\begin{array}{cccc}
AG & FE & NM & CON
\end{array}
$$

$$
M = \left[\begin{array}{cccc}
0.076 & 0.000 & 0.000 & 0.008 \\
0.006 & 0.305 & 0.000 & 0.152 \\
0.005 & 0.010 & 0.101 & 0.530 \\
0.026 & 0.004 & 0.002 & 0.000
\end{array} \right] \begin{array}{l} AG \\ FE \\ NM \\ CON \end{array} \tag{9.1}
$$

Let us say we want to find the cost to the Agricultures (AG) caused by the gross production of the following column matrix X,

$$
X = \left[\begin{array}{c}
2,000 \\
3,000 \\
10,000 \\
15,000
\end{array} \right] \tag{9.2}
$$

which represents: $2,000$ units of AG; $3,000$ units of FE; $10,000$ units of NM; and $15,000$ units of CON. Start with the cost to AG in the production of $2,000$ units of AG. We observe that AG consumes 0.076 units of AG, itself, for each single unit of AG produced. So, if AG produces $2,000$ units, it uses up $0.076 \times 2,000 = 152$ units of AG.

To calculate this cost for each of the other segments, move on to the $3,000$ units of FE. It is $0.000 \times 3,000 = 0$. What about $10,000$ units of NM? It is $0.000 \times 10,000 = 0$. Finally what is the cost to AG in producing $15,000$ units of CON? It is $0.008 \times 15,000 = 120$.

Therefore AG, contributed $152 + 0 + 0 + 120$, which equals 272 units of its output to the gross production of this four-segment economic system.

$$\begin{pmatrix} 0.076 & 0.000 & 0.000 & 0.008 \end{pmatrix} \times \begin{bmatrix} 2,000 \\ 3,000 \\ 10,000 \\ 15,000 \end{bmatrix} = 272$$

Do the same thing to calculate the cost to each of the other segments, FE, and so forth. But what is this "same thing?" you are supposed to be doing? What we did was to multiply the first row $(0.076, 0.000, 0.000, 0.008)$, the AG row by the the system's total production, the column matrix, X.

Now, just as we did with AG, let us find the cost to FE in producing this same gross product, X. We will multiply X by the FE row and we get:

$$\begin{pmatrix} 0.006 & 0.305 & 0.000 & 0.152 \end{pmatrix} \times \begin{bmatrix} 2,000 \\ 3,000 \\ 10,000 \\ 15,000 \end{bmatrix} = 3207$$

We repeat this for each row and what we have done is to multiply the matrix M by X. You may wish to look back at Chapter 2 for details on how this is done.

$$\begin{bmatrix} 0.076 & 0.000 & 0.000 & 0.008 \\ 0.006 & 0.305 & 0.000 & 0.152 \\ 0.005 & 0.010 & 0.101 & 0.530 \\ 0.026 & 0.004 & 0.002 & 0.000 \end{bmatrix} \times \begin{bmatrix} 2,000 \\ 3,000 \\ 10,000 \\ 15,000 \end{bmatrix} = \begin{bmatrix} 272 \\ 3,207 \\ 9,000 \\ 84 \end{bmatrix} \tag{9.3}$$

In words this equation says:

(The System's input-output matrix model) $\times$ (Gross product vector) $=$ (Cost vector). In symbols,

$$MX = C$$

The vector, C is the cost to the entire system in producing X.

To find the net production (the NNP, or the "Net National Product") for the entire economic system, we subtract the cost vector C from the gross product vector X. In symbols, let Y denote the net product then

$$Y = X - C \tag{9.4}$$

Problem: What is the *net product* of the *gross product* vector X for the economic model M defined in Equation (9.1)?

Answer: Using Equation (9.4), we get

$$Y = X - C = \begin{bmatrix} 2,000 \\ 3,000 \\ 10,000 \\ 15,000 \end{bmatrix} - \begin{bmatrix} 272 \\ 3,207 \\ 9,000 \\ 84 \end{bmatrix} = \begin{bmatrix} 1728 \\ -207 \\ 1000 \\ 14916 \end{bmatrix}$$

The negative entry -207 in the second row (ferrous metals) of the net product Y shows that in producing X, the system M consumed more ferrous metals than it produced. It had to depend upon an outside source for this material, in the same way that certain economies, such as Japan and Holland, depend upon outside sources for their petroleum.

In a theoretical problem, we could set, as a target, a gross product X, that will yield net product, Y, with only positive entries.

Thus, we would be required to solve the following equation for X.

$$X - MX = Y$$

or

$$(I - M)X = Y$$

which could easily be done mathematically, by finding the inverse $(I - M)^{-1}$. We use computers to solve 46 component, and larger economies, such as the Leontief's model.

Transition matrices and diagrams

This input-output Leontief matrix is an example of what is called a *transition matrix*; that is, a matrix in which the rows and columns are real events related to each other by some numerical entry in the matrix. A *transition diagram* is a way to depict the same data as the matrix in terms of a *flow chart* or a *graph* that shows the interconnection between the various components by short line segments or curves.

Example: Let **B** be the following matrix connecting three segments X, Y, and Z of an economy:

$$
\begin{array}{c}
 \\
X \\
Y \\
Z
\end{array}
\mathbf{B} =
\begin{array}{c}
\begin{array}{ccc} X & Y & Z \end{array} \\
\begin{pmatrix}
0.18 & 0.31 & 0.25 \\
0.64 & 0.07 & 0.00 \\
0.13 & 0.05 & 0.71
\end{pmatrix}
\end{array}
$$

This says that the transition from X to X is assigned a value of 0.18, and from X to Y the value is 0.31; X to Z it is 0.25, and so forth. In Figure 9.1, below, we sketch the transition diagram for this matrix.

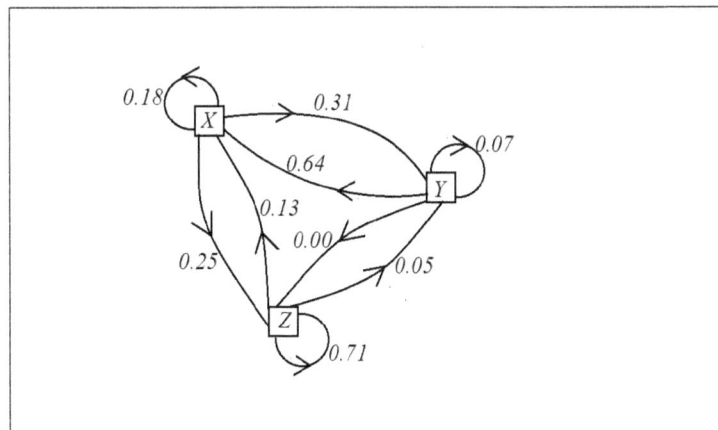

FIGURE 9.1 TRANSITION DIAGRAM FOR **B**

The reader may be inspired to draw a similar diagram for the abbreviated version of the Leontief model we discussed above. It is doable and not too messy since three of the matrix entries are zero and require no arrow to be drawn.

Stockroom inventory

Consider the following stockroom problem in the management of high and low inventories.

A stockroom can be described as being in one of two situations at the end of each month: $L =$ that it is a low inventory and $H =$ it is a high inventory. If, at the end of one month, it is L, the stock clerk will order and receive more supplies before the end of the next month. Then it has an 80% chance that it will be H by the end of that month, (and a 20% chance it will be L again). In other words, it has a probability of 0.8 to transition from L to H. If, on the other hand the stockroom is H at the end of the month, the clerk will not order supplies and it will have a 50% chance to still be in H at the end of that month. That is, it will transition from H to H with probability 0.5.

Problems:

1 Write out the transition matrix for this diagram.

2. Draw the transition diagram illustrating this situation.

Solutions:

1. The matrix for this inventory problem is:

$$\begin{array}{cc} & \begin{array}{cc} L & H \end{array} \\ \begin{array}{c} L \\ H \end{array} & \begin{pmatrix} 0.2 & 0.8 \\ 0.5 & 0.5 \end{pmatrix} \end{array}$$

2. The transition diagram is shown in Figure 9.2.

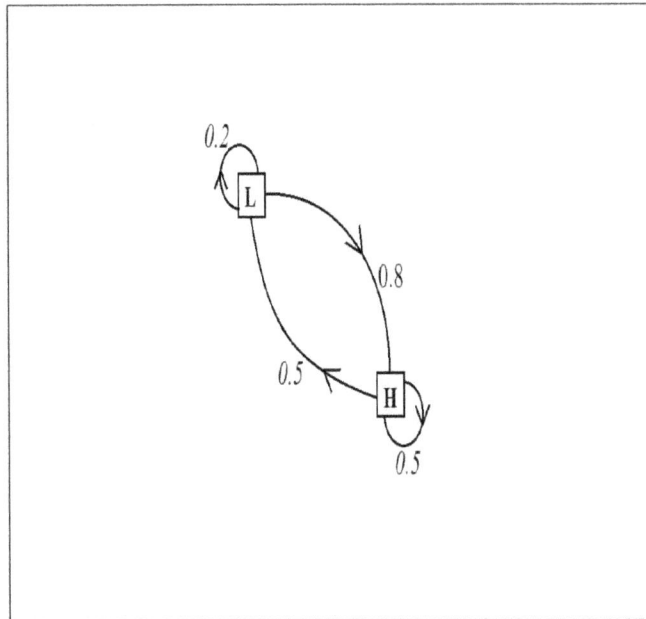

FIGURE 9.2 STOCKROOM TRANSITION

Transition matrices and diagrams were developed by a Russian mathematician, Andrey Markov, in 1880 and they provide us with useful insights and interesting ideas for formulating models in the social sciences. We use them to study transitions from one state to another through probabilities and through random contingencies.

In his theorems, Markov defines a *state space*, which is actually a matrix of random events, which are connected by probabilities. These can be used to make predictions about transitions into new states without knowledge of any previous state. A string of such transitions is called a "Markov Chain." Since each step of the process, as in a series of coin tosses, does not use information about a previous state, it is called a *Memoryless Markov Chain*. Believe it or not, this process is extremely useful in decision making because it shows you what to expect down the line from various paths into the unknown future.

Here is a relatively simple transition matrix model is based upon "brand loyalty."

Brand loyalty matrix

Suppose that a store is selling three brands, *Brand A*, *Brand B* and *Brand C*. A study of customer purchasing patterns shows that a person who purchases A this time will also purchase A the next time with 0.83 frequency, (that customers have a 83% loyalty to A). But there is a 17% chance that a Brand A customer will switch to Brand B and a 0% chance that a Brand A customer will switch to Brand C on the next purchase. This same study shows that Brand B customers switch to Brand A 6% of the time, but will be 92% loyal to Brand B, and switch to Brand C 2% of the time. Brand C customers switch to A 5% of the time and to Brand B 0% but stay with Brand C 95% of the time. In the table we indicate a customer switching from one brand to another by the arrow shown in the upper left corner of the table.

Next Purchase

	Brand A	Brand B	Brand C
Brand A	0.83	0.17	0.00
Brand B	0.06	0.92	0.02
Brand C	0.05	0.00	0.95

Previous Purchase

Table 9.2 Brand Loyalty

Let us convert this table to the following 3 by 3 *Brand Loyalty Matrix L*:

$$L = \begin{pmatrix} 0.83 & 0.17 & 0.00 \\ 0.06 & 0.92 & 0.02 \\ 0.05 & 0.00 & 0.95 \end{pmatrix}$$

Say that, one day, you take a survey of purchases and on that day you find that 3000 customers bought Brand A, 2000 customers bought Brand B and 1000 customers bought Brand C. You will have the 3 by 1 matrix, S_0 that tells you the *market state* on that day.

$$S_0 = \begin{pmatrix} 3000 \\ 2000 \\ 1000 \end{pmatrix} \begin{matrix} \text{Brand A} \\ \text{Brand B} \\ \text{Brand C} \end{matrix}$$

Assuming the loyalty matrix L is true, find the market state, S_1 the next day. That is, you want to find how many customers will purchase Brand A, how many will purchase Brand B and how many will purchase C.

Here is how you do it. Multiply the matrix L times the matrix S_0, this gives you S_1

$$\begin{aligned} S_1 &= L \times S_0 \\ &= \begin{pmatrix} 0.83 & 0.17 & 0.00 \\ 0.06 & 0.92 & 0.02 \\ 0.05 & 0.00 & 0.95 \end{pmatrix} \times \begin{pmatrix} 3000 \\ 2000 \\ 1000 \end{pmatrix} \\ &= \begin{pmatrix} 2830 \\ 2040 \\ 1100 \end{pmatrix} \end{aligned}$$

To get the next market state, multiply L by S_1 to find $S_2 = L \times S_1$. Can you see why Brand A should start to worry?

$$S_2 = L \times S_1 = \begin{pmatrix} 0.83 & 0.17 & 0.00 \\ 0.06 & 0.92 & 0.02 \\ 0.05 & 0.00 & 0.95 \end{pmatrix} \times \begin{pmatrix} 2830 \\ 2040 \\ 1100 \end{pmatrix} = \begin{pmatrix} 2696 \\ 2069 \\ 1186 \end{pmatrix}$$

Predictions

In February 2017, the American Association for the Advancement of Science (AAAS) Journal *Science* published a special section, "Prediction and its Limits." The articles dealt with the possibilities and probabilities in predicting outcomes of events as varied as armed conflicts, new discoveries in science, policy debates, explanations in social systems, intelligent driverless cars, and predicting human behavior. These AAAS articles were *essays*, taking a broad overview of the significant real-life problems facing humanity today. High speed computers with terabytes of memory can give rise to simulation models for studying these problems. What this means is that you can load up the computer with trillions of details, ostensibly sufficient to mimic reality. The resulting computer output could be used to make predictions or to take some kind of action.

For example, with a driverless car, you can program into its memory every conceivable situation that might be encountered in every conceivable traffic pattern. You could design the "awareness" of the vehicle to be able to see any impending hazard and react appropriately. The drawback to this simulation modeling is that there may be more variables than you have accounted for, and it would be difficult to build in instructions for the contingency of a random event that might occur in an *inconceivable* situation. Never-the-less it may happen that a driverless car with a large enough data base in its memory and fast enough computational ability could drive better and safer than any human being.

As early as the 17th century, the mathematical concepts needed to study these modern problems were laid out by two French and one Russian mathematician: Blaise Pascal, Pierre de Fermat, and Andrey Markov.

In 1641, to help his father's tax collecting job, Pascal, at the age of 19, invented a calculating machine that basically mechanized the abacus by using a system of gears that automatically tripped a carryover to a new column in any base ten calculations. This successful device became the mother of calculating machines, that inspired other similar computers in the 17th century. I guess you could say the abacus was the grandmother.

Later, when Pascal was in his twenties, he corresponded back and forth with Fermat through a series of letters discussing a certain probability problem. It involved finding a fair method for assessing how to divide the payoff between two players in an incomplete game that was interrupted and could not be resumed. What came out of that correspondence was the use of the binomial theorem to establish a mathematical basis for calculating probabilities in gambling. Today, applications in gambling, economics, politics and statistical mechanics are based upon their work.

In 1798, Simeon Poisson, at age 18 entered École Polytechnique, and was immediately recognized by his mathematics professors as having superior mathematical talent. He published two papers on finite difference equations by the time he was 20 and in the early 1800's he went on to develop a number of mathematical techniques, including the theorems called the Poisson distributions. These define the occurrence of a discrete event in fixed periods of time. The problems that can be solved by this method include:

- determining how many beds should be always available in a maternity ward in a hospital

- the number of mutations in a strand of DNA over a given time

- the losses and claims in an insurance policy over a given time

- how streets should be laid out to accommodate various traffic patterns and in one interesting case

- "...the number of soldiers in the Prussian army killed accidently by horse kicks..."[2]

Any effort concerned with the science of predictions and decision making is an interesting amalgamation of random events, probability, vectors, matrices, and expected values, together known as *stochastic analysis.* Here is what we mean by these terms.

Probability vectors

A probability vector is any vector whose components are non-negative numbers adding up to 1. It can be a row vector or a column vector.

Example: Probability row vector

$$\mathbf{v} = (0.030, 0.630, 0.001, 0.210, 0.129)$$

is a probability vector because all of its coordinates are non-negative and the sum of the components,

$$0.030 + 0.630 + 0.001 + 0.210 + 0.129 = 1$$

The five coordinates tell you that there are five things that can happen, and the first one can happen 3% of the time, the second one can happen 63% of the time, the third 1/10 of 1% of the time, the fourth happens 21% of the time and the fifth 12.9% of the time.

Example: Also, a column vector is a probability vector if its components are non-negative and sum to 1.

$$\mathbf{w} = \begin{pmatrix} 0.582 \\ 0.201 \\ 0.217 \\ 0.000 \end{pmatrix}$$

This represents a situation in which four things can happen, the first 58.2% percent of the time, which is the same thing as saying *with a probability of* 0.582. The second thing can happen with a probability of 0.201, the third with a probability of 0.217 and the fourth never happens.

The general definition of a probability vector is as follows.

The vector, $\mathbf{v}$

$$\mathbf{v} = (x_1,\ x_2,\ ...,\ x_n)$$

or the vector, $\mathbf{w}$

$$\mathbf{w} = \begin{pmatrix} x_1 \\ x_2 \\ \vdots \\ x_n \end{pmatrix}$$

---

[2] From *Poisson Distibution*–Wikipedia

is a probability vector if and only if

$$x_i \geq 0, \text{ for all integers } i \text{ from 1 to } n.$$
$$\text{and}$$
$$\sum_{i=1}^{i=n} x_i = 1$$

Stochastic matrices

Any matrix whose rows,

$$\begin{pmatrix} p_{k1} & p_{k2} & \cdots & p_{kn} \end{pmatrix}$$

are probability vectors *or* whose columns,

$$\begin{pmatrix} p_{1k} \\ p_{2k} \\ \cdots \\ p_{mk} \end{pmatrix}$$

are probability vectors, is called a *stochastic matrix*.

Examples: The following matrices **A** and **B** are stochastic matrices. **A** is one whose rows are probability vectors and **B** is one whose columns are probability vectors.

$$\mathbf{A} = \begin{pmatrix} 0.83 & 0.17 & 0.00 \\ 0.06 & 0.92 & 0.02 \\ 0.05 & 0.00 & 0.95 \end{pmatrix}$$

$$\mathbf{B} = \begin{pmatrix} 0.7 & 0.833 \\ 0.3 & 0.167 \end{pmatrix}$$

Problem:

A local airport has a history of flight delays caused by various weather conditions. Records show that if the weather is mild, there is no delay, if there is a heavy snow storm, the delay is one and one-half hours, and if the storm is severe there is a three hour delay. We can depict these delay times as a column vector.

$$\begin{bmatrix} \begin{vmatrix} \text{Severity} \\ \text{Mild} \\ \text{Heavy} \\ \text{Severe} \end{vmatrix} \end{bmatrix} \quad \begin{pmatrix} \text{Delay, minutes} \\ 0 \\ 90 \\ 180 \end{pmatrix}$$

One holiday week-end in the year 2018 a four-day weather forecast predicted that on Friday the probability of mild weather was 0.40, of a heavy storm 0.20, and of a severe storm 0.40. But on Saturday conditions improved so that the probability of mild was 0.70, of heavy 0.20, and of severe only 0.10. Sunday was even better with probability predictions of 0.9, mild, 0.06, heavy and only 0.04, severe. By Monday the storm had completely moved on, and the prediction was 100% mild with 0 probabilities of either a heavy or severe storm. We can depict these days and predictions as the following stochastic matrix.

$$
\begin{array}{ccccc}
\text{Day} & \text{Mild} & \text{Heavy} & \text{Severe} \\
\end{array}
$$

$$
\begin{array}{c}
\text{Fri.} \\
\text{Sat} \\
\text{Sun} \\
\text{Mon}
\end{array}
\left(
\begin{array}{ccc}
0.40 & 0.20 & 0.40 \\
0.70 & 0.20 & 0.10 \\
0.90 & 0.06 & 0.04 \\
1.00 & 0.00 & 0.00
\end{array}
\right)
$$

Now, to find the predicted delay times for each of the four days we multiply the given stochastic matrix by the vector of weather predictions.

$$
\left(
\begin{array}{ccc}
0.40 & 0.20 & 0.40 \\
0.70 & 0.20 & 0.10 \\
0.90 & 0.06 & 0.04 \\
1.00 & 0.00 & 0.00
\end{array}
\right)
\times
\left(
\begin{array}{c}
0 \\
90 \\
180
\end{array}
\right)
=
\begin{array}{c}
0 + 18 + 72 \\
0 + 18 + 18 \\
0 + 5.4 + 7.2 \\
0 + 0 + 0
\end{array}
$$

$$
=
\left(
\begin{array}{c}
90 \\
36 \\
12.6 \\
0
\end{array}
\right)
\text{minutes}
$$

If you had a choice, you should buy your ticket for Monday, when there is no delay. This is an example in which stochastic matrices are useful in solving problems involving *random events*, such as weather patterns.

Random events

Stuff happens. When something happens, then that happening is called an *event*. When an event is not deliberately planned, it is called a *random event*. If you are about to cook a meal on your electric range and there is a power outage interfering with your plan, then you have experienced a random event. Could you have expected it? Yes, if the weather outside was a raging electrical storm, with nearby lightening strikes, you might have thought there was a high probability that it would happen. You did not plan it; you did not do something to cause it. Obviously something, some random event caused it. Maybe lightening struck a transformer at the nearest power station; maybe a truck crashed into a power pole.

A random event need not be a surprise; it could just be one of several known outcomes. Your power company is likely to have a computer record of possible outcomes, and the frequency with which power outages, or other events, occur under various weather conditions.

If you are tossing a coin on a table, there are just two possible random events; it will come up heads or tails when it lands on the table. This, of course, ignores the negligible event that it would stand on its edge, and the even less likely event that all of the atoms in the coin would move at the same time in the same random direction sending the coin straight out the window.

A more interesting example was the problem we discussed in Chapter 2, regarding births of lambs. In any given flock of sheep, if you know the probabilities for the three events that a mature ewe could give birth to either zero or one or two lambs, then you know how many lambs to expect in some future year.

For any such collection of random events, if you are given enough historical data you could assign probabilities to them. These types of problems are called *discrete probability* or *finite probability models* because there are only a finite number of such events; they can be counted and assigned a numerical value.

Usually problems in which an *infinite* number of events can occur would be something involving continuous variables like time, distance, area, volume and so forth. For example suppose you are throwing darts at a circular target. It would be impossible to define, as an event, the fact that a dart hit and stuck to a point in the circle. This is because there are infinitely many points in the circle; therefore, the random selection of any one, single, point out of an infinity of points is zero.[3] Thus, all probabilities would be zero and would add up to zero for the whole system when they should add up to 1.

You can turn this problem into a manageable *finite probability* problem by painting rings on the target. If you drew a finite number of concentric circles about the center then you could designate the fact that a dart would hit and stick to a point in the inner circle or to an area between any two such circles, a *ring* or *annulus*, as an event. Then the probability will not be zero, but some positive number computable in terms of the area of the region compared to the total target area. Let us define such a toss of the dart as a *simple event*. This means that the event is unique; no dart can stick in two regions at the same time. Of course, this requires us to have a previous rule or agreement that resolves the case in which a dart sticks on the circumference of a circle. Such a rule could be, for example: "The circumference of a circle is part of that circle." We want the events to be not only unique, but we want them to be conjointly exhaustive.

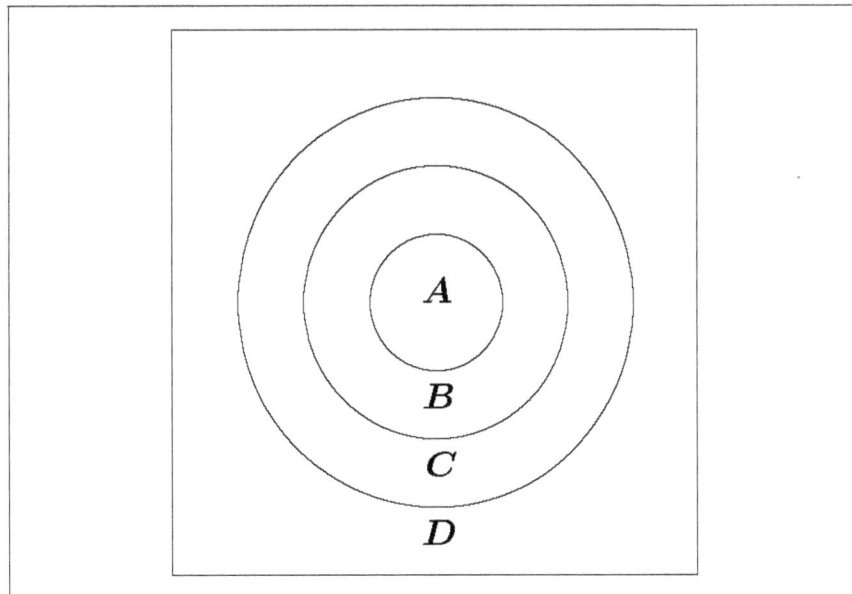

Figure 9.3 FOUR REGIONS OF A TARGET

That is, nothing other than sticking to one of these regions can count as an event.

---

[3] With a finite number like a billion, selecting one of them at random is "one in a billion." If we convert this to a probability it would be $\frac{1}{1,000,000,000}$, or 0.000000001. But the phrase "One in an infinity" does not translate to 1 divided by infinity, but to 1 divided by a number n approaching infinity; that is, $\frac{1}{n} \to 0$ as $n \to \infty$.

For example, a dart that misses the entire target and lands on the floor is not an event. Take as an example, the target depicted above in Figure 9.3.

This target defines a discrete probability space consisting of four simple events, namely throwing a dart that sticks into one of the four regions labeled A, B. C, or D. Next, we need to define the probability that such an event will occur.

One way to do that for any region is to use the area of that region as a fraction of the area of the whole target. For example, suppose the radii, in inches, of the three circles are: 2, 4, and 6, and that the side of the square containing them is 16 inches. Then the area of the square is 256 in^2, and the areas of the 4 regions, A, B, C, and D, respectively, are as shown in the following table.

Region	Area	Rel. Area	Probability
A	4π	$4\pi/256$	0.049
B	$(16-4)\pi$	$12\pi/256$	0.147
C	$(36-16)\pi$	$20\pi/256$	0.245
D	$256-36\pi$	$1-9\pi/64$	0.558
Square	16^2	$256/256$	1.000

Table 9.3 Relative areas of the target regions

Here we are using the idea that we can interpret the relative frequency of a event as the probability that such an event will happen. This is an assumption made in discrete probability models. Thus, if you throw 256 darts, and they are uniformly distributed, then the chances are that 4π of them will hit region A, the "bull's eye". This is because the region A covers 4π in^2, about 12.6 in^2. Since the total target area is 256 in^2, relative frequency of such an event is $12.6/256 \approx 0.049$.

Now that you have defined the random events (the regions) and you have an estimate of the probabilities that they will occur, you can assign a *value* to their occurrence. The number assigned to the event in this last step, is called a *random variable.* Although assigning random variable can be arbitrary, the nature of the problem usually provides a clue as to what would be a reasonable set of numbers. In the target problem discussed here you would want to give fewer points to the event of just hitting the target than you would to hitting bull's eye. Let us assign the values to the regions as follows:

Region	Value
A	10 points
B	7 points
C	3 points
D	1 point

Table 9.4 Random variables

Now here is a surprising question.

Problem: What is the value of the average random dart throw?

After getting over the shock of being seriously asked a question like this, let us think about it for a while.

Solution:

Assume that on any one throw you will get a bull's eye about 4.9% of the time because the area of A is about 4.9% of the total target area   Thus, 4.9% of 10 points is $0.049 \times 10 = 0.49$ of a point. Similarly, we are likely to hit region B, about 14.7% of the time. Therefore, 14.7% of 7 is: $0.147 \times 7 = 1.03$; for region C is it $0.245 \times 3 = 0.735$, and, finally, for region D, we have $0.558 \times 1 = 0.558$. Adding these values the average dart throw would have a value of

$$0.49 + 1.03 + 0.735 + 0.588 = 2.813$$

What we have found is the *expected value* of a probability distribution. It really is an average because we have added up the points multiplied by probabilities, which are actually fractions; we can say we have done a "pre-division" by the number of throws. For this set of events and their probabilities,the expected value of 2.813, is approximately 3. That is, the average dart throw is in the C annulus.

Expected value in a discrete probability space

We define the expected value of a discrete probability distribution as follows.

Definition (Expected value):

If a set S consists of n non-overlapping events: E_1, E_2,...E_n, and the probabilities of these events are, respectively, p_1, p_2, ..p_n, and the random variables of these events are, respectively, the numbers: v_1, v_2, ..v_n, then the expected value, E, of this discrete probability distribution is:

$$E = \sum_{i=1}^{i=n} p_i \times v_i \tag{9.5}$$

Problem

In 2010, an insurance company collected data over a long time with a large number of policies in force. As shown in Table 9.5 below, they found that in one year they could operate a hospital insurance policy as if each policyholder had probability A of filing and collecting a payable claim for benefits B, with an average payment of C per year. What should be the yearly premium on this policy just to cover the probable claims?

A	B	C
Probability	Benefit Claimed	Benefit Cost
0.42	Hospitalization	$10,800
0.02	Outpatient	$440
0.04	Home Care	$2,000
0.18	Extended Special Care	$8,200
0.34	None Claimed	0

Table 9.5 Health care probabilities and payments

Solution:

This problem is asking what is the expected value of such a policy? It is the amount the insurance company must charge each person in order to break even when it fulfills all such claims. Using equation (9.5), we get that the expected value is:

$$\begin{aligned} E &= 0.42 \times 10,800 + 0.02 \times 440 + 0.04 \times 2,000 + 0.18 \times 8,200 + 0.34 \times 0 \\ &= \$6100.80 \end{aligned}$$

This says that on the average every policy holder is getting $6100.80 worth of benefits every year; so, the monthly insurance premiums should be at least

$$\frac{1}{12} \times (\$6100.80) = \$508.40$$

Buying insurance is a way to share the cost of getting sick. It protects anyone who got sick and received a benefit as well as those who did not get sick and received no benefit. But what if all of the "well people" were prescient enough to guess they would not need the insurance. Let us say that these 34%, who claimed no benefits, did not buy this policy. Then the other 66% were the only policy holders. This would require a re-scaling of the percentages, called "normalizing" the data. This means you are making the 66% become 100% of the policy holders. This is done by dividing 1 by 0.66, getting 1.515..., because

$$0.66 \times (1.515...) = 100\%$$

Thus, the monthly cost for this same policy is,

$$\$508.40 \times (1.515...) = \$770.23$$

Games and Artificial Intelligence

Just as so many other competitive animals, we humans play games. Yes, we do it sometimes to win food, territory, and mates, but we often seek more cerebral and abstract rewards, than say a robin or chimpanzee or even a bonobo. Our penchant for abstraction makes us want to understand what a game is and how we can be winners.

Some two-player games such as board games have been "completely solved" which means that the programs have been written that take advantage of modern computers' large memory capacities and their ability to make rapid calculations that let them consistently defeat any human player. Backgammon fell in 1979, checkers in 1995, chess in 2000, and the ancient Chinese game of Go in 2016. These are what are called *perfect information* games. This means that both players know everything about the position and know everything about what a player can do on the next move. In the case of backgammon, however, there is an element of chance in the roll of the dice so that neither player knows before the dice roll what moves are possible

The solutions of these games grew out of a field of mathematics called AI, Artificial Intelligence, which is an amalgamation of probability, psychology and physics, and computer programs that learn. And just how do they do that? Just like we do: repeating a task, retracing old paths until they run into a previously failed branch, changing paths and keeping records of previous successes Also, a human overseer might intervene from time to time with a surgical replacement of part of the code. Eventually the program will be weaned from its human baby-sitter because the human is no longer a good enough player to make any meaningful changes

Poker

A different type of game, one with imperfect information, such as any version of poker, is more difficult to crack. The information you have is imperfect because the other players are privy to some information, namely their own hidden cards, not available to you. Also, the involvement of money creates a new wrinkle. A player may gain some leverage by bluffing, thereby confronting other players with the extra problem of possibly losing money on the chance that the bettor really does have superior cards.

A program called *DeepStack*, written in 2017 seems to be able to eke out wins against the world's best Texas hold'em poker players. This program builds in bluffing buffers as well as other Artificial Intelligence algorithms. It learned how to handle bluffing by playing thousands of games against thousand of players with different simulated bluffing personalities.

Matrix games

To understand how game theory got started before it evolved into the sophisticated topic we know as Artificial Intelligence, we will give examples of two-player (or two-team) games in which each side has a restricted number of choices. These will be perfect information games because the available choices for each side are clearly displayed. Each side will be able to calculate their own and their opponents possible gains and losses depending upon the choices each side makes. Each side reaps a reward or suffers a loss according to a mutually agreed upon prearranged scheme. If one side wins exactly what the other side loses, then it is a *zero sum* game, otherwise a *non-zero sum* game.

The word "game" is used here as a catch-all term. It can be used to actually represent a conflict between two parties, competing for some dominance, or for some prize. It could also, be a negotiation, in which each side is attempting to get the best possible deal. It could simply be looking at a list of options and a list of their concomitant consequences suggesting or even dictating some action on the given options. It was probably in this sense of decision making that, in 1947, John von Neumann, a physicist, and Oskar Morgenstern, an economist, wrote their book *Theory of Games and Economic Behavior* in which they developed the basic assumptions for "Game Theory."

The most fundamental ones are:

1. The desire to win postulate, and
2. The Intelligent player postulate.

We can easily compare how the play of real games and the play of theory games use these two postulates.

Real games vs Theory games

• *Clarity of goal–the desire to win postulate*

In real games, players may have fuzzy, unclear, or even contradictory, motivation to win a "unit of utility." They may wish to "win by losing." In playing a weaker opponent or a child, the stronger players may let the weaker player win; or they may not try to take every point. This could happen in a teaching situation.

In Theory games, the players always try to optimize their gain.

• *Knowledge of available strategies–the intelligent player postulate*

In real games, players may not be mindful of the opponents' possible moves, or they may misjudge the opponents, assuming that they will not make the best move because they are not playing intelligently.

In theory games players are aware of each others' possible strategies, and assume that their opponents are intelligent players.

A game as a decision-making tool

Game theory may be applied to a decision making problem, in which your opponent is not a person or some other contentious entity trying to outwit you. It could be the weather, or the future, or some set of possible random events you must anticipate in order to choose your best possible plan of action. You may not have perfect information, but you could have a good idea about the most serious and most probable challenges you will be facing.

   Example:
   Let's say that a high school student is trying to decide among three colleges that have invited her to apply. She sets up a matrix game in which the row "player" is her decision to select one of the three colleges, and the column player is the future, that is, a set of contingencies that determine her future job interests, based upon her evolving, self-assessed inclinations and aptitudes.

   She partitions the three colleges into

$$r_1 \quad : \quad \text{Technological Institute}$$
$$r_2 \quad : \quad \text{Business College}$$
$$r_3 \quad : \quad \text{Liberal Arts College}$$

The student's future job interests, the column player, is partitioned into

$$c_1 \quad : \quad \text{Engineer}$$
$$c_2 \quad : \quad \text{Mathematician}$$
$$c_3 \quad : \quad \text{Musician}$$
$$c_4 \quad : \quad \text{Business Manager}$$

She constructs Table 9.6 in her attempt to make a decision. This is a typical matrix game; the table, itself, is called a *payoff matrix;* the numbers in the table are called utilities (or values). When a row is selected, and a column is selected then the payoff is the utility in the intersection of that row and that column.

Column Player

	c_1	c_2	c_3	c_4
r_1	14	6	2	4
r_2	4	3	1	12
r_3	10	8	9	9

Row Player (labeled to the left of the matrix)

Table 9.6 Matrix Game

   If she selects r_1, the Tech Institute, *and* her future career choice is engineering, c_1, row 1, column 1, (r_1, c_1), then the utility is high, 14. But, if her future choice is mathematics, c_2, then the Tech Institute gives her a lower utility, 6. If she ends up in music, c_3, then the Tech Institute gives her an even lower payoff, 2, found in (r_1, c_3). Finally, the Tech Institute, Business Career, (r_1, c_4) pay off is 4.

   Problem:

   1. Suppose the high school student selects r_2, show that her best career choice would

be as a business manager.

2. If she chooses r_3 show that her best future career choice is to become an engineer.

3. If her career leads toward becoming a mathematician, what is her best college choice?

4. What is the smallest number in each row and what does it mean for future careers?

5. Is there a number that is the smallest in its row and the largest in its column?

Answers:

1. Look at r_2, then going across that entire row, looking at the numbers column by column and you find the greatest utility for this row is in c_4, business manager.

2. For r_3, the largest value is in c_1, engineer.

3. In this case we look at the column c_2 and try to find the college that gives her the largest utility, and that is r_3.

4. The smallest value is any row is the least amount of utility you could get from that college for the career defined in that column.

5. Yes, the number 8 in Row 3, Column 2 is a number that is both its row minimum and its column maximum.

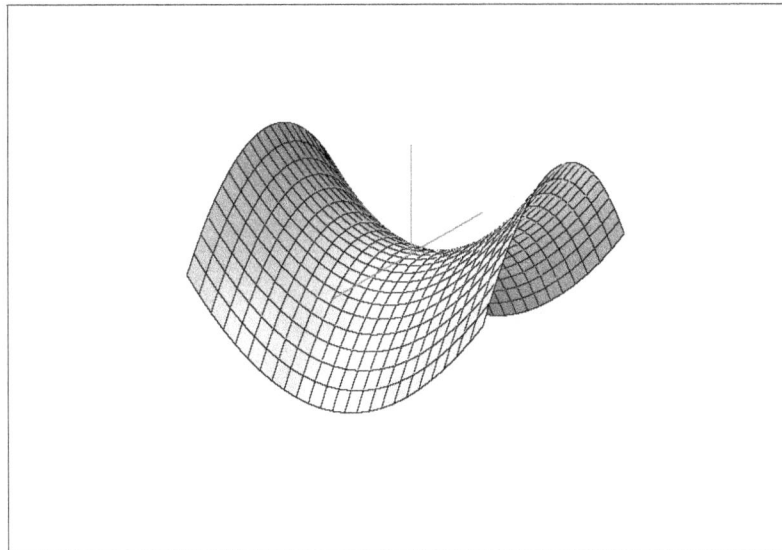

Figure 9.4 A SADDLE POINT = MINIMAX STRATEGY

Under the uncertain conditions about a future career, the row r_3 is the *safest* choice in terms of career and college. There are future careers in which some other college has a greater reward, but *all* colleges, other than r_3, have some future career with a smaller reward.

The selection of a number that is the minimum in its row and the maximum in its column is called the *minimax strategy*. The number in that position is called a "saddle point" for the game.   See Figure 9.4.   It depicts a surface that has a point that is simultaneously the minimum in one direction and the maximum in the perpendicular direction. It is the best that can be done for that matrix.

Table 9.7, is another example of a saddle point matrix game.

Column Player

		c_1	c_2	c_3	c_4
	r_1	R wins 1	C wins 3	C wins 1	C wins 2
Row Player	r_2	R wins 5	R wins 10	C wins 2	C wins 4
	r_3	R wins 7	R wins 5	R wins 9	C wins 1

Table 9.7

This game is unlike the college selection game in which the column player was a fixed entity, the "future career." In that game the column player did not deliberately switch around selecting various columns hoping to trip up the row player. The game in Table 9.7 has two aggressive active players. Let's call them the Raven and the Crow, respectively, for Row and Column. In order to maintain gender neutrality in our discussion, we will refer to a team on either side as *it*.

Let's imagine that the following takes place. Raven (R) and Crow (C) walk into the room. They sit down and are shown the payoff table for the first time. R's task is to *secretly* pick a row to play; the whole row is R's pick. Now, C's task is to *secretly* select a column to play, again, the whole column. Neither player knows its opponent's choice. They simultaneously reveal their choices and are rewarded or penalized according to what is in the intersection of this row-column combination.

Although neither player knows what the other will play, they can easily figure out what the other player *should* play because they both can see the entire payoff matrix.

Problems:

1. Show that there is a column that C can play and never lose.

2. Show that R does not have a no-lose row

3. When the game is to be played over and over again, what row is safest for R?

4. Why should C never play c_1, nor c_2, nor c_3?

Solutions:

1. If C selects column c_4, it will never lose no matter what R plays.

2. All three of the rows have at least one loss to C, so R does not have a no-lose row.

3. R, being an intelligent player, knows that C will play its no lose column c_4, so R must minimize its losses by *always* playing r_3. This is its safest row. Suppose R thinks about playing some other row, say, r_2, hoping C is going to play c_2 giving R a large 10 point win, or column c_1 giving R 5 points, but this is an unintelligent play by R because C, by continuing play to c_4 makes R suffer a 4 point loss. Similarly, R should not think about playing r_1.

4. Because C is intelligent and knows that R will always play r_3, making C lose 7, 5, or 9 points, if it tried to play c_1, c_2, or c_3.

It should be clear that the matrix game in Table 9.7 is biased in favor of the C player. It turns out to also be a game which has a saddle point; we will rewrite it in a standard matrix form, that lets us recognize such games and immediately apply a special strategy for solving them.

The payoff matrix in the previous game can be expressed as follows.

Crow

	c_1	c_2	c_3	c_4
r_1	+1	−3	−1	−2
r_2	+5	+10	−2	−4
r_3	+7	+5	+9	−1

Raven

This is the same matrix as the one in Table 9.7, written entirely from Raven's point of view. The *pluses* mean Raven *wins* and the *minuses* mean Raven *loses* (Crow wins). So, for example the +5 in the second row, first column means that R wins 5 (and C loses 5), while the −4 in the second row, fourth column means R loses 4 and C wins 4.

The boxed in number $\boxed{-1}$ in the third row fourth column is the saddle point. It is a number that is the smallest in its row and largest in its column. This means that when R selects r_3, then 1 is the amount that R loses if C plays correctly. This is the best R can do, in this biased game. Also, when C plays column c_4, 1 point is the maximum that C can win if R plays correctly.

Saddle point strategy

As we mentioned above, the saddle point strategy is also called the minimax strategy. Not all games have a saddle point, but in any that do, this is the best strategy for both players, any other attempts to select a different row by R will be worst for R. Similarly, if C tries to select a column other that the one containing the saddle point, then C will fare worse. If a game has more than one saddle point, playing for any one of them is also the best each player can do. When there is no saddle point, we will play for a mixed strategy.

Probability strategy in matrix games

What is the best way to play a game that does not have a saddle point? This question has a practical answer only if one anticipates playing the game several times. This is because the best strategy has the players switching back and forth randomly among the rows and columns. This strategy requires the players to have previously calculated the probabilities with which they must select their rows and columns to insure a beneficial outcome independent of the opponents choices.

Example:

A door-to-door salesman from Walla Walla, Washington devises the following two person game. His sales pitch is the row player and his customers constitute the column players. He partitions his sales pitch into two factors r_1= "Glad-Hand" approach and r_2= "Serious" approach. His customers are of two types: c_1= "Naive" customer and "Wary" customer. Here are the payoffs according to his first encounter with any customer. He finds that when he applies the glad-hand approach and the customer turns out to be Naive, he makes an $16 sale. If the glad-hand approach is used against a Wary customer, he makes a $8 sale. A serious approach to a naive customer earns him $9 and a serious approach to a wary customer yields him $13.

Customers

		Naive	Wary
Sales Pitch	Glad Hand	16	8
	Serious	9	13

When he first encounters customers, he has no idea whether they are Naive or Serious. He notices that there is no saddle point since no number is the row minimum and column maximum. So, he cannot use the saddle point strategy. What other strategy should he adopt to get a consistently decent sale? He could mix his approaches in hopes of getting something between $16 sale and a $8 dollar sales. Here is how to do that. He assumes that there is some optimal expected value to his contacts, regardless of what type of customer he is encountering. He wants to find out what combination of random approaches he can use to insure him at least that value.

What he will do is find out what he can expect when he assumes that the customer is Naive. Then find out what to expect if every customer is Wary. He will then set these two expected values equal to each other and find out what proportion of Glad Hand vs Serious approaches will result in this *same expected value* no matter what kind of customer he is facing. He will settle for this amount (between 8 and 16) as his game value, regardless of the nature of the customer.

Starting with the assumption that all customers are Naive, he wants to know what fraction, p, (when $0 \leq p \leq 1$), of his approaches should be Glad Hand. Which is the same thing as saying that we will make $(1 - p)$ be the fraction of Serious approaches. Thus, playing Glad Hand with probability p gives him $p \times 16$ against a Naive player $(1 - p) \times 9$ against a serious player. Hie expected value is E_1.

$$E_1 = 16 \times p + 9 \times (1 - p) \tag{9.6}$$

On the other hand, if he assumes all customers are Wary. Again, assume he uses Glad Hand with probability p and Serious with probability $(1 - p)$ against Wary. Then the expected value which we denote as E_2, is

$$E_2 = 8 \times p + 13 \times (1 - p) \tag{9.7}$$

To take care of having to encounter either Naive or Wary customers, he can find what probability p will make the two expected values be the same; that is, $E_1 = E_2$. This is a one variable one equation problem:

$$
\begin{aligned}
16p + 9 - 9p &= 8p + 13 - 13p \\
7p + 9 &= -5p + 13 \\
12p &= 4 \\
p &= \frac{1}{3}, \text{ and} \\
1 - p &= \frac{2}{3}
\end{aligned}
$$

So,

$$
\begin{aligned}
E_1 &= 16 \times \frac{1}{3} + 9 \times \frac{2}{3} = \$11.33, \text{ and} \\
E_2 &= 8 \times \frac{1}{3} + 13 \times \frac{2}{3} = \$11.33
\end{aligned}
$$

Therefore, this game has a value of $11\frac{1}{3}$ to the salesman, no matter what kind of customer he encounters..

But, wait just a cotton picking minute! What about the customers? Couldn't they appear in some random or even contrived way, say with probabilities q and $(1-q)$? What would be the expected game value, from their point of view, in that case?

Let us check that out. Assume the approach is always Glad Hand then the customer will be Naive with probability q and Wary with probability $(1 - q)$. This is switching from Naive to Wary against Glad hand, so the expected value is

$$
F_1 = 16 \times q + 8 \times (1 - q) \tag{9.8}
$$

Now switching from Naive to Wary against Serious, we get the expected value

$$
F_2 = 9 \times q + 13 \times (1 - q)
$$

To find the value of q needed to make these two expected values the same, set $F_1 = F_2$.

$$
\begin{aligned}
16q + 8 - 8q &= 9q + 13 - 13q \\
8q + 8 &= -4q + 13 \\
12q &= 5 \\
q &= \frac{5}{12} \\
1 - q &= \frac{7}{12}
\end{aligned}
$$

So the best customer strategy is to be Naive 5/12 of the time and Wary 7/12 of the time, making the game values be

$$
\begin{aligned}
F_1 &= 16 \times \frac{5}{12} + 8 \times \frac{7}{12} = \$11.33, \text{ and} \\
F_2 &= 9 \times \frac{5}{12} + 13 \times \frac{7}{12} = \$11.33
\end{aligned}
$$

Thus, the mixed strategy of the repeated game gives the salesman his best compromised pay off.

Optimal strategies for all two by two games

The formula for two-by-two games can be depicted in the following flowchart.

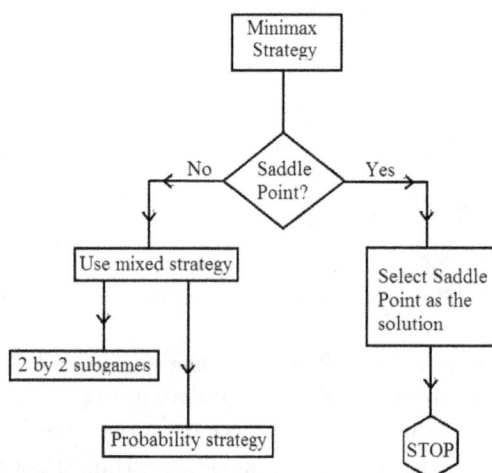

FIGURE 9.5 GAME STRATEGY

1. If there are one or more saddle points, play for any one of them.

2. If there is no saddle point play a mixed strategy

If a two by two game

$$\begin{bmatrix} a & b \\ c & d \end{bmatrix}$$

has a saddle point, both players must play it, since neither player can get a better payoff by playing other rows or columns. If the game does not have a saddle point, then the best play for the row player is the following probability strategy:

$$\text{Play } r_1 \text{ with probability } p_1 = \frac{|d - c|}{|a - b| + |d - c|}$$

$$\text{Play } r_2 \text{ with probability } p_2 = \frac{|a - b|}{|a - b| + |d - c|}$$

and for the column player:

$$\text{Play } c_1 \text{ with probability } q_1 = \frac{|d - b|}{|a - c| + |d - b|}$$

$$\text{Play } r_1 \text{ with probability } q_2 = \frac{|a - c|}{|a - c| + |d - b|}$$

We call these four formulas the probability strategy for any two by two game with no saddle point. You need to fully compute only p_1, because $p_2 = 1 - p_1$, and the game value found by the row player, using p_1 and p_2 is the same as that found by the column player using q_1 and q_2. The door-to-door salesman problem given above is an example of this probability strategy, both sides came up with \$11.33 as the game value.

Optimal strategies for all three by three games

If a game has 3 rows and 3 columns with no saddle point, it is a bit more complicated, but you can still set up probabilities for playing the three rows, finding the expected value of the game, and playing for that value as the best either player can do.

Subgame strategy for n by 2 or 2 by n games

An interesting configuration for a game is one that has 3 rows and 2 columns or one that has 2 rows and 3 columns. In general, all 2 by n or n by 2 games can be solved for any, $n > 2$ by a method called the *subgame strategy*. Let's say, for example, you have a 2 by 3 game. Three different two-by two matrices can be created by "dropping" the columns one at a time[4]. Then the values for these 2 by 2 games can be calculated. Using these computed game values the side that has control over which subgame to play can select the one that is best for them.

Example: In a game of "Flag Football," Navy is the row player and has two defenses, r_1 and r_2. Army has three offensive plays c_1, c_2, and c_3.

Here is a 2 by 3 game matrix for one play. Notice that it has no saddle point, but one of the subgames does have a saddle point. Using the subgame strategy find the game value for this matrix.

	Army		
Navy	c_1	c_2	c_3
r_1	3	2	1
r_2	-1	0	2

We find the three 2 by 2 subgames by dropping out the columns one at a time, then find the value of each subgame.

Drop column 3 to get Game A_1

$$A_1 = \begin{pmatrix} 3 & 2 & \dots \\ -1 & 0 & \dots \end{pmatrix} = \begin{pmatrix} 3 & 2 \\ -1 & 0 \end{pmatrix}$$

Game A_1 has a saddle point, a row minimum and column maximum, 2. It is in row 1, column 2. The game value is

$$V_1 = 2$$

Drop column 2 to get the next game

$$A_2 = \begin{pmatrix} 3 & \dots & 1 \\ -1 & \dots & 2 \end{pmatrix} = \begin{pmatrix} 3 & 1 \\ -1 & 2 \end{pmatrix}$$

---

[4]"Dropping" a column means that the column player decides, at the outset, to play that column with probability 0.

Game A_2 has no saddle point, use the 2 by 2 probability, q, for column 1.

$$q_1 = \frac{|2-1|}{|2-1| + |-1-3|} = \frac{1}{5}$$

the game value is

$$V_2 = \frac{1}{5} \times 3 + \frac{4}{5} \times 1 = \frac{7}{5}$$

Drop column 1 to get Game A_3

$$A_3 = \begin{pmatrix} \cdots & 2 & 1 \\ \cdots & 0 & 2 \end{pmatrix} = \begin{pmatrix} 2 & 1 \\ 0 & 2 \end{pmatrix}$$

In game A_3, the 2 by 2 strategy formula for the column players is to play column 1 with probability

$$\frac{|2-1|}{|2-1| + |2-0|} = \frac{1}{3}$$

and the game value is

$$V_3 = \frac{1}{3} \times 2 + \frac{2}{3} \times 1 = \frac{4}{3}$$

The three game values, $V_1 = 2$, $V_2 = \frac{7}{5}$, and $V_3 = \frac{4}{3}$ are positive so they all favor the row player, the Navy. What the Army wants is the smallest one, namely, V_3. Since the Army has the choice as to which column to drop, it is the side that decides which subgame to play, so it picks the one with the smallest game value, A_3, in order to minimize its loss.

Non-zero sum games

To learn what a non-zero sum game is, we need to know how the matrix for a non-zero sum game differs from the one for a zero sum game. In a zero sum game, such as the ones we have been discussing, the payoff in a row and column is a single positive or negative number favoring the row player, if positive and the column player if negative One player wins exactly the same amount the other player loses.

In a matrix for a non zero sum game, however, the payoffs displayed are *pairs* of positive or negative numbers, (x, y) indicating a payoff of x to the row player and y to the column player

Example: The following matrix game is a non-zero sum game. It is a battle between Earth Space Ship (EARTH1) and the space ship, ZHAKA1, from a nearby exo-planet, Zhaka. The rows and columns are *actions*, which may include negotiations, or the use of a variety of weapons. The ships meet somewhere in the Milky Way, with EARTH1 muttering something like, "This Galaxy ain't big enough for the both of us." In this, their first, and perhaps last, meeting, they play the following payoff table.

$$\text{Zhaka} \quad \begin{array}{c} \\ r_1 \\ r_2 \\ r_3 \end{array} \begin{pmatrix} \overset{\textstyle \text{Earth}}{} \\ \overset{c_1 \qquad\quad c_2 \qquad\quad c_3}{} \\ (5,0) & (2,7) & (6,-1) \\ (4,5) & (9,24) & (4,4) \\ (-8,-8) & (8,2) & (11,4) \end{pmatrix}$$

Observations:

1. If row plays r_1 and column plays c_3 Zhaka wins by $+7$ because the score is 6 to -1. (The first term, 6 is the row's payoff and the second term, -1, is the column's payoff)

2. If row plays r_2 and column plays c_2 then Zhaka gets clobbered by a deficit of 15. Earth wins 24 to 9.

3. If they played r_2, c_3 they both win 4; this is the *win-win* outcome, a tie with both sides winning points.

4. If they play r_3, c_1, they both lose 8, a tie with both sides losing points. It is a *lose-lose* move.

Looking at the whole column c_3, Earth realizes that the total is unfavorable because it never gains more than it loses. Similarly, Zhaka would probably avoid r_2 because over-all it's payoffs are not greater than Earth's. The problem is that row r_2 and column c_3 is the payoff where neither side would lose. Actually, it would be best for them to agree to cooperate by choosing the win-win option, r_2, c_3. But there is no incentive for the planets to play this combination. The only way it would ever be played would be through cooperation. It is a negotiated play.

After sizing up their matrix, the players, not only agree to play the r_2, c_3 combination but they also stipulate that if one side did play the agreed to move, but the other side cheated by making an alternative choice in order to try to win, then a *third party*, the Intergalactic Police, would destroy both planets, resulting in another type of lose-lose scenario.

The main problem in a non-zero sum game is: Who makes up the difference when both sides get a positive payoff? Or, in a lose-lose move, who gets the amount lost by the two sides? The answer is that there *is always a hidden player in any non-zero sum game.* That player might be, say, the sponsors of an event in which the games are played. These guys would have to make good on the payoffs when both players win something, and they would pocket the amount lost by one side and not paid to the other side. Thus, for example, the third player could be "the house" in a gambling casino.

In a negotiated settlement of a management *vs* labor dispute, the third party could be the consumer who would experience a price increase that management passes along because of the cost of the settlement. Sometimes, the hidden player is the tax payer in government assistance for national disasters, or a volunteer group in a local disaster, and so forth.

Management *vs* Labor

Example: A company has two choices to either *accede to* or *deny* the demands of its striking employees. And, independently, the labor union has two choices to either *end* or *continue* the strike. It is clear that each side has possible payoffs depending upon their own and their opponents choices. Generally these types of confrontations are not zero sum games. Let us make up a possible such game. Here is an example of a possible management *vs* labor strike.

Suppose there is a workers' strike for wage raises and safer working conditions.

$$
\text{Labor} \begin{array}{c} \\ \text{End} \\ \text{Continue} \end{array}
\begin{array}{c} \text{Management} \\
\left(\begin{array}{cc}
\text{Accede} & \text{Deny} \\
(10, 10) & (-10, 5) \\
(-15, -2) & (-10, -10)
\end{array} \right)
\end{array}
$$

As usual, the order pair (x, y) payoffs means x goes to the row player and y goes to the column player. If management accedes to the demands and Labor ends the strike, then both sides win. But who pays? Labor wins 10 in wages and conditions, Management wins 10 by getting to restart business, but probably has to raise prices on their products, so the consumer is the hidden player. If Labor ends the strike but Management still denies the demands, Labor loses big, because its payoff is -10 and Management regains production with out having to make changes; it wins $+5$. If Management accedes, but Labor continues the strike, Labor loses income and public support and management survives with a small loss. Finally, if Labor continues the strike and Management continues to deny the demands, both lose big time on salaries, conditions and production and sales.

Here Labor is afraid to independently play row 1 because they would lose -10 to 5 if Management played column 2. And Management is afraid to independently play column 1, because Labor may play row 2 and cause management to look weak by acceding to demands and still not settle the strike. They look at the payoff matrix and realize that row 1, column 1 is the win-win option, but they must agree to both play that one and promise not to cheat at the last minute by making another choice. It is a negotiated play.

Operations research

In the next chapter we see a mathematical concept called *Operations Research* or *Linear Programming*. This concept is intimately related to game theory. It is widely used in business, hospitals, military, communication, and in any other field in which it is important to manage large networks to optimize performance, subject to constraints. This could mean maximizing profits, minimizing costs while simultaneously conserving resources and satisfying demand. It is undoubtedly the most, other than calculus, wide spread "use" of mathematics in the anthropocene age.

CHAPTER 10

OPERATIONS RESEARCH

Assuming that mathematics is created by attempts to solve problems, we must conclude that certain branches of mathematics are more suitable for some types of problems, than they are for others. At the risk of oversimplifying this idea, we are inclined to say that those industries dealing with management of resources, communication, and assignment problems tend to use probability and finite mathematics, which includes *operations research*, a major component of which is linear programming. On the other hand, arenas involving medicine, engineering, construction and space exploration tend to use calculus and group theory.

We can get some idea of the wide-spread use of linear programming by noticing the amount of computer time consumed in working problems in this area. We quote the Hungarian computer scientist Laslo Lovasz, who said in 1980:

> *If one would take statistics about which mathematical problem is using up most of the computer time in the world the answer would probably be linear programming.*

In the same year Eugene Lawler, a computer scientist at the University of California at Berkeley, said about linear programming:

> *It is used to allocate resources, plan production, schedule workers, plan investment portfolios, formulate marketing and military strategies. The versatility and economic impact of linear programming in today's industrial world is truly awesome.*

Specifically some of the really big linear programming problems are:

1. Oil refinement problems–A company can access quantities and *qualities* of crude oil, and must meet demands by producing various products. The objective being either to maximize profits or minimize costs.

2. Internet problems–How to construct a network which allows the maximal flow of information across the connections.

253

3. Transportation and shipping–These are related to internet problems, often involving finding the shortest possible routes.

4. Military problems–These include rapid transportation of supplies, to meet varying demands, and efficient deployment of troops.

Most of these problems involve solving large systems of n linear equations in m variables. It would not be unusual for fractional distillation of crude oil to produce as many as $n = 30$ products requiring as many as $m = 30$ steps. Other industrial and manufacturing or communication systems would require many times that many variables and equations.

In this book, we will severely restrict the sizes of m and n in our examples. We can illustrate the important aspects of such problems, by discussing fairly modest ones, such as 2 or 3 equations in 4 or 5 variables.

The City Manager's problem

Example

A city manager has a budget that must be shared among various departments: *law enforcement*, *fire protection*, *parks and recreation*, *street maintenance*, *sanitation*, etc. The costs in these departments are variables with a range of values, but because the budget is finite, the range is restricted. Each cost influences and competes with the others. For this reason, they are said to *constrain* each other. The budget can be described as a *function of several independent variables subject to constraints*. The constraints are given in the form of *inequalities* with the property that when one of the variable increases its share of the budget, at least one of the others is forced to decrease its share.

Let us consider two line-items: *law enforcement* and *park services*, that need to be budgeted by the city manager. Money spent on recreation reduces the amount available for law enforcement, which is justified because the existence of recreational facilities helps to reduce the actual cost of law enforcement.

Suppose that the money available in some year for both law enforcement and recreation is $A = \$150,000$. Assume that equipment, facilities, fixed costs, and legal requirements, makes it turn out that the park budget plus \$600 cannot exceed $\frac{1}{10}$ of the law enforcement budget. We will put these constraints into mathematical notation as follows:

$$
\begin{aligned}
x_1 &= \text{Law enforcement} \\
x_2 &= \text{Parks and recreation} \\
x_1 + x_2 &\leq 150,000, \text{ total available for both items} \\
x_2 + 600 &\leq \frac{1}{10}x_1 \text{ facilities constraint}
\end{aligned}
$$

We rewrite these as the system of inequalities with the variables on the left hand side:

$$
\begin{aligned}
x_1 + x_2 &\leq 150,000 \\
-\frac{1}{10}x_1 + x_2 &\leq -600
\end{aligned}
$$

To clear fractions, multiply the second inequality by -10 (which reverses it from $\leq$ to $\geq$) we get:

$$x_1 + x_2 \quad \leq \quad 150,000 \tag{10.1}$$

$$x_1 - 10x_2 \quad \geq \quad 6000 \tag{10.2}$$

We also have, in this and in every linear programming problem, the constraint that none of the variables can have a negative value. This is called the *feasibility* constraint. Neither line-item can be less than zero, so both $x_1 \geq 0$ and $x_2 \geq 0$. A negative value for a variable is *infeasible*.

Suppose that a study shows that for every dollar spent on *parks*, the city accrues a benefit of 0.007 rating points as determined by tourist surveys and for each dollar spent on *law enforcement* the city accrues a benefit of 0.001 rating points. Thus, the rating points equation is:

$$R = 0.001x_1 + 0.007x_2 \tag{10.3}$$

The city wants to consider the the rating points in budgeting these two line-items. The objective is to maximize the rating points, subject to the constraints. We call Equation 10.3 the *objective equation*. So now we have the problem:

Problem:

Find solution to the Inequalities 10.1, and 10.2 that maximizes the objective function R.

Note:

We need to interrupt the attempt to solve this problem right now to point out that the system 10.1 and 10.2 does not consist of two *equations*, but rather two *inequalities*. Therefore, we cannot use row reduction to solve it as a system of two equations in two unknowns as we did in Chapter 2. Look back at Chapter 2. What can we do about that? Let us introduce to you *slack variables*.

Slack and surplus variables

How can we turn an inequality into an equation? Let us say that you know that some non-negative number y is less than or equal to 100

$$0 \leq y \leq 100$$

Let s denote the difference between 100 and y, that is $s = 100 - y$. This number is called a *slack variable* because it takes up the slack between y and 100, so it can be added to y in order to bring it up to 100. That is,

$$\text{If } y \quad \leq \quad 100 \text{ then there is a number, } s \geq 0, \text{ such that}$$
$$y + s \quad = \quad 100$$

Or, if $y \geq 100$, then there is a number $t \geq 0$, such that $y - t = 100$. The number t, describes the amount in excess of 100. It is sometimes called a *surplus variable*. So, an inequality such as $x_1 + x_2 \leq 150,000$ can be written as an equation: $x_1 + x_2 + s = 150,000$, and the inequality $x_1 - 10x_2 \geq 6000$, can be written as $x_1 - 10x_2 - t = 6000$. Now, back to the city manager problem.

We will re-write the system 10.1 and 10.2 using slack and surplus variables, and we will still use Equation 10.3 as the objective function, but we will include the slack and surplus variables and ascribe to them 0 rating points in the equation for R. This is tantamount to using the slack and surplus variables as a catalyst in solving the problem. They give us enough variables in a system of equations, and they drop out of the solution without affecting the outcome.

ORIGINAL VERSION OF THE PROBLEM USING INEQUALITIES

Maximize R, where:

$$R = 0.001x_1 + 0.007x_2$$

subject to

$$x_1 + x_2 \leq 150,000 \text{ and}$$
$$x_1 - 10x_2 \geq 6000$$

NEW VERSION OF THE PROBLEM USING EQUATIONS

Using a slack variable s and a surplus variable t, we we change the objective formula R so that it will include these two new variables, but with coefficients of 0. This tells us that the slack and surplus variables do not contribute to the rating. points. The problem now is to maximize R, where

$$R = 0.001x_1 + 0.007x_2 + 0s + 0t$$

subject to

$$x_1 + x_2 + 1s + 0t = 150,000 \tag{10.4}$$
$$x_1 - 10x_2 + 0s - 1t = 6000 \tag{10.5}$$

This is a 2 by 4 system, two equations with four variables. And, we are trying to find only the non-negative values of x_1, x_2, s, and t that makes R the largest.

Well, have we really made any progress? Can we solve a system that has four variables and only two equations? Yes, we can and will; here is how. Solve all of the possible 2 by 2 systems that we can create by letting two of the variables at a time be zero[1]. In each system, the remaining two variables will have solutions, provided the coefficients do not have a zero determinant (see Chapter 2), but we must reject them if any variable is negative. This would be an infeasible solution.

On the other hand, if a system has a feasible solution we compute R. Then we look for the feasible solution with the largest value of R in maximizing problems and the smallest value of R in minimizing problems. We have solved the problem of finding the optimal system.

Basic and non-basic variables

Any of the variables that we deliberately set equal to zero are called *non-basic* variables. We are leaving them out of the system and solving for the other variables, those that we left in the system. These latter are called the *basic* variables. There are six ways we can select two of the four variables to keep basic. This is the binomial coefficient $\binom{4}{2} = 6$ that we studied in Chapter 6.

---

[1]This should remind you of the subgame strategy we used in solving 2 by n games.

Solution to the city manager's problem

Here are solutions to the 6 different systems. We are going to list the six possible systems in the following table. The headings are: **S** = System; **N** = non-basic variables; **B** =Basic variables; **Equations** = the equations of the basic variables; **Solutions** = Solutions of the equations; **F?** = Is the solution feasible?; **R** = value of R for the feasible solutions. We put an **X** in places where R is not to be computed.

S	N	B	Equations	Solutions	F?	R
1	x_1, x_2	s, t	$\begin{pmatrix} s = 150K \\ -t = 6K \end{pmatrix}$	$\begin{pmatrix} s = 150K \\ t = -6K \end{pmatrix}$	No	X
2	t, x_2	x_1, s	$\begin{pmatrix} x_1 + s = 150K \\ x_1 = 6K \end{pmatrix}$	$\begin{pmatrix} x_1 = 6K \\ s = 144K \end{pmatrix}$	Yes	$R = 6$
3	x_2, s	x_1, t	$\begin{pmatrix} x_1 = 150K \\ x_1 - t = 6K \end{pmatrix}$	$\begin{pmatrix} x_1 = 150K \\ t = 144K \end{pmatrix}$	Yes	$R = 150$
4	x_1, t	x_2, s	$\begin{pmatrix} x_2 + s = 150K \\ -10x_2 = 6K \end{pmatrix}$	$\begin{pmatrix} s = 150,600 \\ x_2 = -600 \end{pmatrix}$	No	X
5	x_1, s	x_2, t	$\begin{pmatrix} x_2 = 150K \\ -10x_2 - t = 6K \end{pmatrix}$	$\begin{pmatrix} x_2 = 150K \\ t = -1,506K \end{pmatrix}$	No	X
6	s, t	x_1, x_2	$\begin{pmatrix} x_1 + x_2 = 150K \\ x_1 - 10x_2 = 6K \end{pmatrix}$	$\begin{pmatrix} x_1 = 136,909 \\ x_2 = 13,090 \end{pmatrix}$	Yes	$R \approx 228.5$

Table 10.1 Solving the city manager's problem by trial and error

The method we just used is not the best way to solve these problems. It wastes too much time computing infeasible solutions and does not provide a systematic way to get to the feasible ones. A better method, *the simplex algorithm*, was invented in the 1940's by the American mathematician George Dantzig and, independently, by the Russian mathematician, Leonid. V. Kantorovich. It is one of the two most used formulas in solving linear programming and Operations Research problems. The origin of large problems involving efficient movement of equipment and military personnel as well as the economical use of resources can be traced back to around the time that electronic computers were beginning to emerge in the late 1930's and early 1940's.

Much of the work done on these types of problems between 1940 to 1960 was developed by Dantzig and other mathematicians. The simplex method was in widespread use during these years and expanded hand-in-hand along with the development of faster and more powerful computers. This was the type of mathematics that the world needed. Increasing population and increasing demand coupled with limited resources and crowded and complex conditions required efficient management. This is what attracted communication, transportation, other industries, large and small, as well as government to linear programming. It took until 1963 for Dantzig's book; *Linear Programming and Extension* (Princeton University Press, 1963) to appear. This book gives a comprehensive history of linear programming and is the classical reference source for various methods of solution.

Another method for solving linear programming problems, the Karmarkar Algorithm appeared in 1984. While working for I.B.M., a mathematician from India, Narenfra Karmarkar developed a linear programming algorithm that uses properties from the field of *projective geometry*. It is a satisfactory method for solving linear programming problems. Karmarkar moved on to Bell Laboratories and his algorithm, evolved into a computationally practical method. Both the simplex method and modern versions of Karmarkar are currently in use to solve linear programming problems.

Since it is easier to describe the geometry of Dantzig's method, we will confine our discussion to the Simplex Algorithm. Here is how to interpret it. In the above city manager problem, the four variables x_1, x_2, s, and t could be designated as a point (x_1, x_2, x_3, x_4) in 4 dimensional space. Now, what would a geometric entity confined by linear equations in these variables look like?

Let us back up and start by looking at a two dimensional space. Begin with linear constraints in the plane and build up to higher dimensional simplices. For example, start with the *x-axis*, whose equation is $y = 0$ and the *y-axis* whose equation is $x = 0$, and the following two linear equations, (10.6) and (10.7), in two variables, as sketched in Figure 10.1

$$
\begin{aligned}
x_1 + 3x_2 &= 15 && \text{(10.6)} \\
x_1 + x_2 &= 10 && \text{(10.7)}
\end{aligned}
$$

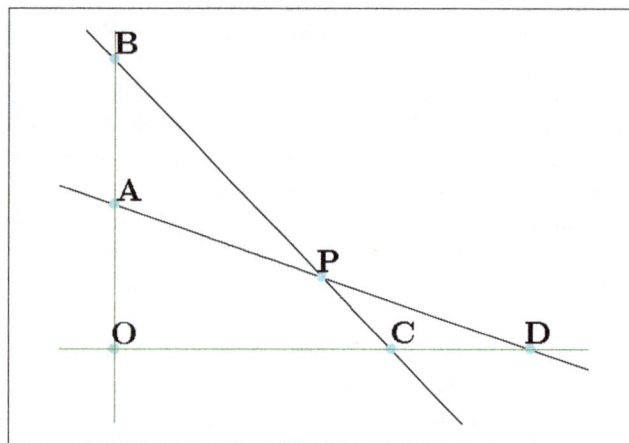

Figure 10.1 LINES INTERSECT IN POINTS

If these had been *inequalities*, instead of *equations*, then, we might have

$$
\begin{aligned}
x_1 + 3x_2 &\leq 15 && \text{(10.8)} \\
x_1 + x_2 &\leq 10 && \text{(10.9)}
\end{aligned}
$$

And, including the constraints, $x_1 \geq 0$ and $x_2 \geq 0$, we get the interior to the the polygon $APCO$, the shaded area in Figure 10.2.

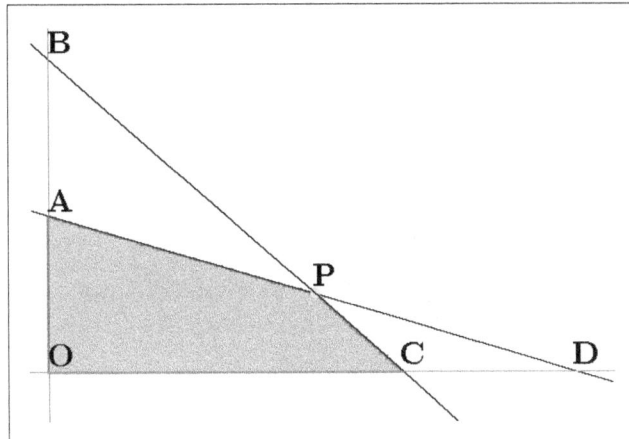

Figure 10.2 A 2-DIMENSIONAL SIMPLEX

If, however, the inequalities had been:

$$x_1 + 3x_2 \leq 15 \tag{10.10}$$

$$x_1 + x_2 \geq 10 \tag{10.11}$$

then the points would have been in the interior of triangle PCD as in Figure 10.3, and this would be the simplex defined by the inequalities, (10.10) and (10.11).

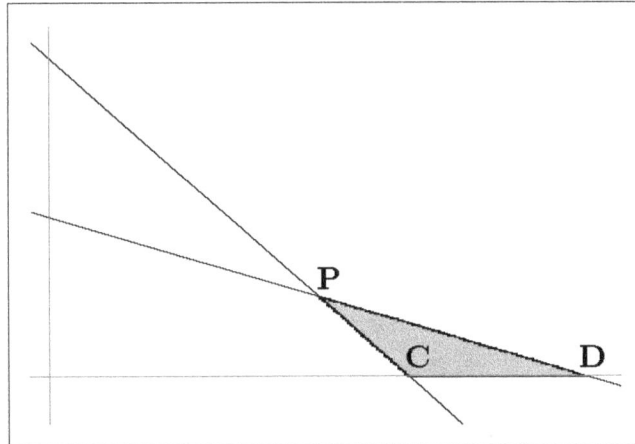

Figure 10.3 ANOTHER 2-DIMENSIONAL SIMPLEX

What about a 3 or higher dimensional simplex? It is defined by interior points of a convex polyhedron defined by points bound by n dimensional planes, or "walls", instead of lines. This means that the points satisfy linear inequalities in three or more variables.

What would a higher dimensional (4 or 5 or more) simplex look like? A whimsical answer is that a whole system would look like a kid's giant ramshackle cardboard "fort" in your living room.

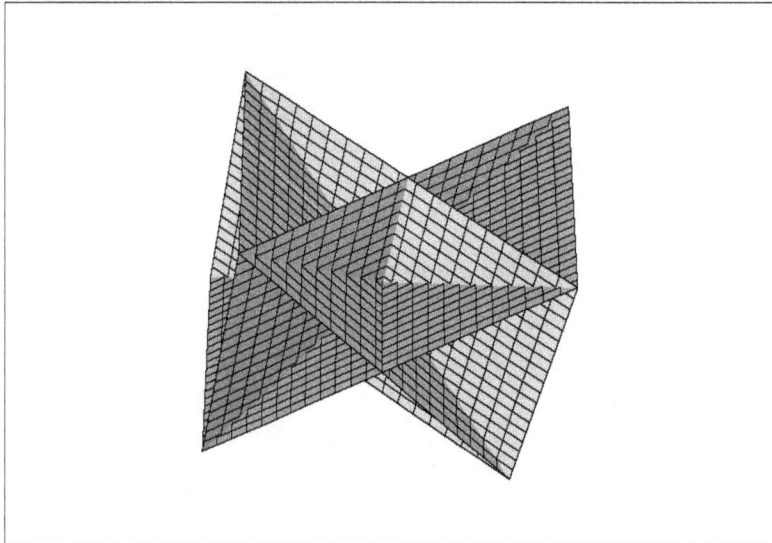

Figure 10.4 AN N-DIMENSIONAL SIMPLEX

The objective function distributes a prize at each of the corner points, the feasible solutions, of this structure. So, the kids just have to start at some *convenient* corner point, called the initial feasible solution, inside the structure, then move from that corner to another feasible solution, at some other corner. At each point they can evaluate the extant prize, and pivot to another corner, if any, that offers a better prize. This continues until they find the best prize. This will be known to them because, at each corner, they can test the *opportunity cost* of either staying put or moving on to a new adjacent corner point.

Mathematically, the whole procedure is called the simplex algorithm, and works for any n by n simplex, and for either maximizing or minimizing objective functions. In the next section we will work on a problem and describe all of the algorithm rules required to optimize a solution to an objective function under both less than and greater than constraints.

Programming the simplex algorithm

If we have a system of m linear inequalities in n variables, (where n is, usually, larger than m) and we have a linear objective function of the n variables then we can apply the following algorithm to find a point in n space, if any, that furnishes the optimal value of the objective function.

PREPARATION FOR THE PROGRAM

First we will assume that all of the variables are non-negative. That is $x_1 \geq 0$, $x_2 \geq 0$, ...$x_n \geq 0$. Second make sure that the right hand sides of all of the inequalities are also non-negative.

Example:

If you are given the following system of constraints in four variables

$$
\begin{array}{rcl}
2x_1 - 7x_2 + 13x_3 + 0x_4 & \leq & 237 \\
-6x_1 + 10x_2 - 17x_3 + 44x_4 & \geq & 88 \\
8x_1 + 190x_2 - 19x_3 + x_4 & \geq & -1170 \\
x_1 - 1x_2 - 10x_3 - 39x_4 & \leq & 1000
\end{array}
\tag{10.12}
$$

You notice that the third inequality has a negative number, -1170, on the right hand side. You need to change this to one that has a positive number on the right hand side. This is easily done just by multiplying both sides by (-1), changing this constraint to

$$-8x_1 - 190x_2 + 19x_3 - x_4 \leq 1170$$

It is important to notice that this act of multiplying both sides by (-1) not only changed the signs on all of the coefficients, but it also *reversed* the inequality from $\geq$ to $\leq$. Now the system can be written as:

$$
\begin{array}{rcl}
2x_1 - 7x_2 + 13x_3 + 0x_4 & \leq & 237 \\
-6x_1 + 10x_2 - 17x_3 + 44x_4 & \geq & 88 \\
-8x_1 - 190x_2 + 19x_3 - x_4 & \leq & 1170 \\
x_1 - 1x_2 - 10x_3 - 39x_4 & \leq & 1000
\end{array}
$$

A system written this way is said to be written in *standard*, or *canonical* form. If any $\geq$ inequality, in the standard form, has only negative coefficients, and a positive right hand side, then the solution does not exist. This is a good check point. If some $\geq$ inequality has at least one variable with a positive coefficient, then continue to work on the problem. If, however, all of the coefficients in such a constraint are negative, when the right hand side is positive, then stop and report *no solution.*

STEPS FOR SOLVING A LINEAR PROGRAMMING PROBLEM

Once we have put the constraints in standard form, convert the constraints from a system of *inequalities* into a system of *equations* as follows. For each $\leq$ inequality introduce a *slack* variable, with $(+1)$ as its coefficient, to convert the inequality to an equation and for each $\geq$ inequality introduce a *surplus* variable with a coefficient of (-1) in order to convert the constraint into an equation. Keep in mind that for any solution to be feasible, all of the slack and surplus variables must be non-negative

Example: In the above system we will introduce slack and surplus variables x_5, x_6, x_7, x_8 for the four inequalities, getting:

$$
\begin{array}{rcl}
2x_1 - 7x_2 + 13x_3 + 0x_4 + 1x_5 + 0x_6 + 0x_7 + 0x_8 & = & 237 \\
-6x_1 + 10x_2 - 17x_3 + 44x_4 + 0x_5 - 1x_6 + 0x_7 + 0x_8 & = & 88 \\
-8x_1 - 190x_2 + 19x_3 - x_4 + 0x_5 + 0x_6 + 1x_7 + 0x_8 & = & 1170 \\
x_1 - 1x_2 - 10x_3 - 39x_4 + 0x_5 + 0x_6 + 0x_7 + 1x_8 & = & 1000
\end{array}
$$

In *maximizing* problems, change the objective function by adding all of the slack variables, including the surplus variables, with coefficients equal to a negative number $-M$, with a large absolute value, *"minus a bazillion."* On the other hand, in *minimizing* problems, let the coefficients of the slack and surplus variables be $+M$," *plus a bazillion."*

Note:

The use of this silly nonsensical word, "bazillion" is to emphasize that we want the slack variables in the objective function to have a really *large negative coefficient in maximizing problems* and *large positive coefficients in minimizing problems.* This is to preclude them from ever being a permanent member of the feasible solutions. The number M is presumed to be much larger, say a million times larger, than any positive constraint or any positive coefficient in the original problems.

Now, here let us suppose that the objective function F is to be *maximized*, where F is

$$F(x_1, x_2, x_3, x_4) = 31x_1 - 1x_2 + 50x_3 + 7x_4$$

Taking M to be any number larger than say 1170 times a million, re-write the objective function, using M, as follows

$$\begin{aligned} F(x_1, x_2, x_3, x_4) \;=\; & 31x_1 - 1x_2 + 50x_3 + 7x_4 \\ & -Mx_5 - Mx_6 - Mx_7 - Mx_8 \end{aligned}$$

Then, even if you had used x_5 or x_7, or x_8 as part of your initial variables, the $(-M)$ coefficients would quickly weed them out as candidates for solving a maximizing problem.

Similarly, for an objective function you want to *minimize*,

$$G(x_1, x_2, x_3, x_4) = -6x_1 + 100x_2 - 13x_3 + 37x_4$$

use $+M$ as the coefficients for the slack and surplus variables, getting

$$\begin{aligned} G(x_1, x_2, x_3, x_4) \;=\; & -6x_1 + 100x_2 - 13x_3 + 37x_4 \\ & +Mx_5 + Mx_6 + Mx_7 + Mx_8 \end{aligned}$$

In either case, the system now has too many variables for the number of equations. We are going to make some of the variables have a zero value in order to leave us n variables for the n equations giving us a square (n by n) system. The variables in the square system are called *basic variables* and the variables deliberately set aside (made zero) are the *non-basic variables.* As we solve the square n by n system for the basic variables, we will keep an eye on the non-basic variables to see whether or not it would be better to have one of them in the system of basic variables, instead of one that is in the system. It is like maintaining an understudy, off-stage, for a current performer in order to keep the show going without a hitch. Only, in this case we want to replace the current performer with a superior one, if possible.

Here is the procedure for doing that

PROGRAM

1. **Put the system into standard form:**   Change inequalities to equations and change the objective function.

2. **Get a feasible initial solution:**   Use slack or other variables with positive coefficients as your first basic variables.

3. **Evaluate the objective function:** Evaluate the objective function for this solution.

4. **Test the opportunity cost:** Improvement Possible? **YES:** Go to **Step 5**. **NO:** Go to **Step 9**.

5. **Identify in the better non-basic variable:** Select this one as the candidate to be made basic.

6. **Determine the outgoing basic variable:** Find one whose removal yields a new feasible solution.

7. **Solve the new system:** Evaluate the objective function for this solution.

8. **Go to Step 4.**

9. **END**

Let us apply this algorithm to a problem we have already solved, the city manager problem concerning the budgeting of Law Enforcement *vs.* Park Services.

City manager's problem revisited

Restatement of the problem:

A city manager wants to maximize the city's score on a survey in which tourists rate the city as a vacation spot.

The objective function is the equation:

$$R = 0.001x_1 + 0.007x_2 \qquad (10.13)$$

Where x_1 and x_2, are the amounts in dollars that the city assigns to two line-items:

$$x_1 = \text{Law enforcement}$$
$$x_2 = \text{Parks and Recreation}$$

We are given the following constraints on the budget.

$$x_1 + x_2 \leq 150,000$$
$$x_1 - 10x_2 \geq 6,000$$

These constraints are in standard form because the right-hand sides of the inequalities are positive. We add a slack variable, x_3, with coefficient $(+1)$ to the *less than* inequality, subtract a surplus variable, x_4, from the *greater than* inequality.

$$1x_1 + 1x_2 + 1x_3 + 0x_4 = 150,000$$
$$1x_1 - 10x_2 + 0x_3 - 1x_4 = 6000$$

This is a maximizing problem, so letting M be a large positive number, and use coefficients of (-1) for the slack and surplus variables making the objective function

$$R = 0.001x_1 + 0.007x_2 - Mx_3 - Mx_4$$

We can find a feasible solution by finding two variables in this system that have coefficients,

$$\begin{bmatrix} a & c \\ b & d \end{bmatrix}$$

whose determinant $ad - bc \neq 0$. Let's start with the 2 by 4 system

$$\boxed{1x_1} + 1x_2 + \boxed{1x_3} + 0x_4 \;=\; 150,000 \tag{10.14}$$

$$\boxed{1x_1} - 10x_2 + \boxed{0x_3} - 1x_4 \;=\; 6000 \tag{10.15}$$

The boxed in terms tell us that the basic variables in the system are x_1, x_3 and the non-basic ones are x_2, x_4. Here is the determinant of the coefficient matrix for the basic variables,

$$\det \begin{bmatrix} 1 & 1 \\ 1 & 0 \end{bmatrix} = -1$$

Working with the whole system at once, use *row reduction*[2]. Keep row 2 and replace row 1 by row 1 minus row 2:

$$\boxed{0x_1} + 11x_2 + \boxed{1x_3} + 1x_4 \;=\; \boxed{144,000}$$

$$\boxed{1x_1} - 10x_2 + \boxed{0x_3} - 1x_4 \;=\; \boxed{6,000}$$

Interchange rows.

$$\boxed{1x_1} - 10x_2 + \boxed{0x_3} - 1x_4 \;=\; \boxed{6,000} \tag{10.16}$$

$$\boxed{0x_1} + 11x_2 + \boxed{1x_3} + 1x_4 \;=\; \boxed{144,000} \tag{10.17}$$

The solution is: $\{x_1 = 6000,\ x_2 = 0,\ x_3 = 144000,\ x_4 = 0\}$.

Using K to stand for 1000, the objective function for this solution is

$$\begin{aligned} R &= 0.001 \times 6K + 0.007 \times 0 - M \times 144K - M \times 0 \\ R &= 6 - 144K \times M \end{aligned}$$

Could this be the maximum value for R? No at all. In the first place R cannot be negative because there is no budget distribution that can make that happen, and here $-144K \times M$ has an extremely negative value less than negative 144 million. Remember M, itself, has been assigned a very large number. Actually, in terms of the original variables this solution is $x_1 = 6000$ and $x_2 = 0$, so the value of R is:

$$R = 0.001 \times 6000 + 0.007 \times 0 = 6$$

It should be pointed out that we apply the row reduction process to the entire system, Equations (10.14) and (10.15) to get to the system with Equations (10.16) and (10.17). This keeps the non-basic variables "in the loop," that is to say, upgraded. This is unlike the solution of the 6 systems obtained when we arbitrarily made two of the variables

---

[2]See chapter 2.

basic and two non-basic at a time and kept returning to the original equations each time. The advantage in working on the equations with all four of the variables is that we can test which basic variable is causing an *opportunity cost* to the system because there is a non-basic variable capable of contributing more value to the objective function.

This might be compared to a major league baseball team continuing to challenge one of the pitchers on its farm team, while evaluating one of the pitchers on its national team. When its major team pitcher is not performing well, they may feel that they are suffering an opportunity cost, losing games they should have won. So they want to swap him out with the minor league player.

We are now going to reveal the steps in the simplex algorithm, that tell us when and how to perform these swap-outs. I feel that this is extremely interesting stuff, but I want to warn the reader that the next few pages involve some heavy details.

To start, Equations(10.16) and (10.17) have been entered into the following table.

C_i	0.001	0.007	$-M$	$-M$		
v	x_1	x_2	x_3	x_4	z_0	c_0
x_1	1	-10	0	-1	$6K$	0.001
x_3	0	11	1	1	$144K$	$-M$
z_i	0.001	$-0.01 - 11M$	$-M$	$-0.001 - M$	R	
$C_i - z_i$	0	$11M + 0.017$	0	$+0.001$		

Now, let us explain the other entries.

1. The C_i's in the top row are the coefficients for the objective function for all of the variables.

2. Horizontally, the **v** in second row shows all of the variables. Vertically, **v** shows the *currently* basic variables.

3. The column c_0 are the coefficients of the objective function for the current basic variables.

4. The z_0 column is the solution to the system when using the current basic variables.

5. R is the value of the objective function for the current basis. It is c_0 written as a row vector times z_0 written as a column vector.

$$R = c_0 \cdot z_0$$
$$= (0.001, -M) \cdot \begin{pmatrix} 6K \\ 144K \end{pmatrix}$$
$$6 - 144K \times M$$

6. In the next to the last row, for any integer i, the z_i stands for the value that the objective function would have if the *currently* non-basic variable, x_i were currently basic. In other words, when a new x_i is in the system, z_i's would be the new c_0, written as a row times the new z_0 column. We want to see how each currently non-basic variable would effect objective function if it had been in the basis. How do we compute the z_i's? Writing c_0 as a row matrix $(0.001, -M)$ and each of the x_i's as a column matrix, we define z_i as: $z_i = c_0 \cdot x_i$. Just try calculating one of the z_i's, for example, z_2.

$$z_2 = (0.001, -M) \times \begin{pmatrix} -10 \\ 11 \end{pmatrix} = -0.01 - 11M$$

Each entry in the last row $C_i - z_i$ is the *opportunity cost* obtained by subtracting z_i from the objective function coefficient of x_i. When x_i is non-basic then $C_i - z_i$ measures what value x_i would have contributed to the objective function had it actually been a basic variable. Notice that if we had computed z_1 or z_3 for the two variables x_1 and x_3 that are in the system, then you get $C_1 - z_1 = 0$, and $C_3 - z_3 = 0$, indicating there is no opportunity cost for variables that are already in the system. It makes sense.

If the opportunity cost is negative for every non-basic variable, then the maximum value of the objective function has been reached and no other variable will increase the value. Stop. The maximization problem is solved.

Otherwise, if $C_i - z_i$ is positive for some non-basic x_i, then that variable is beckoning you to bring it in. By doing so you must be prepared to remove one of the current basic variables, which is tantamount to pivoting to a new corner point and get a better prize (that is, improve the value of the objective function).

In this initial solution we see that the opportunity cost, $C_2 - z_2 = 11M + 0.017$ is the largest positive number in the last row; that means that we can get a better value for R if we bring in x_2. We will make x_2 the incoming column. Looking at the x_2 column, which variable x_1 or x_3 should come out? Intuitively, we can easily see that it is x_3 that is doomed for removal, but we need a mathematical justification for this decision in case some other situation may be less obvious.

There is a rule, called the *replacement rule* and it goes like this: Let b_1 and b_2 be the entries, in the *incoming* column, here they are (-10), and 11. Now look at the entries of the current solution, the z_0 column, call them a_1 and a_2, here they are: $6K$ and $144K$.

For every *positive* coordinate in the incoming column (the b_i's) find the *replacement ratio* a_i/b_i. Pick the smallest such ratio as the one that identifies the outgoing variable. Here since b_1 is negative, we cannot use the replacement ratio a_1/b_1. Here we use b_2 and compute a_2/b_2. This is the smallest and only positive replacement ratio and it appears in row x_3.

OK, since the smallest such positive ratio is in row x_3, BINGO! x_3 is the outgoing variable It will be replaced by x_2, making (x_1, x_2) the new basis What this replacement ratio really does is to make the next solution *feasible* because it insures that you will be subtracting smaller numbers from larger numbers.

Before solving for this basis, the new table would be:

C_i	0.001	0.007	$-M$	$-M$		
v	x_1	x_2	x_3	x_4	z_0	c_0
x_1	1	-10	0	-1	$6K$	0.001
x_2	0	11	1	1	$144K$	0.007

Use row reduction to solve this system; first, replace Row 1 by Row $1 + \frac{10}{11} \times$ Row 2.

C_i	0.001	0.007	$-M$	$-M$		
v	x_1	x_2	x_3	x_4	z_0	c_0
x_1	1	0	10/11	$-1/11$	$136.9K$	0.001
x_2	0	11	1	1	$144K$	0.007

Then divide row 2 by 11.

C_i	0.001	0.007	$-M$	$-M$		
v	x_1	x_2	x_3	x_4	z_0	c_0
x_1	1	0	10/11	$-1/11$	136.9K	0.001
x_2	0	1	1/11	1/11	13.09K	0.007

Now that we have solved the system for these two variables, we can return to a full table with the opportunity costs entries.

C_i	0.001	0.007	$-M$	$-M$		
v	x_1	x_2	x_3	x_4	z_0	c_0
x_1	1	0	10/11	$-1/11$	136.9K	0.001
x_2	0	1	1/11	1/11	13.09K	0.007
z_i	0.001	.007	ϵ	$-\epsilon$	228.5	
$C_i - z_i$	0	0	$-M - \epsilon$	$-M + \epsilon$		

The ϵ, is a small positive negligible amount, such as $\frac{0.017}{11}$, indicating that $-M - \epsilon$ or $-M + \epsilon$, are essentially just $-M$. In any case, no opportunity cost is positive, therefore, there is no better incoming variable and the solution is $\{x_1 = 136,900,\ x_2 = 13,090,\ R = 228.5\}$. This is the same answer that we got when we worked this problem by trial and error.

Simplex Algorithm Flow Chart

Here is a summary of the simplex method, and in Figure 10.5, below, we display a flow chart that depicts the steps in this algorithm. You may wish to look back and forth between this list of steps and the flow chart.

1. Put in standard form with slack and surplus variables.

2. Is this a maximizing problem?

  (a). If No, skip to step 7. (b) If Yes, continue.

3. Set objective coefficients of the slack and surplus variables $= (-M)$

4. Compute z, z_i, $C_i - z_i$.

5. Is $C_i - z_i$ positive for some x_i?

  (a). If No, STOP. Optimal solution found. (b) If Yes, continue.

6. For the most positive $C_i - z_i$, x_i enters the basis; go to step 11.

7. Set objective coefficients of the slack and surplus variables $= (+M)$.

8. Compute z, z_i, $C_i - z_i$.

9. Is $C_i - z_i$ positive for some x_i?

  (a). If No, STOP. Optimal solution found. (b) If Yes, continue.

10. For the most negative $C_i - z_i$; x_i enters the basis; go to step 11.

11. Does the incoming variable have a positive replacement quotient?

  (a). If No, go to step 16. (b). If Yes, continue.

12. Compute replacement quotients for the incoming variable.

13. Denote by x_k the row with the smallest nonnegative replacement quotient. Replace x_k by x_i.

14. By row-reduction, solve the new system for these basic variables.

15 . Is this a maximizing problem?

    (a). If No, loop back to step 8 .   (b). If Yes loop back to step 4 .

16 . Analyze the problem.

A Note on this algorithm

Step 16 happens because the incoming variable has no possible outgoing variable. This means that we cannot get a better solution for this incoming variable. It may turn out that there is none because the simplex may not be confined to a finite space. It could be unbounded, or there may not be a unique solution.

    The city manager problem is so small that the use of the simplex algorithm is an awkward way to solve it. But in systems with more equations and variables, the trial an error method would be worse. It turns out that the simplex method is the perfect tool for these *optimization under constraints* types of problems.

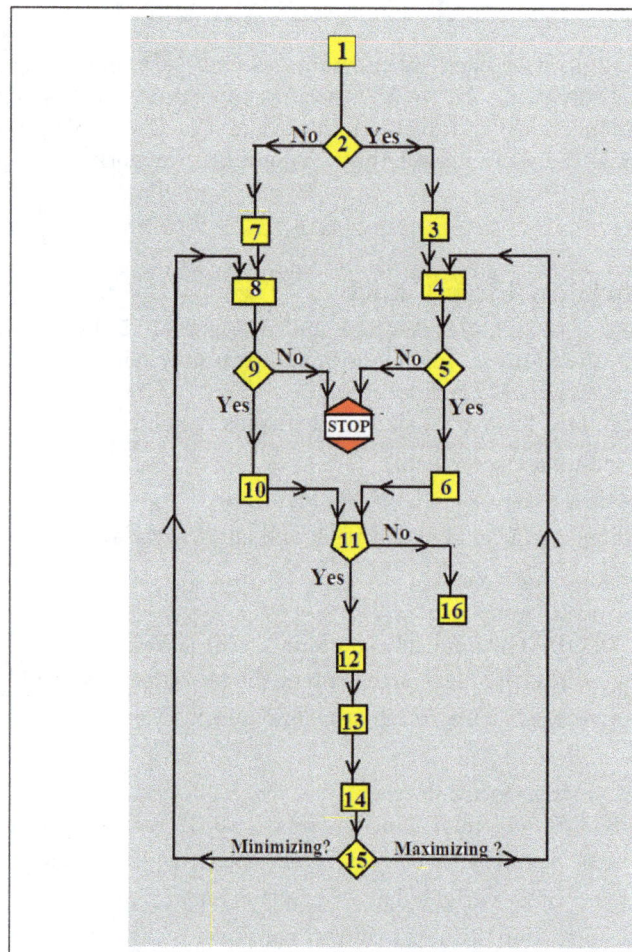

FIGURE 10.5 SIMPLEX ALGORITHM

CHAPTER 11 _____

THE CALCULUS

En plein air is something artists, especially painters, do fresh, and spontaneously. It is done in real time, on the spot. You might even say "on the fly." Mathematicians do something like this when they use calculus to solve a problem dealing with instantaneous changes in force, velocity or direction. A baseball outfielder does the same thing watching the trajectory of a fly ball and adjusting his or her own path just in time to arrive at the right spot at the right time.

Apparently, other living creatures behave in this same kinesthetic dance we call calculus.

Do animals use calculus?

The worms

Of course, we human animals do, but what about dogs and worms? In 2008, University of Oregon biologist, Shawn Lockery did an experiment in which he was able to cause round worms to change their path to new directions by changing the intensity of the smell of food. Like the baseball outfielder, the worms altered their path in order to get to the ultimate goal. Their reaction, taking an alternate path, was interpreted as evidence that the worm's brain computed the derivative[1], causing them to change their path relative to the change in food's odor. The worm's behavior actually is an application of calculus. A mechanical analog of the worm experiment uses mindless steel balls and magnets causing the balls to alter their path by changing the magnets positions and strengths.

The dog

Another interesting story is found in an article, pp 178 – 184 of the May 2003 issue of *The College Mathematics Journal*, Vol. 34. The American mathematician, Tim Penning from, Hope College in Holland, Michigan, has done several experiments in which his dog solved the calculus problem of finding the path through two different media (land and

---

[1]The derivative of a function is a computation that tells you the rate of change in a function derived from a change in its variable.

water) to a target in the minimum time. A typical result of these experiments, idealized here, for discussion, is the following.

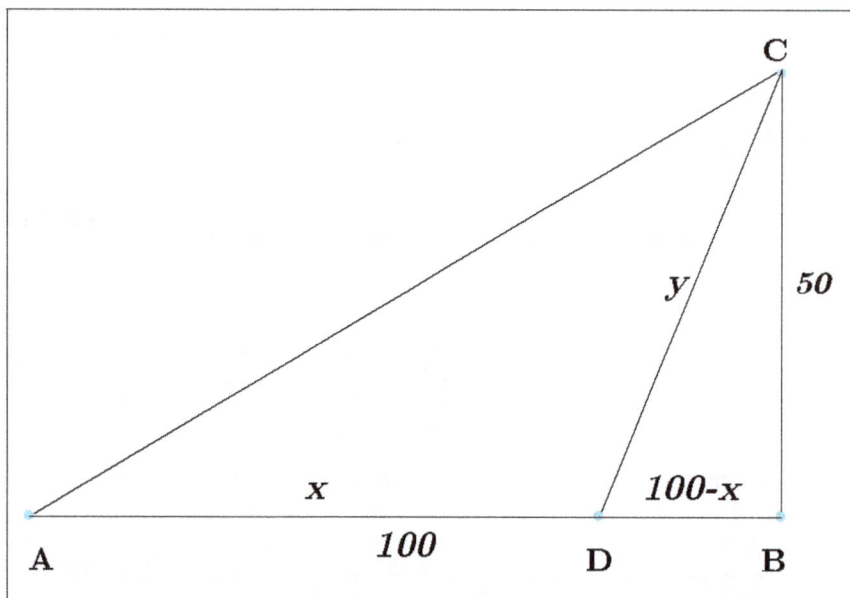

FIGURE 11.1 QUICKEST RECOVERY OF THE BALL

Problem:

Professor Pennington takes his dog, Elvis, to a flat linear beach on Lake Michigan. Elvis can run faster than he can swim (say about 3.25 times as fast).   Pennington, standing at point A, throws a tennis ball to a point C in the water. The line from A to B is land at the water's edge and is 100 ft. long. The points A, B, and C form a right triangle with BC being 50 ft. long. Most of the time Elvis does not jump in the water at point A, and swim the whole 112 ft. to C, but rather runs as fast as he can, on land, and jumps into the water at a point D on his own volition and swims as fast as he can toward the ball and recovers it. The problem Elvis solves is: At what point D must he enter the water so that the time it takes to run from A to D plus the time it takes to swim from D to C is the *shortest* possible? This is not a problem of finding the shortest distance AC because he cannot swim as fast as he can run. This is a problem of finding what value of x will make retrieving the ball fastest?

Solution:

Let us review the formulas for speed.

$$r = \frac{d}{t}$$

Where r = rate, d = distance, and t = time:

So the formula for time is

$$t = \frac{d}{r}$$

Say that Elvis' rate on land (running) is r feet per second and his swimming rate is $\frac{r}{3.25}$. This is because we assumed that he could run 3.25 times as fast as he could swim.

In Figure 11.1 Elvis runs a distance x and swims a distance y. We do not yet know what these distances are, but here are the times taken up by these modes of transportation:

$$t_{\text{run}} = \frac{x}{r}$$

$$t_{\text{swim}} = \frac{y}{\frac{r}{3.25}} = \frac{3.25y}{r}$$

These are true no matter when Elvis jumps in the water, because the variables x and y are just part of the geometry. Notice that, in the right triangle, DBC, y is the length DC, so, by the Pythagorean theorem, y is a function of x.

$$y = \sqrt{(100 - x)^2 + 50^2}$$

What we are doing now is trying to find the value of x (and consequently y) that will make the sum of the running time, t_{run} and the swimming time, t_{swim} be the least it can be. The formula for the *total* time, $t_{\text{run}} + t_{\text{swim}}$ is actually a function of x; call it $T(x)$ as we show here.

$$
\begin{aligned}
\text{Total time} \quad &= \quad \text{running time } + \text{ swimming time} \\
T(x) \quad &= \quad t_{\text{run}} + t_{\text{swim}} \\
T(x) \quad &= \quad \frac{x}{r} + \frac{3.25}{r}y \\
T(x) \quad &= \quad \frac{x}{r} + \frac{3.25}{r}\sqrt{(100 - x)^2 + 50^2}
\end{aligned}
$$

We will assume that Elvis can run $10ft.$ per second. So, $r = 10$. This makes the formula for the total time:

$$T(x) = \frac{x}{10} + \frac{3.25}{10}\sqrt{(100 - x)^2 + 50^2} \qquad (11.1)$$

The graph of Equation 11.1 is shown in Figure 11.2

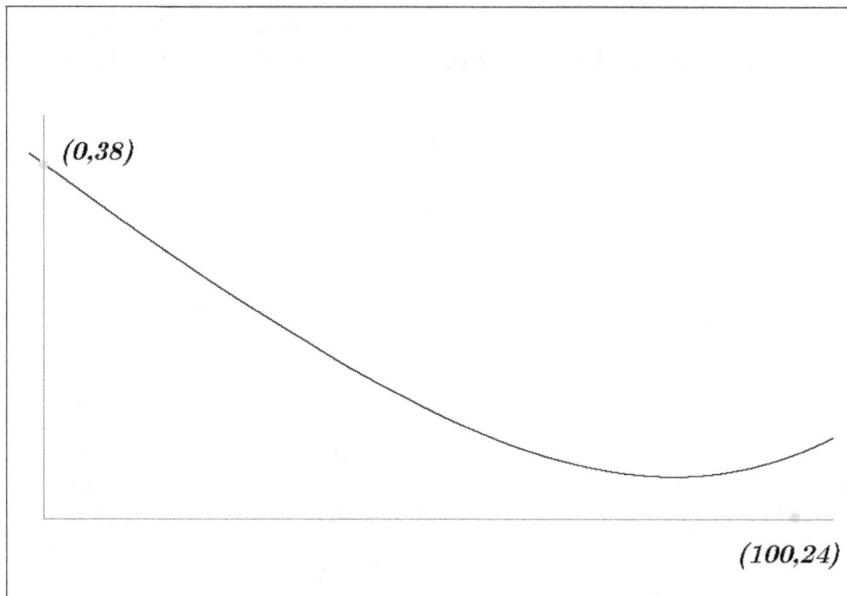

(0,38)

(100,24)

FIGURE 11.2 SECONDS ELVIS USES TO GET THE BALL

From this graph, we can see that if Elvis enters the water where x is zero, it takes him a little more than 36 seconds to get the ball. If, instead, he runs the whole 100 feet to B, where $x = 100$, and jumps in then, according to Figure 11.2, it only takes him about 27 seconds. But if he jumps in at about $x = 85$, it takes him about 26 seconds, which is close to the shortest possible time. We did not use calculus to get this answer, we used a graph of the function of x in Figure 11.2, that gave us the time for every value of x, then we just used our "eyeball" measurement by looking at the graph and finding what looked like it should be the lowest point. It turns out to be a very good answer, but not the absolutely best answer. To get the best answer we need to use the calculus answer.

What is the calculus answer?

Starting with Equation 11.1, we can use calculus to find its *derivative*, which is as follows:

$$T'(x) = \frac{1}{10} + \frac{3.25}{10}\left[\frac{-2(100 - x)}{2\sqrt{(100 - x)^2 + 50^2}}\right] \tag{11.2}$$

where $T'(x)$ is read "T prime of x" and it is equal to an expression that tells you the slope at every point $(x, T(x))$ of the graph in Figure 11.2. The beauty of that is you can find the smallest value of $T(x)$ by setting the derivative $T'(x) = 0$, because the lowest point of the graph of $T(x)$ has slope zero, so the wonderful formula Equation (11.2) lets you find the value of x for that lowest point; it is precisely

$$x = 100 - \frac{50}{\sqrt{(3.25^2 - 1}}$$

But we finally admit, because of the square root, that we have to resort to an approximation, after all.

$$x \approx 83.8$$

making $T \approx 25.46$ seconds.

In the next section we will show you how to find the wonderful derivative formula for Equation (11.1).

But how did the dog solve the problem?

I don't know. But I am willing to guess that when he saw his master throw the ball he could see, generally, that it was headed down beach and he preferred running to swimming. So, he ran along the beach roughly in the direction the ball went, after running a while he realized that he was not getting much closer to the ball, and, in fact, that it seemed to be fading away more and more across the water. He decided to jump in the water and start swimming after it. Exactly what I would have done too, as a dog, since we dogs don't have pockets to carry around a calculator.

What is calculus?

Calculus is the mathematics of continuity and infinity. We use it to understand the relationship between various quantities in real life. It is used to explain motion across surfaces and through space. It lets us examine how infinitessimal changes in a variable can cause changes in a function of that variable. It lets us study what happens to a function when one of the variables "gets close to" or "approaches" some fixed number. For example what happens to $\frac{1}{x-100}$ when x approaches 100? And what do we even mean by saying x approaches 100. Most of the time, fortunately, we can get by with an intuitive idea as to what this means. The notation we use is $x \to 100$, which reads "x approaches 100," or "100 is the limit of x." We will use these ill-defined words interchangeably. They are not precise, but are suggestive.

Archimedes' time

In 250 BCE, Archimedes first used trigonometry and the areas of n sided polygons (squares, pentagons, hexagons, etc.) that, he said, *verged* to the area of a circle. He was anticipating the ideas of limits, infinite series, and other calculus concepts. In 1906, a 1000 year-old document was found that had originally started out as a copy of a mathematics book written by Archimedes, but was over-written as a book of prayers by church scribes. It is known as the Archimedes *palimpsest,* which is a document that has been effaced by scraping off the original text and writing over the old text at a later time. In 2008, new x-ray techniques were developed that could detect the iron content in the ink that was used to make this copy. Enough of the hidden writing has been revealed for researchers to be able to read the underlying mathematical treatise. The library at the Walters Art Museum in Baltimore, Maryland, has this document in its archives. Much of its content is currently available on line on the Internet. Nothing more was done with calculus until the beginning of the *Age of the Enlightenment*, a period of time around 1600 CE.

Modern times

So, it took nearly two thousand years after Archimedes for calculus to reappear. At that time, mathematics was being driven by problems in geometry, navigation and astronomy. Then sometime in the early 17th century, problems in science, especially, physics and chemistry, began to emerge that stimulated new ideas needed to solve new problems. These were calculus ideas about continuity and limits. Thus, calculus was "in the air" and being used piece-wise throughout the world by mathematicians like René Descartes (coordinate systems, 1630), Pierre de Fermat (slopes, and derivatives, 1635), and Bonaventura Cavalieri (areas and integrals, 1635). Finally, the basic principles became organized into a coherent discipline in a book by Isaac Newton (1660) and, almost simultaneously by Gotfried Leibnitz (1670). Then, we started to see a more accurate use of the words, "approaches" and "limit".

Intuitive vs rigorous

We will illustrate what an intuitive *vs.* a rigorous idea of limit looks like by an example.

Example:

Question: If $y = 3x$, what happens to y as x "gets close to" 8?

Answer: y "gets close to" 24 as x "gets close to" 8.

Symbolically, a good looking mathematical expression of this would be: If $x \to 8$, then $3x \to 24$ or

$$\lim_{x \to 8} 3x = 24$$

But really, what do we mean by "close to" or by "limit"? The secret answer is in the inequality symbol, $<$, "less than." The words "close to" can mean "less than a given distance from." Thus 24 is the limit of $3x$ as x approaches 8 means that: If ϵ is any positive number, then there exists a positive number δ, such that when x is any number within a distance of δ from 8, then $3x$ will be within a distance of ϵ from 24.

In symbols, this says that if $|x - 8| < \delta$ then $|3x - 24| < \epsilon$.

Example: If $\epsilon = 0.009$, then we can take δ to be equal to 0.003, or less, because if x is any number within a distance of 0.003 from 8, then $3x$ will be within 0.009 of 24 proved as follows[2]:

$$
\begin{aligned}
|x - 8| &< 0.003 \\
-0.003 &< x - 8 < 0.003 \\
-0.009 &< 3x - 24 < 0.009 \\
|3x - 24| &< 0.009
\end{aligned}
$$

The general statement would be: "The number 24 is the limit of $3x$ as x approaches 8" means that it is true that: "No matter how small a tolerance ϵ you want between 24 and $3x$, you can confine x to a small enough distance δ from 8 so that $|3x - 24|$ is smaller than your prescribed tolerance ϵ."

If you can understand this concept of limits, you have gone a long way to mastering calculus.

It may not be brain surgery, but it really *is* rocket science. Physicist are interested in the distance that an object travels as a function of time, called *displacement*, how fast it travels, (distance divided by time) called *velocity* and how fast the velocity changes over time, called *acceleration* (velocity divided by time or distance divided by time squared). Engineers are interested in how fast acceleration changes over time, called *jerk*, acceleration divided by time or distance divided by time cubed.

---

[2]The symbols | | are indicating the *absolute value* of the expression between them. Thus, when $b > 0$, $|x - a| = b$ means $x - a = \pm b$, or $x = a \pm b$. And $|x - a| < b$ means $-b < x - a < b$ or $b > \pm(x - a)$.

One new application in satellite technology is *snap*, the instantaneous rate of change in the *jerk*. This is used in the control system for pointing the Hubble Telescope. It is the fourth derivative of displacement.

But these folks don't just want average speed or average acceleration, over long periods or even short periods of time. They need to know the instantaneous values of these quantities. That is why they need to know about limits and measurements that meet specific tolerances. One reason they need to know these things is so that they can find out what the *escape velocity* a rocket needs to reach in order to overcome the gravitational pull of the Earth. Assuming we have not yet achieved infinitely continuous *photonic propulsion*, we will continue to need to achieve escape velocity or there will be no space probes.

Falling objects in a gravitational field

In the next example, we will give you the instantaneous velocity, $v(t)$, of a falling object at any time t, and show how to get the change in this velocity well enough to compute the *instantaneous acceleration* at any time t. From this example we will see that we can also step back from the velocity and find the displacement equation.

Example:

If an object is thrown with a downward velocity of $-v_0$ at an initial time $t_0 = 0$, and at all times $t > 0$, it also experiences gravitational pull, adding $-32t$ to its velocity, then its velocity, $v(t)$, at any time t, in feet/second is:

$$v(t) = -32t - v_0 \tag{11.3}$$

Both of the minus signs in this formula indicate that the body is moving *downward*, in the *opposite direction* of distance, measured as positive from the Earth upward. That is, the body is coming down, thereby *shrinking* its distance from the Earth down to zero when it hits the ground. If we had originally thrown the object upward with positive velocity v_0, the sign on that term would have been $+$, not minus, and if we had just dropped the body without throwing it all, v_0 would not even appear in the equation, because then $v_0 = 0$. Also, in this discussion we are assuming *no* air resistance. If we had included air resistance it would have been a positive term because it would work against the downward fall of the body.

Still using Equation (11.3), after $t + h$ seconds the velocity is:

$$v(t + h) = -32(t + h) - v_0$$

If we want to know how much the velocity has changed over the time interval from t seconds to $t + h$ seconds, we just find the difference in the two velocities $v(t + h) - v(t)$, which we denote as Δv, also in ft./sec.

$$\begin{aligned} \Delta v &= -32(t + h) - v_0 - (-32t - v_0) \\ &= -32t - 32h - v_0 + 32t + v_0 \\ &= -32h \text{ ft./sec.} \end{aligned}$$

Now let Δt be the change in time during the given time interval, that is:

$$\Delta t = (t + h) - t = h \text{ sec.}$$

In this problem we were asked to find the *acceleration*, how fast the velocity changes over time. Specifically, we will want to find the instantaneous acceleration.

This is asking for $\frac{\Delta v}{\Delta t}$, the rate of change in velocity with respect to time at any instant. The units of measurement for this quantity are $\frac{\text{feet per second}}{\text{second}} = (\text{ft./sec.})/\text{sec.}$ and is sometimes written: ft./sec.^{2}

$$\frac{\Delta v}{\Delta t} = \frac{-32h\frac{\text{ft.}}{\text{sec.}}}{h \text{ sec.}} = \frac{-32h}{h}\frac{\text{ft.}}{\text{sec.}^2} = -32 \text{ ft./sec.}^2$$

This is the instantaneous acceleration at any time.

Here we started with a formula, Equation (11.3) for the velocity of a falling object, and found a formula for acceleration in a field of gravity. This is one of the types of problems that calculus can solve. Amazingly, another type of problem is just the opposite. From the velocity formula we can give you the *position formula,* called the "displacement," $x(t)$. Where $x(t)$ is the distance the object moved in t seconds.

$$x(t) = -16t^2 - v_0t + x_0 \tag{11.4}$$

where x_0 is the height from which the object is thrown at time $t = 0$, and v_0, as before, is the velocity with which it is thrown downward. The initial height is the displacement at time t equal zero:

$$x(0) = -16 \times 0^2 - v_0 \times 0 + x_0$$

We did *not* tell you yet how we got this displacement formula, Equation (11.4) but, as you will see, later, it depends upon knowing how to find a function that is the *anti-derivative* of a given function. Here, we use the known speed equation to determine the displacement equation. This is an important calculus operation called *integration* of a function. It is not always possible to find exact functions whose derivatives are known, but the use of modern computers often allows you to find approximations, when exact answers are not possible or not practical.

Problem:

In order to find out if Equation (11.4) is really the displacement equation for Equation (11.3), find the velocity of an object whose displacement (or position if you wish) at any time t is $x(t)$ as given in Equation (11.4).

Solution:

We start by finding the displacement at two different times t and $t + h$:

$$\begin{aligned} x(t) &= -16t^2 - v_0t + x_0 \\ x(t+h) &= -16(t+h)^2 - v_0(t+h) + x_0 \end{aligned}$$

Then the difference in the heights, Δx, divided by the difference in the times: Δt, gives you the average velocity during the given time interval So let us find Δx and Δt :

$$\begin{aligned} \Delta x &= x(t+h) - x(t) \\ &= -16(t+h)^2 - v_0(t+h) + x_0 - (-16t^2 - v_0t + x_0) \\ &= -16t^2 - 32ht - 16h^2 - v_0t - v_0h + x_0 + 16t^2 + v_0t - x_0 \\ &= -32ht - 16h^2 - v_0h \tag{11.5} \end{aligned}$$

To get Equation (11.5) we cancelled out the terms that added up to zero in the previous equation. That is

$$-16t^2 + 16t^2 = 0$$
$$-v_0 t + v_0 t = 0, \text{ and}$$
$$x_0 - x_0 = 0$$

Finding Δt is much easier, it is:

$$\Delta t = (t + h) - t = h.$$

Divide Δx in Equation(11.5) by Δt, and you have

$$\frac{\Delta x}{\Delta t} = \frac{-32ht - 16h^2 - v_0 h}{h}$$
$$= \frac{(-32t - v_0 - 16h)\, h}{h}$$

Here, we factored h out of the numerator and re-arranged the terms. Note that the term $-16h^2$ had two factors of h so we could only factor out one of them. Now cancel h, from the numerator and denominator, getting

$$\frac{\Delta x}{\Delta t} = -32t - v_0 - 16h$$

This is the average speed over the whole time interval from t to $t + h$. How do you think we would get the exact instantaneous speed at t? By assuming that the time interval is not h seconds in length but *zero* seconds in length. This is where calculus gets the reputation of dealing with "ghosts of departed quantities." It is inaccurate to say that we let the time interval have length $h = 0$. Instead, we say that we have computed the velocity over an interval of length h, and we find the *limit* of that average velocity as h "approaches" zero. That is, $h \to 0$.

Thus, the final step is to say that the instantaneous velocity $v(t)$ is defined as

$$
\begin{aligned}
v(t) &= \lim_{h \to 0} \frac{\Delta x}{\Delta t} \\
v(t) &= \lim_{h \to 0} \frac{x(t + h) - x(t)}{h} \\
v(t) &= \lim_{h \to 0} -32t + v_0 - 16h \\
v(t) &= -32t - v_0
\end{aligned}
\tag{11.6}
$$

The derivative

Derivative is instantaneous rate of change

That's it! The instantaneous rate of change in a quantity is called the *derivative* of that quantity and its definition is Equation (11.6). Usually we don't switch the notation from x to some other variable like v as we did here. Ordinarily, we keep the same notation for the function $x(t)$ and its derivative: sometimes written $\frac{dx}{dt}$, or $x'(t)$. Not every function has a derivative at every point. And not all functions use the same notation. For example, a function might be denoted f and the independent variable denoted x, and h could be denoted Δx, then the definition would look like this:

$$\frac{df}{dx} = \lim_{\Delta x \to 0} \frac{f(x + \Delta x) - f(x)}{\Delta x} \tag{11.7}$$

In this gravity problem the velocity $v(t)$ is the derivative of the displacement function $x(t)$, so in most physics applications, the velocity is defined as,

$$v(t) = \frac{dx}{dt} \text{ or as } x'(t)$$

Problem:

Given the function

$$f(x) = x^3$$

Use Equation (11.7) to find the derivative $\frac{df}{dx}$ at x

Solution: First find $f(x + \Delta x)$

$$\begin{aligned} f(x + \Delta x) &= (x + \Delta x)^3 \\ f(x + \Delta x) &= x^3 + 3x^2\Delta x + 3x(\Delta x)^2 + (\Delta x)^3 \end{aligned}$$

We got this by the formula for the third power of a binomial:

$$(a + b)^3 = a^3 + 3a^2b + 3ab^2 + b^3$$

Now subtract $f(x)$ from $f(x + \Delta x)$, getting

$$\begin{aligned} f(x + \Delta x) - f(x) &= x^3 + 3x^2\Delta x + 3x(\Delta x)^2 + (\Delta x)^3 - x^3 \\ f(x + \Delta x) - f(x) &= 3x^2\Delta x + 3x(\Delta x)^2 + (\Delta x)^3 \end{aligned}$$

Now divide by Δx, getting

$$\begin{aligned} \frac{f(x + \Delta x) - f(x)}{\Delta x} &= \frac{(3x^2 + 3x\Delta x + (\Delta x)^2)\Delta x}{\Delta x} \\ \frac{f(x + \Delta x) - f(x)}{\Delta x} &= 3x^2 + 3x\Delta x + (\Delta x)^2 \end{aligned}$$

Now find the limit as $\Delta x \to 0$. Thus,

$$\begin{aligned} \lim_{\Delta x \to 0} \frac{f(x + \Delta x) - f(x)}{\Delta x} &= \lim_{\Delta x \to 0} \left(3x^2 + 3x\Delta x + (\Delta x)^2\right) \\ \lim_{\Delta x \to 0} \frac{f(x + \Delta x) - f(x)}{\Delta x} &= 3x^2 + 0 + 0 \end{aligned}$$

$$\frac{d(x^3)}{dx} = 3x^2$$

Derivative is slope

In mathematics the derivative has the geometric interpretation as the slope of a line tangent to a graph at a given point.

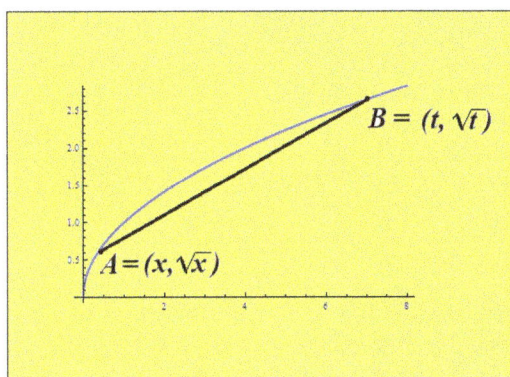

FIGURE 11.3 THE SECANT LINE AB

Example: Figure 11.3 is the graph of the equation

$$y(x) = \sqrt{x} \tag{11.8}$$

Since the points A and B are points on the graph then their coordinates are:

$$
\begin{aligned}
A &= (x, \sqrt{x}) \text{ and} \\
B &= (t, \sqrt{t})
\end{aligned}
$$

Recall the slope of any non-vertical line through two given points is defined as the difference in their ordinates (y values) divided by the difference in their abscissas, (x values). Sometimes, the slope is called the *rise* to the *run*:

$$\text{slope} = \frac{\text{rise}}{\text{run}} = \frac{\Delta y}{\Delta x}$$

Question:
So, in Figure 11.3 what is the slope of the line AB?
Answer

$$\text{slope of } (AB) = \frac{\sqrt{x} - \sqrt{t}}{x - t} \tag{11.9}$$

This gives you the average slope over the whole interval from x to t. A line like AB is called a "secant line," *cutting line*, of the graph. It cuts the graph at two points. But we want the slope of a line at one point. When the two points that define the secant line are brought closer together, and eventually become one point, then the secant line becomes the *tangent line* and the slopes of all of such secant lines become closer and closer to the slope of the tangent line. It is a limiting process. Let's see if we can do the algebra on Equation (11.9).

What happens if we leave the point A fixed, that is make x a constant, and let B "approach" A, by letting t approach x.

But, wait! we cannot let $t = x$ because then when we divide by $x - t$, we would be dividing by zero. What to do? Is there some algebraic expression the is equal to $\frac{\sqrt{x}-\sqrt{t}}{x-t}$, that does not have the problem that the denominator is $(x - t)$? Yes! as long as $t \neq x$, we can write Equation (11.9) as

$$\text{slope of } (AB) = \frac{\sqrt{x} - \sqrt{t}}{x - t} \times \frac{\sqrt{x} + \sqrt{t}}{\sqrt{x} + \sqrt{t}} \tag{11.10}$$

because $\frac{\sqrt{x}+\sqrt{t}}{\sqrt{x}+\sqrt{t}} = 1$, and we can always multiply an expression by 1 without changing the expression. Simplifying Equation (11.10), we get

$$\begin{aligned} \text{slope of } (AB) &= \frac{(x - t)}{(x - t) \times (\sqrt{x} + \sqrt{t})} \\ &= \frac{1}{(\sqrt{x} + \sqrt{t})} \end{aligned} \tag{11.11}$$

This is not yet the slope of the line tangent to the graph at the point A. But we are getting there. It is a limiting process, just as it was in finding the derivative in the falling body problem. We let $t \to x$, (meaning we let the point B approach the point A, and the secant lines that cut the graph at two points that are getting closer and closer together and they become the one point A. Thus, the secant lines becomes the line T as shown in Figure 11.4.

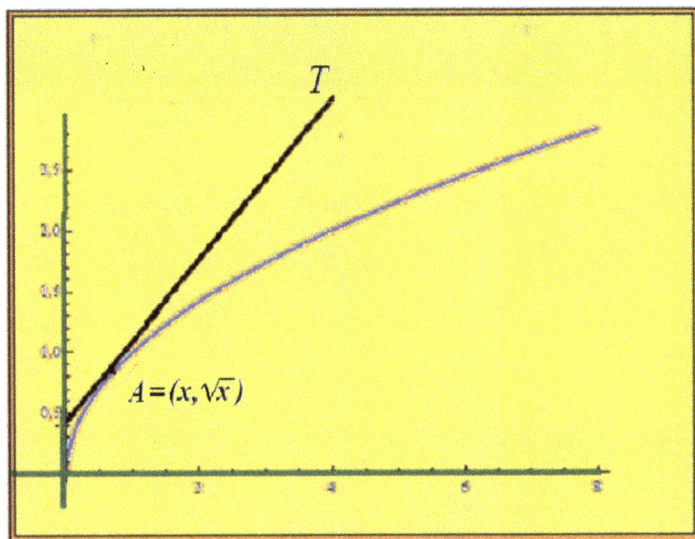

FIGURE 11.4 LINE T IS TANGENT TO THE GRAPH AT A

The slope of that tangent line (the "touching line" rather than the cutting line) is

$$\frac{1}{2\sqrt{x}}$$

We got this result by making algebra and geometry work together:

Questions and answers:

Q: What is the slope of the graph at the point for which $x = 1.5$?

A: It is:

$$\frac{1}{2\sqrt{1.5}} \approx \frac{1}{2 \times 1.2247..} = 0.4082$$

Q: What is the slope of the graph when $x = 100$?

A: It is 0.05

Q: Will the slope ever be zero?

A: No because, there is not any real number x such that $\frac{1}{2\sqrt{x}} = 0$

Q: Will the graph ever have a negative slope?

A: No, $\frac{1}{2\sqrt{x}}$ can never be negative for any real number x.

Q: Does this graph have a tangent line at $x = 0$?

A: Yes, it is a vertical line, the y-axis.

Q: Does this graph have a slope at $x = 0$?

A: No, vertical lines do not have slope.

Geometric derivative

In general, the geometric version of the derivative is as follows:

If $f(x)$ is a function whose derivative $\frac{df}{dx}$ exists then,

$$f'(x) = \frac{df}{dx} = \lim_{t \to x} \frac{f(t) - f(x)}{t - x} \tag{11.12}$$

Problem:

If C is any constant and n is any positive integer, and

$$f(x) = Cx^n$$

use the definition of derivative in Equation (11.12) find the derivative $f'(x)$.

Solution:

Find: $f(t) - f(x)$ and divide by $t - x$

$$
\begin{aligned}
\frac{f(t) - f(x)}{t - x} &= \frac{Ct^n - Cx^n}{t - x} \\
&= \frac{C(t^n - x^n)}{t - x} \\
&= \frac{C(t - x)(t^{n-1} + t^{n-2}x + t^{n-3}x^2 + \ldots + tx^{n-2} + x^{n-1})}{(t - x)} \\
&= C(t^{n-1} + t^{n-2}x + t^{n-3}x^2 + \ldots + tx^{n-2} + x^{n-1})
\end{aligned}
$$

We know $t^n - x^n = (t - x)(t^{n-1} + t^{n-2}x + t^{n-3}x^2 + \ldots + tx^{n-2} + x^{n-1})$ because it is factoring the difference of two nth powers.

Now, finding the limit as $t \to x$, we have

$$\lim_{t \to x} \frac{f(t) - f(x)}{t - x} = \lim_{t \to x} C(t^{n-1} + t^{n-2}x + t^{n-3}x^2 + \ldots + tx^{n-2} + x^{n-1})$$

Taking the limit as $t \to x$, we get

$$f'(x) = Cnx^{n-1}$$

Example

$$\frac{d(5x^4)}{dx} = 20x^3$$

Derivatives of sums and products of functions

Q1: If each of two functions, $f(x)$ and $g(x)$, has a derivative on the same domain, what is the derivative of their *sum* $f(x) + g(x)$?

A1:

$$\frac{d(f(x) + g(x))}{dx} = \frac{df(x)}{dx} + \frac{dg(x)}{dx}$$

Expected and true! Prove it.

Q2: If each of two functions, $f(x)$ and $g(x)$ has a derivative on the same domain what is the derivative of their *product* $f(x) \times g(x)$?

A2:
$$\frac{d(f(x) \times g(x))}{dx} = f(x) \times \frac{dg(x)}{dx} + g(x) \times \frac{df(x)}{dx}$$

Surprising, but true! Prove it.

Higher derivatives of functions

If $f(x)$ is function that has a derivative $\frac{df(x)}{dx}$, then the derivative of that derivative, if it exists, is

$$\frac{d\frac{df(x)}{dx}}{dx}$$

is called the *second derivative* of $f(x)$ and is denoted by either

$$f''(x), \text{ or}$$
$$\frac{d^2 f(x)}{dx^2}$$

Problem: Given:
$$f(x) = 17x^2 + 8x^5$$

Find $f'(x)$ and $f''(x)$.
    Solution:

$$f'(x) = 34x + 40x^4, \text{ and}$$
$$f''(x) = 34 + 160x^3$$

The third derivative is the derivative of the second derivative, and so forth. Thus,

$$f'''(x) = 0 + 480x^2 = 480x^2$$

because the derivative of the constant 34, $\frac{d(34)}{dx} = 0$

Why do we need higher derivatives of function? For one thing, certain concepts in physics: force, work, and acceleration, as well as concepts in economics: cost of living, and inflation are defined in terms of derivatives and second derivatives.

Velocity, momentum, and force as derivatives

When we are given a displacement function in physics,

$$x(t) = f(t), \text{ } displacement$$

where t is time and x is distance, we learned that *velocity* $v(t)$ is the derivative of $x(t)$.

$$v(t) = \frac{dx(t)}{dt}, \text{ } velocity$$

The derivative of velocity is acceleration $a(t)$

$$a(t) = \frac{dv(t)}{dt} = x''(t), \text{ } acceleration$$

What is the *mass* of an object? This term in physics is one that could be taken as primitive or "undefined," much in the same sense that the word *set* is in mathematics. Mass is usually characterized as resistance to movement when a force is applied. It is not the same thing as weight because the same mass will weigh different amounts on different planets. An astronaut standing on the earth does not weigh the same as she would standing on the moon.

A pretty good rough definition of the mass of a body is the number of atoms in that body. Usually the mass of a body does not change unless the body changes. For example, if you consider a rocket plus all of its fuel as being a "body," and you launch it then, as the fuel burns, the number of molecules in the body changes, so the mass changes. But if you tow your car with a tow-truck the mass of you car does not change, unless something happens like the motor dropping out.

Another concept involving mass is *momentum*. The momentum of your car is its mass times how fast you tow it, or how fast it is going when you push it off a cliff. More precisely, if you multiply the velocity $v(t)$, of an object by its mass, m you get the quantity called *momentum*

$$mv(t) = m\frac{dx(t)}{dt}, \; momentum$$

What happens when you try to find how fast momentum changes? Well, for example if mass, m, is constant then the derivative of momentum with respect to time is mass times acceleration. May this always be with you, because it is force.

$$\frac{dmv(t)}{dt} = m\frac{dv(t)}{dt} = ma(t), \; force$$

Force is what happens as the car is falling off the cliff because the mass is being multiplied by the acceleration due to gravity.

In many applications, such as *missile launching science* there are more complicated definitions of force. These systems require that we use mass, $m(t)$, that varies with time. Two formulas for force in launching missiles are:

$$\begin{aligned} F_1 &= m^2(t) \times \frac{d}{dt}\left(\frac{v(t)}{m(t)}\right) \; \text{or} \\ F_2 &= \frac{d}{dt}\left(m(t)v(t)\right) - v'(t)m'(t) \end{aligned}$$

Notice that both of these reduce to the usual $F = m \times a(t)$ when $m(t)$ is constant, m.

Weight and force

The weight of an object on the surface of any planet can be thought of as a force because it is the mass times the acceleration due to gravity of that planet. If you apply the force needed to move such a body then that force must be greater than the weight.

Work

Another important physics concept is *work*. When you apply a force to an object and move it a distance x, then the quantity you are defining is *work*. Work is force times distance. More precisely, work is defined as the product of a force applied to a body along a given straight line causing that body to move some distance along that straight line.

Work = force times distance

In the English system we use the pound and ounce to measure weight, the avoir dupois system. The non-metric system of length, is the foot and the yard. When we use these units, F is force in pounds, lbs., and the distance x is in feet, so work is in foot-pounds, ft.-lbs. In the metric system the units of work are measured in meters and kilograms.

The thing about work, other than its being "the curse of the drinking classes[3]," is that force is not always a constant. A force may be applied to a short part of a distance and the force changes at a different part, or the force might just continuously change over some distance, or the force could be one that is changing direction at any time, and so forth. Also, force could be changing because mass or acceleration are changing. For these reasons, we will need to consider *instantaneous changes* in force over various distances. This brings us to the calculus use of limits, and quantities approaching other quantities.

Assume that the aerodynamics of the object and the wind resistance to its motion are negligible and can be ignored. Ignore, also, the friction and the coefficient of rolling resistance that would be a factor reducing how much force you need to apply to push any such object that is on wheels. This assumes that the work done on rolling an object of 50 lbs. 2 feet is exactly 100 ft.-lbs. It is in this highly over-simplified sense that we propose the following problem.

Problem (Constant force)

A man pushing a grocery cart weighing 40 lbs. leaves the store and goes to his car in the parking lot, 20 ft. from the door.

How much work did he do in this little trip?

Solution:

We take the weight of the cart to be the force; here it is $F = 40$ lbs., and the distance is 20 feet. The work is

$$\begin{aligned} W &= \text{40 lbs.} \times \text{20 ft.} \\ &= \text{800 ft.-lbs.} \end{aligned}$$

We want to set this problem up in a way that lets us generalize it to a more complicated problem. Let us break up the 20 ft. trip into a collection of short pieces, say 20 one-foot long subintervals. We will compute the work done over each sub interval as 40 lbs. times 1 ft., or 40 ft.-lbs. To get the total work done over the twenty intervals we add them up.

$$\begin{aligned} W &= \text{40 ft.-lbs.} + \text{40 ft.-lbs.} + .. + \text{40 ft.-lbs.} \qquad (11.13) \\ &= \text{800 ft.-lbs.} \end{aligned}$$

Let's say the whole distance from the store to the car starts with $x = 0$ and goes to $x = 20$. Now, starting at zero, create twenty subintervals using the twenty-one end points:.

$$\left\{ x_0, x_1, x_2, ..., x_{19}, x_{20} \right\}$$

as shown below in Figure 11.5.

---

[3] Oscar Wilde (1880)

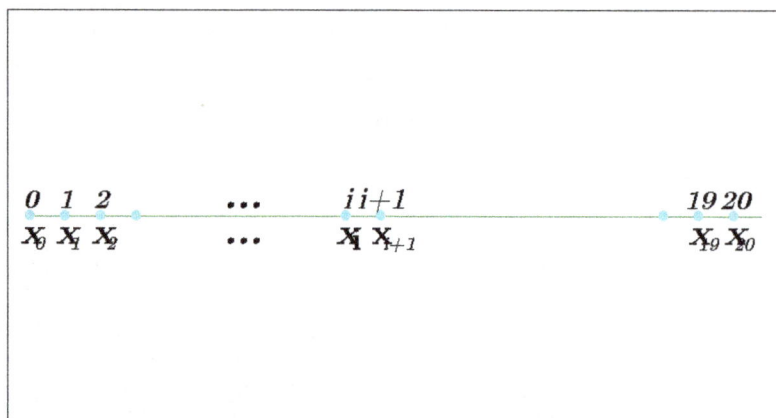

Figure 11.5 A PARTITION OF THE INTERVAL $[0, 20]$

Let Δx_i be the length, in feet, of the *ith* subinterval, where i is an integer between 0 and 19. That is,

$$\Delta x_i = x_{i+1} - x_i$$

Now, by Equation (11.13), we can write work as

$$W = \sum_{i=0}^{i=19} 40 \times \Delta x_i \text{ ft.-lbs.} \tag{11.14}$$

$$= 800 \text{ ft.-lbs.}$$

where the constant force, $F = 40$, can be thought of as the force at any point, ξ_i in the *ith* subinterval.[4] Thus, for ξ_i, we can write Equation (11.14) as

$$W = \sum_{i=0}^{i=19} F(\xi_i) \times \Delta x_i \tag{11.15}$$

In a real situation, we will be required to deal with a force, $F(x)$, that is *not constant* over a given distance. When that is the case, we set up the problem by pretending that the variable force *is* actually constant over short subintervals, using one of the fixed values of the $F(\xi_i)$ at some point, ξ_i in the short subinterval, Δx_i. Thus, we partition the interval into m pieces, and calculate the work done over each little piece Δx_i. In the first interval, where $i = 1$, the work done is

$$F(\xi_1) \times \Delta x_1$$

Where ξ_1 is some number between x_0 and x_1. Then moving over to the next subinterval, where $i = 2$, compute the force as if it were *another* constant $F(\xi_2)$ where ξ_2 is between x_1 and x_2, and so forth. In the long run the work done over the whole distance from a to b is approximated by the sum of all of these little pieces of work. This means that the work done by a variable force over a distance $b - a$, where $x_0 = a$, and $x_m = b$, is approximated by

---

[4]We are using the Greek letter ξ, "Xi," as x.

$$W = \sum_{i=0}^{i=m-1} F(\xi_i) \times \Delta x_i \text{ ft.-lbs}$$

This idea of turning a long-run variable force into a collection of short-run constant forces is the basic tool for setting up problems in calculus. Let us use it in the grocery cart problem, this time with a variable force.

Problem: (Variable force)

Now, the man is pushing a grocery cart that weighed 40 lbs. including a ten pound bag of flour. Unknown to him, when he left the store and started to cross the parking lot, the bag of flour started to leak. He pushed the cart at a rate of 2 feet per second going to his car, 20 feet from the store. During this trip, the bag leaked flour out at the rate of 1 pound per 2 feet leaving what looked like a painted dashed highway stripe leading straight to his car. How much work did he do during this ill-fated trip?

Solution:

The distance is 20 feet, which, we will break up into, say, ten two-foot intervals; and the force is not constant over all these intervals. In fact, the force started out as 40 lbs. at $x = 0$, when he left the store and dwindled down to 30 lbs when he reached the car, because during that 20 foot trip he lost 10 lbs of flour.

Here, the force at any distance x, was $40 - \frac{1}{2}x$ lbs. At the beginning, when $x = 0$ and the force was 40, two feet later, that is, one second later, the force was $40 - 1 = 39$ lbs. The weight at 2 seconds was 38 lbs etc. To simplify the problem, suppose that one pound drops, all at once, every two feet. So that in any 2 foot interval, the weight is constant throughout that short distance, but in the next interval the weight is one pound less and so forth.

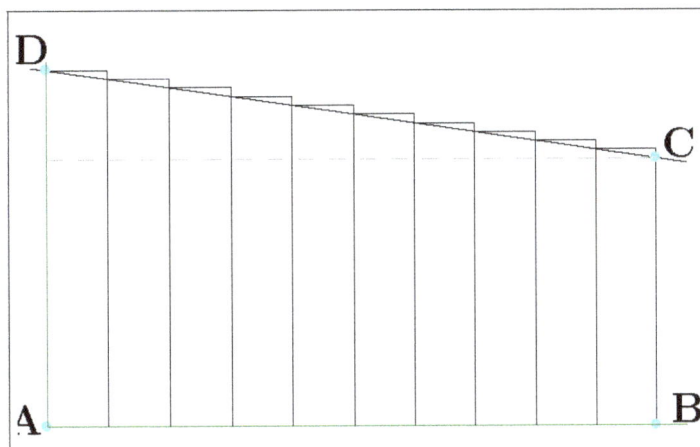

Figure 11.6 WORK DONE WITH LEAKING FLOUR BAG

Therefore, in the nth two-foot interval the weight was $(40 - n)$ lbs, and remained at that weight for 2 feet. The work done was $2 \times (40 - n)$ ft.-lbs. To find an approximation to the total work done, we add up all of these short two-foot pieces of work, from $n = 0$

to $n = 9$. In this approximation $\Delta x_i = 2$

$$W \;=\; \sum_{n=0}^{n=9} (40 - n) \times 2 \text{ ft.-lbs.}$$
$$W \;=\; 710 \text{ ft.-lbs.}$$

See the graph in Figure 11.6. The area of the *nth* rectangle is the work done over two feet, by a force of $40 - n$. The sum of the rectangles in this figure *over-estimates* the work done, because in each foot of the trip we assumed the weight remained constant and then dumped the flour all at once. But it is more likely that the flour dribbled out continuously over each foot. This stair step graph is an approximation to the area under the curve. Thus, the actual (instantaneous) work done is the area *under* the line CD, above the x-axis and between the two vertical lines at A and B. That is, the trapezoid, $ABCD$.

$$20 \times \frac{40 + 30}{2} = 700 \text{ ft.-lbs}$$

rather than the 710 ft.-lbs, sum of the rectangles.

Summary: In the leaking flour bag problem above, we got an approximation of the work done by a variable force over a given distance.

$$F(x) = 40 - \frac{x}{2}$$

We approximated the total work by constructing a set of ten approximating rectangles and adding up their areas. A better approximation would come from increasing the number of rectangles and making them "skinnier," say partitioning the interval $[0, 20]$ into 20 subintervals as in Figure 11.7. Compare this to Figure 11.6, which has ten subintervals.

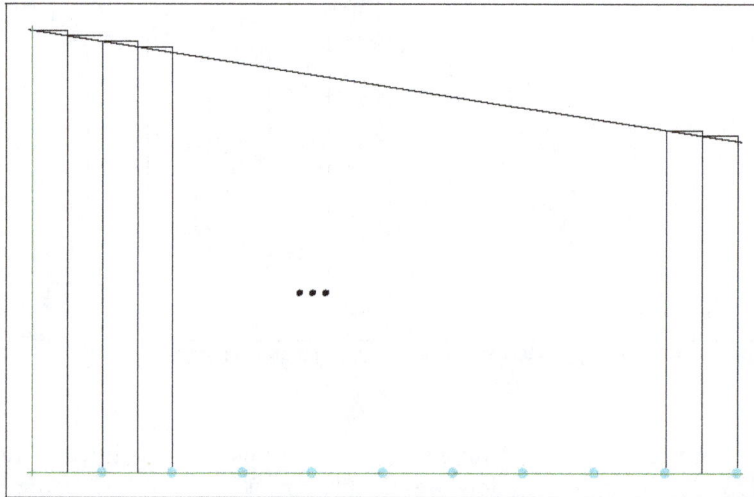

Figure 11.7 MORE AND SKINNIER RECTANGLES

The sum of the 20 rectangles, where $\Delta x_i = 1$ in Figure 11.7 is

$$W = \sum_{n=0}^{n=19} (40 - \frac{n}{2}) \times 1 = 705 \text{ ft.-lbs.}$$

which is better approximation than the 710 ft.-lb. approximation we got for Figure 11.6.

How can we get the exact, actual work? We need to divide the interval $[0, 20]$ into even smaller subintervals, Δx, and increase the number, n, of approximating rectangles even more. In the limit we need to consider letting $\Delta x_i \to 0$ and $n \to \infty$. This process leads to what is called the *integral of* $F(x)$ from 0 to 20, denoted as follows:

$$W = \int_0^{20} (40 - \frac{x}{2}) dx = 700 \text{ ft-lbs.}$$

Which is called "the *definite integral* of $(40 - \frac{x}{2})$ from 0 to 20 with respect to x." Say What? Alright, be patient, we are ready to tell you exactly what the definite integral is.

The definite integral

Here are a few preliminary definitions and concepts we need in order to define and explain the definite integral of a function. You will be delighted, if not surprised, when you find out what the definite integral of a function has to do with the derivative of a function.

Definitions:

The **interval** $[a, b]$ is the set of all real numbers x, such that $a \le x \le b$.

Uniform Partition of an interval:

If m is a positive integer, then the uniform partition $P_m[a, b]$ of the interval $[a, b]$ is a collection of $(m + 1)$ points,

$$\{x_0, x_1, x_2, ... x_m\}$$

with the following properties:

1. $x_0 = a$ and $x_m = b$
2. For any integer i between 1 and m, $x_{i-1} < x_i$
3. The length, Δx_i, of the *ith* subinterval is

$$\Delta x_i = x_i - x_{i-1-} = \frac{b-a}{m}$$

The partition is called *uniform* because all of the lengths, Δx_i, are the same number.

Example:

The points shown in Figure 11.5 is the partition $P_{20}[0, 20]$. Each Δx_i is $\frac{20-0}{20} = 1$.

Definition: (The definite integral of $F(x)$ from a to b.)

Let a and b be real numbers with $a < b$, and let $F(x)$ be a continuous function whose domain is the interval $[a, b]$. If m is a positive integer and $P_m[a, b]$ is a uniform partition of $[a, b]$ and r_i is the midpoint of the ith subinterval, then

$$A(b - a) = \lim_{m \to \infty} \sum_{i=0}^{i=m} F(r_i) \Delta x_i$$

is a number, called the definite integral of $F(x)$ with respect to x from $x = a$ to $x = b$.

This number, $A(b-a)$, is the consequence of letting the number, m, of approximating rectangles increase without bound whilst their widths approach zero. That is $m \to \infty$ and $\Delta x_i \to 0$. It's all about limits. The usual notation for $A(b-a)$ is

$$A(b-a) = \int_a^b F(x)dx$$

Geometrically, the integral of a function from a to b is a number that is the area bound by the x-axis, the graph of the function and the vertical lines at a and b.

Important properties of the definite integral.

We can formally prove the following properties of integrals by using the axioms of the real number system, especially the least upper bound axiom as stated in Chapter 4. But we prefer to use our intuition about areas, such as: areas can be added, areas of parts of a figure are less than the area of the whole figure. The area of a zero figure or the area of a straight line is zero. The areas computed in one direction are the negatives of the areas computed in the opposite direction, and so forth.

1. $A(b) = A(b-0)$ means

$$A(b) = \int_0^b F(x)dx$$

2. $A(0) = 0$ means

$$A(0) = \int_0^0 F(x)dx = 0$$

3. $A(b-a) = A(b) - A(a)$ means

$$\int_a^b F(x)dx = \int_0^b F(x)dx - \int_0^a F(x)dx$$

4. $A(b-a) = -A(a-b)$ means

$$\int_a^b F(x)dx = -\int_b^a F(x)dx$$

We will take as an example, the graph of a specific function $F(x)$ to show the geometric properties of the area under its graph. This special case will give us a good intuitive idea about the definite integral of a function and how it is related to a derivative.

Example:

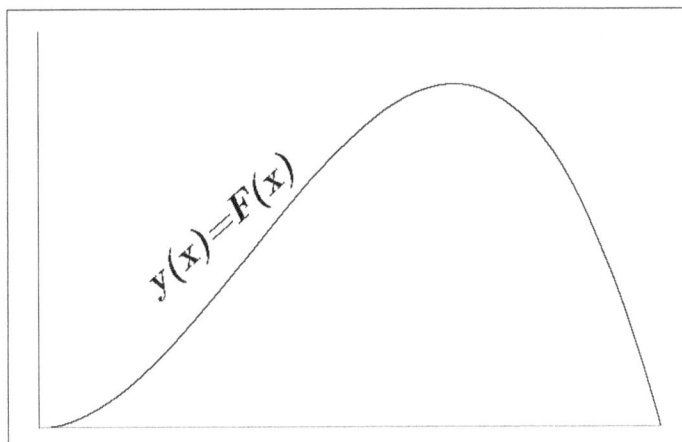

Figure 11.8 GRAPH OF $y = F(x)$

If $F(x)$ is the function:

$$F(x) = 0.06x^2 - 0.001x^3$$

then the graph of F is as shown in Figure 11.8.

The area under this graph is shown, below, in Figure 11.9. We have divided this region into some parts; for example the area from 0 to t is $A(t)$ and the area between t and s is $A(s) - A(t)$.

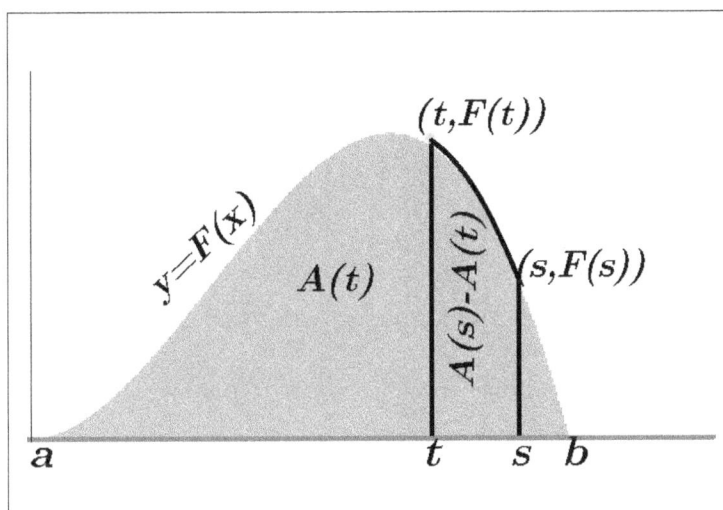

Figure 11.9 THE AREAS $A(t)$ AND $A(s) - A(t)$

$$A(t) = \int_0^t F(x)dx = \int_0^t (0.06x^2 - 0.001x^3)dx$$

$$A(s) - A(t) = \int_0^s F(x)dx - \int_0^t F(x)dx$$

Look at the geometry of Figure 11.9. The region under the graph of $y = F(x)$ and between $x = t$ and $x = s$ (and above the x-axis) is bound on three sides by straight lines and on the fourth by a curve. The lengths of the straight line pieces are $s - t$, (the piece on the x-axis), $F(t)$ and $F(s)$ the two vertical pieces.

Can we construct a rectangle whose area is the same as this region? This is exactly the kind of problem that Archimedes was trying to solve when he tried to find a square whose area is the same as a given circle. If we have a rectangle of a known area, we can find its height just by dividing the area by the base. So let us try to find the height of a rectangle whose base is $s - t$ and whose area is $A(s) - A(t)$.

$$\frac{Area}{base} = height$$

$$\frac{A(s) - A(t)}{s - t} = h \qquad\qquad (11.16)$$

See Figure 11.10. This number h is the height of an actual rectangle that would have the same area as the shaded region.

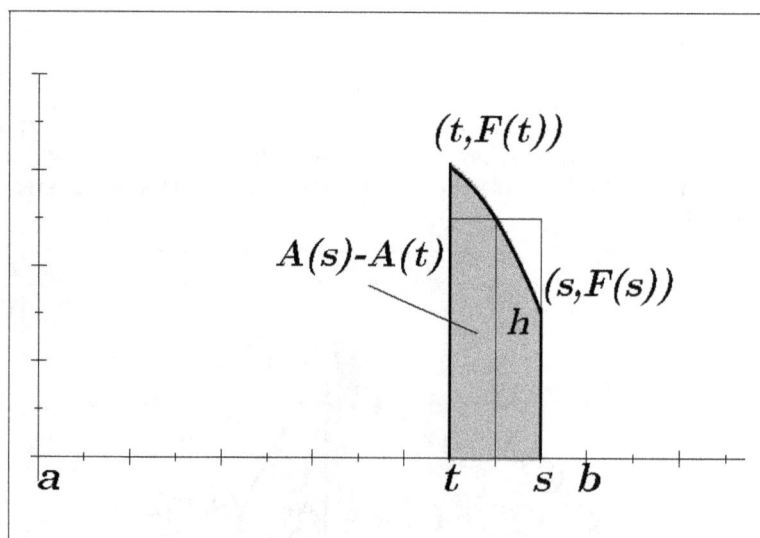

Figure 11.10 THE REGION WITH AREA A(S)-A(T)

Notice that height h is less than $F(t)$, but greater than $F(s)$. So

$$F(t) \geq h \geq F(s)$$

Use Equation 11.16 to replace h, getting

$$F(t) \geq \frac{A(s) - A(t)}{s - t} \geq F(s) \qquad\qquad (11.17)$$

Now, here is a limit problem. What happens to every part of the Inequality 11.17 (the smallest part, the middle part and the largest part) when we let $s \rightarrow t$?

Here is a "quick and dirty" answer.[5]

---

[5] A more accurate treatment is one in which we prove, by the least upper bound axiom of Chapter 4, that $h = F(r)$, for some r in $[t, s]$, and $F(t) \geq F(r) \geq F(s)$, as $s \rightarrow t$, and so forth.

$$F(t) \text{ stays } F(t), \text{ when } s \rightarrow t.$$

$$\frac{A(s) - A(t)}{s - t} \text{ approaches } \frac{dA(t)}{dt}, \text{ when } s \rightarrow t$$

$$F(s) \text{ approaches } F(t) \text{ when } s \rightarrow t$$

Therefore, we can conclude that, in the limit

$$\frac{dA(t)}{dt} = F(t)$$

In other words, the derivative of the integral of $F(t)$ is the function $F(t)$, itself.

The Fundamental Theorem of Calculus

This pithy statement is called *The Fundamental Theorem of Calculus*. Another way to say it is that the integral of a function is an anti-derivative of that function.

$$\frac{d}{dx} \int_a^x F(t)dt = F(x)$$

Usually we evaluate the integral by evaluating its antiderivative for the end points. Thus if $G(x)$ is an antiderivative of $F(x)$, then

$$\int_a^x F(t)dt = G(x) - G(a)$$

Example:

The equation for the graph in Figure 11.8 is

$$F(x) = 0.06x^2 - 0.001x^3$$

So $A(t)$, the area from 0 to t, is

$$A(t) = \int_0^t (0.06x^2 - 0.001x^3)dx$$

$$= (0.02x^3 - \frac{0.001}{4}x^4 + C)|_0^t$$

Where the little symbol $|_0^t$ means substitute t in for x and then 0 in for x and subtract the results of the substitutions.

$$= 0.02t^3 - \frac{0.001}{4}t^4 + C - (0 + C)$$

$$= 0.02t^3 - \frac{0.001}{4}t^4$$

In particular, if t is a specific number like 30, then

$$
\begin{aligned}
A(30) & = \int_0^{30} (0.06x^2 - 0.001x^3)dx \\
A(30) & = 0.02 \times (30)^3 - \frac{0.001}{4}(30)^4 \\
& = 337.5
\end{aligned}
$$

This is the end of our discussion on calculus and we will leave it to the interested reader to look at more varied and deeper results, involving a myriad of applications. There are thousands of calculus books available in bookstores and on–line for reasonable prices. You can print up your own copy of "e-books," some of which are available for no cost at all.

Calculus is an example of how things go very smoothly in mathematics, but now let us turn to some things that were surprisingly not so "cut and dried," as is sometimes erroneously said about mathematics.

CHAPTER 12

A CRISIS IN MATHEMATICS

Shoring up the foundations of mathematics

In the second half of the nineteenth century, mathematicians were struggling with inaccurate ideas about irrational numbers, infinity, and flimsy ideas about the use of "infinitessimals" in calculus and differential equations. Various proposals were made to try to fix this problem and in 1872, the German mathematician Richard Dedekind proposed an axiom which came to be known as the *Dedekind Cut*, or the least upper bound axiom. Prior to that, the concept of functions, continuity, and differentiability seemed to be handled in a piecemeal and *ad hoc* fashion. A few years before that, in 1850, a British mathematician, George Boole, had written about rules of logic and rules of *set theory* that seemed to be needed to fill in gaps in sorting out mathematical proofs. In the 1880's, Georg Cantor, a Russian-German mathematician, introduced the notion of higher orders of infinity. For example, the irrational numbers are more numerous than the rational numbers. This was a shocking idea that needed to find its place in the scheme of things.

David Hilbert

All of this end-of-the-nineteenth-century activity arose because of the rapid growth in applications to industry, technology and scientific growth. Scientists and mathematicians began to run into paradoxical results and contradictions in their headlong rush to make claims and produce results that did not have sufficient underpinning. Euclidean Geometry came under close scrutiny for its imprecise use of undefined terms, definitions and unconscious assumptions that should have been included among the axioms. In 1899, the great German mathematician, David Hilbert, published his book, *The Foundations of Geometry*. In place of Euclid's five axioms, Hilbert used 21 axioms, which included making explicit some of Euclid's hidden assumptions, such as what it means to say that a point is *between* two other points. By rigorous use of precise statements, Hilbert and others, believed that he had cleansed Euclidean Geometry of all of its short-comings and redundancies. In 1900, however, an 18 year old American mathematician, Robert L. Moore, published a one-page proof that Hilbert's twenty-first axiom was redundant and could be proved as a theorem from the first 20 axioms.

Not only did Hilbert succeed in axiomatizing geometry, but physicists, mathematicians, and other scientists were beginning to do the same thing in their fields.   For years, Hilbert had been making contributions to partial differential equations, calculus of variations, and, in his 1920 collaboration with Emmy Noether, on mathematical physics.

Hilbert was inspired to try to develop a program to axiomatize all of mathematics. With his attempt to achieve this goal, he began what is known as the "formalist school" of mathematics. It was clear to him that he could not do all this alone; so in 1900, when he was 38 years old, he gave a hefty homework assignment to all of the mathematicians of the world.   This was done when he presented a lecture, entitled **Mathematical Problems**, before the International Congress of Mathematicians in Paris.

Here is the introduction to his lecture.

> *"Who of us would not be glad to lift the veil behind which the future lies hidden; to cast a glance at the next advances of our science and at the secrets of its development during future centuries?   What particular goals will there be toward which the leading mathematical spirits of coming generations will strive?   What new methods and new facts in the wide rich field of mathematical thought will the new centuries disclose?"*

After a brief discussion about the nature of mathematical problems, he stated 23 problems. They were, subsequently, published in English in the *Bulletin of the American Mathematical Society*, Volume 8 (1902) pp 437 – 479.   The interested reader can find them on the Internet.   Did the mathematicians of the twentieth century work on these problems? Oh yes.   And even today in the second decade of the 21*st* century they still do.

Every so often, a rumor, sometimes true, sometimes partially true, will run wild among the mathematical community. "Did you hear that So & So has just solved part of Hilbert's 4*th* problem?" Or, "Somebody just got a negative result to a question in Hilbert's 10*th* problem." It is very nerve wracking when you are waiting anxiously to see whether or not you will be "scooped" by someone else solving your favorite problem before you do; worse yet, you may have heard that So & So is getting a result opposite to the one you are getting!

Gottlob Frege

Going back to the nineteenth century, in 1884, a German mathematician, Friedrich Ludwig Gottlob Frege, who was the founder of mathematical logic and the world's leading mathematical logician, published a massive book called *Die Grundlagen der Arithmetik*, the *Foundations of Arithmetic*.   This work really anticipated the rise of Hilbert's formalism.   It was read and understood by very few mathematicians and by, practically, no mortal person.   It combined ideas about set theory and logic started by Dedekind and Boole, and tried to formally axiomatize all of arithmetic.   It developed an abstract system of undefined terms, axioms and theorems about numbers and sets.   So when Frege published his work, it was supposed to put mathematics back on track as being perfectly logical and having a straight and narrow path to the truth.

According to the Encyclopaedia Britannica:

> *The Grundlagen was a work that must on any count stand as a masterpiece of philosophical writing. The only review the book received, however, was a devastatingly hostile one by Georg Cantor, the mathematician whose ideas were the closest to Frege's who had not bothered to understand Frege's book before subjecting it to totally unmerited scorn.*

Frege continued to write beautiful papers on the subject of mathematical logic and foundations, but had very few readers. Nevertheless, he did have one admirer who had read all of his works, a young British philosopher and mathematician by the name of Bertrand Russell.

Bertrand Russell

In 1903, just when Frege had sent off the second edition of the Foundations of Arithmetic, to the printer, he received a letter from Russell. The letter related how much Russell admired Frege's work and agreed with his philosophy, but

> *"There is just one point where I have encountered a difficulty. You state (p. 17) that a function, too, can act as an indeterminate element. This I formerly believed, but now this view seems doubtful to me because of the following contradiction. Let w be the predicate: to be a predicate that cannot be predicated of itself. Can w be predicated of itself? From each answer its opposite follows...."*

This observation established the fact that self-referent paradoxes existed in Frege's system of axioms. Frege responded with a letter containing the following sad sentence,

> *"... Your discovery of the contradiction caused me the greatest surprise and I would almost say consternation, since it has shaken the basis on which I had intended to build arithmetic...."*

Frege had the book published but appended the letter from Russell and his own response indicating that he had no way to salvage this monumental work. This incident ended Frege's career as a mathematical logician; he never made another significant contribution to the field that he had practically invented.

A few feeble attempts were made by Russell and other mathematicians and philosophers to repair the system by adding restrictive axioms in defining what is meant by a set, what is meant by elements of a set and whether or not subsets can be elements of a set. None of these worked because these restrictive axioms often tried to limit what it means to say something was true, thereby contradicting themselves. For example if you tried to add the axiom: *No statement can assert anything about its own truth.* Then that axiom itself is false because it *is* asserting something about its own truth.

In 1904, Russell and Alfred North Whitehead then started a project called *Principia Mathematica*. They published the first volume in 1910 and two others in 1912 and 1923. In this work, they started from scratch and tried to define everything before even stating an axiom or theorem.[1] They wanted to produce a system that was not contradictory and at the same time complete enough to answer any questions that could be asked about sets, subsets and elements of sets in the most general way. The logic had to be fool-proof.

---

[1] Look back at Figure 2.1 in Chapter 2, to see a page of *Principia*.

Russell, himself, said this attempt took a lot out of him, and he was unsatisfied, finally admitting that he had not succeeded in his goal. And, indeed, it was a failure. The reason for this became apparent within the next decade, when Kurt Gödel startled the world with *his* astonishing discovery. Which we will discuss later.

Mathematical structure

Meanwhile, as mathematicians were trying to solve Hilbert's problems and Whitehead and Russell were working on the foundations of mathematics, some French mathematicians in the 1920's were dealing with several vexing problems having to do with very fundamental questions. Some of which resulted from Russell's paradoxes. Others dealt with selecting elements from each set in an uncountable collection of sets. Still others were related to Hilbert's formalism. And what about the problems of logic and paradoxes we discussed back in Chapter 7? Recall the Hanging Bridge paradox from Don Quixote, in which the traveler told the gate keepers that his destination was to be hung by them. If they did hang him, then he spoke the truth so they had to let him cross the bridge, but if they let him cross he spoke falsely, so they had to hang him. What do we really mean by the truth or falsity of a mathematical statement?

Bourbaki

In the midst of these troubles, a secret society was formed in France. Around 1935, several French mathematicians, came together and started an intellectual cadre, called the *Bourbaki Group*. They were intent upon giving *full axiomatic structure* to all mathematics. They wanted to make sure that rigor was introduced at every level of mathematics. Using the name of a deceased French General, they created a fictitious mathematician, "Nicholas Bourbaki." Between 1940 and 1980 they began publishing small mathematical pamphlets, monographs, under this pseudonym. They had some influence on the textbooks in analysis and set theory, but their rigorous approach was so extreme, that it was difficult to make much progress. One prominent French mathematician, René Thom, claimed that, "... in their search for rigor, the Bourbaki group was suffering from *rigor mortis*."

In 1948, "he" produced a document which was, pejoratively, known as the *Bourbaki Manifesto*, laying out his new standards for mathematical papers. When the English version came out in 1950, it helped influence mathematics teaching by changing the curriculum in American Elementary Schools. It was called the "new math" and was pervasively baffling to the parents of elementary school children. This seminal paper was: "The Architecture of Mathematics" by Nicholas Bourbaki, *The American Mathematical Monthly*, Vol. 57 No.4, 1950. Here is what Bourbaki meant by *full axiomatic structure*.

Elements of mathematical structure

What are the essential ideas in the architecture of mathematics? They are a collection of objects, detailed descriptions of these objects, arbitrary assumptions about these objects, and the logical conclusions you can draw from these assumptions. We enumerate them as follows:

1. Primitive Terms, (the undefined terms)
2. Definitions
3. Axioms

4. Theorems

The *primitive terms* are flexible, unspecified, concepts that can be assigned any meanings what-so-ever. *Defined terms* are specific conditions limiting the scope of the undefined terms and adding new terms which are defined in terms of the undefined ones. *Axioms* are arbitrary assumptions about the defined and the undefined terms. The *Theorems* are statements that can be proved from the axioms or from previously proved theorems.

The axioms have only three limits on them: *consistency, independence and completeness.*

1. *Consistency*: No two axioms can contradict each other and no set of axioms can be used to prove statements that contradict each other.

2. *Independence*: No axiom can be derived as a consequence of combined use of the other axioms. If the axioms in a system are not independent, then the system is said to be *redundant.*

3. *Completeness*: Any statement using the undefined terms or defined terms must be either provable or disprovable from only the axioms.

In effect, Bourbaki was advocating Hilbert's Formalism. Wonderful. Now let us get back to what it was that Russell and Whitehead were trying to do. They didn't really worry about *independence*. If an axiom set is not independent, then no harm is really done, except the system is not as elegant as one that is independent. If someone finds that one of your axioms can be proved by using the other axioms, that's OK, just remove the redundant statement from your list of axioms and add it to your list of theorems.

But Russell and Whitehead *did* want *completeness* and *consistency*. In everyday language, completeness means that there are no unanswered questions, and consistency means there are no contradictory answers. It is analogous to a legal system: it is complete if the laws and case histories are sufficient to settle *any* case as either legal or not legal. If it is consistent, then it *cannot* be true that there is some particular case that can be deemed both legal and illegal at the same time.

Look at what happened to the idea that mathematics could be complete *and* consistent—have its cake and eat it too.

Kurt Gödel

Kurt Gödel was born in Czechoslovakia in 1906, four years after Bertrand Russell wrote his fateful letter destroying Frege's life's work on the foundations of arithmetic. In 1931, Gödel showed that the Russell-Whitehead monumental attempt to completely derive all mathematics from logic and the integers could *never* succeed.

In his article "On Formally Undecidable Propositions of *Principia Mathematica* and Related Systems," published in 1931 in the German journal: *Monatshefte für Mathematik*, Gödel proved the following two theorems, now called the Gödel Incompleteness Theorem.

Theorem 1. In any consistent formal system containing a minimum of arithmetic, a formally undecidable proposition can be found; that is, a closed formula A such that neither A nor not A can be deduced within the system.[2]

---

[2] If a system is consistent, it must be incomplete.

Theorem 2. If certain natural completeness conditions are met, one can take this formula A (as described above) to be the formula which expresses the consistency of the system.

What Gödel's proof of Theorem 1 did was to show that if a system is consistent, then there is some statement, A, that cannot be proved nor disproved within that system. In other words there is a question that can't be answered. So a consistent system is incomplete.

And by Theorem 2, he proved that it is impossible to justify consistency in a complete system. In other words, if every question can be answered then there is going to be some question that will have contradictory answers.

How did Gödel prove this?

The way that Gödel proved both of these theorems was to show that, in any system, if you tried to list *all* of the statements that could be made, you would run into a paradox. Eventually you would have to make the following statement:

$$A = \textit{This sentence cannot be proved by the current set of axioms.}$$

1. If A is true, then it is a true statement that cannot be proved, so the system is *incomplete* because it is incapable of proving every true statement.

2. On the other hand, if A is false, then it *can* be proved, but that would say the system can be used to prove a false statement. So the system is inconsistent.

What can we do about undecidability in mathematics?

So, what do we do? We opt for incompleteness. We would rather have a system that is incomplete than one that is inconsistent. In his book, *The Mathematical Tourist*, 1988, Ivars Peterson says that what Gödel did was to introduce an *uncertainty principle* into mathematics:

> "...*Gödel proved that any* (consistent) *collection of axioms leads to a mathematical system containing statements that can neither be proved nor disproved on the basis of those axioms. A theorem, then, can be undecidable: adding more axioms wouldn't help because some of the theorems would still slip through. Furthermore, Gödel's work implies, that in some cases, mathematicians wouldn't even be able to decide whether a theorem can or cannot be proved. Uncertainty strikes mathematics!*"

This does not mean that, all of a sudden there aren't correct answers anymore in mathematics We can't have students all over the world telling their math teachers that any old answer would be OK because mathematics is uncertain anyway. But it does mean that mathematics may not be any better off than physics or any other empirical science. That is, we will not be able to derive every mathematical statement from logic. And we may not be able to solve every problem.

There are a number of unsolved problems that mathematicians have been working on for many years that we may be required to accept as undecidable. In computer science, *complexity theory* essentially proves Gödel's theorem by showing that it is impossible for a program to be written that can check to see whether its input results in an infinite loop or a well defined halt. A consistent system is shown to be incomplete, which implies that a complete system must be inconsistent.

The power and the beauty

What does Gödel's proof do to the theme of this book? I wrote it to explain why mathematics is, ostensibly, the premier human problem-solving tool. It has the reputation that when all practitioners in some field, such as physics, biology, business, economics, or engineering showed up with a problem on the door step of the house of mathematics they were always welcomed with a friendly solution, or at least a willingness to work with the supplicants to find a solution. But, what is there for *mathematicians* who have arrived at the same door with their hats in their hands? They wanted the satisfaction of knowing that they could have a consistent system that was capable of settling all questions, but it was not to be. We can offer them the solace that incompleteness may not be *too* high a price to pay for consistency. This is our grand compromise, and we will tolerate it.

This is a good time for me to move on to a vacation home. But the old house is still here and will continue be here, serving in a more restricted way, for who knows how long? Young string-theorists, economists, biologists, astrophysicists, artists, chess players, musicians, historians, geologists, and practitioners in yet unknown fields, will continue to visit here. I will excuse myself and step out of the back door, softly singing this little song:

Oh, beautiful for abstract space
Where quantum physics rules
For fixed prime gaps and i-phone apps
And engineering tools
Mathematics, Oh Mathematics,
Gauss shed his grace on thee
And hide thy goofs
With Gödel proofs
From here to infinity!

Clement Falbo was born and raised in San Antonio,Texas. After serving four years in the U. S. Navy, he enrolled in the University of Texas at Austin where he earned his Ph.D. in mathematics Dr. Falbo is a Professor Emeritus at Sonoma State University, where he taught mathematics for thirty-five years. Upon his retirement in1999, he and his wife Jean Ann Falbo, a biologist, served in the U. S. Peace Corps in the African nation of Zimbabwe. Together, they taught African high school students mathematics and science. They now live in Joseph, Oregon. Falbo is the author of several mathematical articles and three mathematical textbooks: Finite Mathematics-Applied, Math Oddyssey 2000, and First Year Calculus-An Inquiry-Based Learning Approach. He has three sons, six grandchildren and four great grandchildren, Jean has a son and daughter and two grandchildren. She helped him write this book.

www.ingramcontent.com/pod-product-compliance
Lightning Source LLC
Chambersburg PA
CBHW081804200326
41597CB00023B/4147